Springer Series in

SOLID-STATE SCIENCES 146

Springer Series in
SOLID-STATE SCIENCES

Series Editors:
M. Cardona P. Fulde K. von Klitzing R. Merlin H.-J. Queisser H. Störmer

The Springer Series in Solid-State Sciences consists of fundamental scientific books prepared by leading researchers in the field. They strive to communicate, in a systematic and comprehensive way, the basic principles as well as new developments in theoretical and experimental solid-state physics.

126 **Physical Properties of Quasicrystals**
Editor: Z.M. Stadnik

127 **Positron Annihilation in Semiconductors**
Defect Studies. By R. Krause-Rehberg and H.S. Leipner

128 **Magneto-Optics**
Editors: S. Sugano and N. Kojima

129 **Computational Materials Science**
From Ab Initio to Monte Carlo Methods. By K. Ohno, K. Esfarjani, and Y. Kawazoe

130 **Contact, Adhesion and Rupture of Elastic Solids**
By D. Maugis

131 **Field Theories for Low-Dimensional Condensed Matter Systems**
Spin Systems and Strongly Correlated Electrons. By G. Morandi, P. Sodano, A. Tagliacozzo, and V. Tognetti

132 **Vortices in Unconventional Superconductors and Superfluids**
Editors: R.P. Huebener, N. Schopohl, and G.E. Volovik

133 **The Quantum Hall Effect**
By D. Yoshioka

134 **Magnetism in the Solid State**
By P. Mohn

135 **Electrodynamics of Magnetoactive Media**
By I. Vagner, B.I. Lembrikov, and P. Wyder

136 **Nanoscale Phase Separation and Colossal Magnetoresistance**
The Physics of Manganites and Related Compounds
By E. Dagotto

137 **Quantum Transport in Submicron Devices**
A Theoretical Introduction
By W. Magnus and W. Schoenmaker

138 **Phase Separation in Soft Matter Physics**
Micellar Solutions, Microemulsions, Critical Phenomena
By P.K. Khabibullaev and A.A. Saidov

139 **Optical Response of Nanostructures**
Microscopic Nonlocal Theory
By K. Cho

140 **Fractal Concepts in Condensed Matter Physics**
By T. Nakayama and K. Yakubo

141 **Excitons in Low-Dimensional Semiconductors**
Theory, Numerical Methods, Applications By S. Glutsch

142 **Two-Dimensional Coulomb Liquids and Solids**
By Y. Monarkha and K. Kono

143 **X-Ray Multiple-Wave Diffraction**
Theory and Application
By S.-L. Chang

144 **Physics of Transition Metal Oxides**
By S. Maekawa, T. Tohyama, S.E. Barnes, S. Ishihara, W. Koshibae, and G. Khaliullin

145 **Point-Contact Spectroscopy**
By Y. Naidyuk and I.K. Yanson

146 **Optics of Semiconductors and Their Nanostructures**
Editors: H. Kalt and M. Hetterich

147 **Electron Scattering in Solid Matter**
A Theoretical and Computational Treatise By J. Zabloudil, R. Hammerling, L. Szunyogh, and P. Weinberger

Volumes 1–125 are listed at the end of the book.

Heinz Kalt Michael Hetterich (Eds.)

Optics of Semiconductors and Their Nanostructures

With 164 Figures

Springer

Professor Dr. Heinz Kalt
Dr. Michael Hetterich
Institut für Angewandte Physik and
DFG-Center for Functional Nanostructures
Universität Karlsruhe (TH)
76128 Karlsruhe, Germany
E-mail: Heinz.Kalt@physik.uni-karlsruhe.de
michael.hetterich@physik.uni-karlsruhe.de

ISSN 0171-1873

ISBN 3-540-22068-2 Springer Berlin Heidelberg New York

Library of Congress Control Number: 2004105248

Springer is a part of Springer Science+Business Media
springeronline.com

Printed in Germany

Typesetting by the editor/authors
Cover concept: eStudio Calamar Steinen
Cover production: *design & production* GmbH, Heidelberg

Printed on acid-free paper SPIN: 10953136 57/3141/ba - 5 4 3 2 1 0

Preface

The history of semiconductor optics started with the first experimental observations of light emission in the 19th century. Einstein developed his quantum description of the interaction of light with matter in the early 20th century and the famous Elliott formula for the optical response of the exciton was published in the 1960s. The solar cell celebrates its 50th birthday this year and the laser diode will turn 35 in 2005. Semiconductor optical devices are nowadays found in every household and are the goods of a multi-billion dollar market. Hence one might expect that semiconductor optics has reached a quite mature state. However, in particular the decade around the transition from the 20th to the 21st century has demonstrated that the latter is still a highly active field constantly bringing up new, surprising breakthroughs and even technical revolutions.

The reasons for this development are manifold and it is the aim of this book to highlight some of the most important ones. Firstly, there is the evolution of new experimental techniques to uncover the microscopic basics of semiconductor optics as well as new concepts to manipulate these processes for novel applications. The theory of semiconductor optics follows new paths to describe or predict such phenomena. The progress in nanotechnology brings along a wealth of new physics resulting from confinement and quantization of both light and carriers. Finally, the advent of new semiconductor materials and the enormous advance in the epitaxial growth of well known compounds extends the applications of semiconductor optics from the ultraviolet through the whole visible range to the near infrared.

To be more specific, ultrafast optical spectroscopy now explores the timescale of, and even before, the first scattering processes of optically excited carriers with their environment. This can be ultimately extended to the timescale of one cycle of light leading to effects of extreme nonlinear optics. Ultrafast laser spectroscopy utilizes and tests the coherent nature of the optical excitations via methods of coherent spectroscopy such as four-wave mixing, quantum-beat spectroscopy, and coherent control. These as well as resonant light scattering (both time-resolved and continuous wave) including the analysis of speckle reveal the coherent contributions to the optical response and the processes of dephasing. Such studies now increasingly focus also on the coherence of spin states in semiconductors. The studies in the traditional frequency range of optics are supplemented by terahertz spectroscopy exploring the intersubband transitions of excitonic or carrier excitations. New approaches in near-field and confocal microscopy give access to a length scale well below the wavelength of light which is governed by quasi-ballistic transport of carriers and excitons. Temporally and spatially resolved phonon-sideband spectroscopy directly monitors exciton populations and their spatio-temporal dynamics.

This evolution of experimental techniques is accompanied by the elaboration of new theoretical concepts. Quantum kinetics theory, which describes e.g. the temporal build-up of correlation effects and screening, has to include memory terms to describe the optical response on a timescale where collisions between carriers can not be viewed as instantaneous. The standard semiconductor Bloch equations are extended to include correlation terms which requires an intelligent truncation of the hierarchy of higher-order effects in a self-consistent way. Including the quantization of the light field, which is obviously necessary for the description of systems with strong carrier–photon coupling like microcavities, also leads to a new interpretation of the optical spectra of excitons and the electron–hole plasma in semiconductors with slow carrier relaxation and weak excitonic binding like (Ga,In)As. Finally, effects related to localization find an elegant description by treating the excitonic center-of-mass motion in a fluctuating potential.

Semiconductor optics has gained enormous momentum from the development of nanostructure technology. One finds atom-like properties in selforganized quantum islands and dots, namely discrete states rather than bands which are unperturbed by inhomogeneous broadening. New fundamental excitations of the semiconductor such as trions are found and multi-particle states like the biexciton are strongly enhanced due to confinement. These novel properties are currently being transferred into applications such as quantum-dot lasers and optical amplifiers. They are also hot candidates for the realization of quantum computing.

New concepts for optoelectronic applications arise from the combination of semiconductors with photonic structures. An illustrative example is the normal-mode coupling, i.e. the splitting of modes due to a coupling between exciton and photon modes, established in microcavities. The ultimate goal is to place a single quantum dot in a photonic nanocavity providing entanglement or single photons on demand. No less innovative is the emerging field of spintronics which uses the spin rather than the charge of carriers for information technology. Experiments triggered by theoretical predictions show that quantum interference of transition amplitudes of one and two-photon absorption can be used to control independently spin and charge currents. And even in the relatively mature field of photovoltaics there is still room for optimization, as is shown by rigorous thermodynamical considerations.

Finally, the last decade has seen a major breakthrough in commercial optoelectronics including lighting, displays, data storage, as well as optical sensing in chemistry and biology. This breakthrough was caused by the development of efficient blue and white light-emitting diodes (LEDs) and the blue diode laser on the basis of group-III nitrides. Alternative wide-gap semiconductors for optoelectronics in the ultraviolet and blue are diamond and even more important ZnO. In the field of telecommunications new infrared emitters based on GaInNAs have been realized with a performance superior to existing technology and unusual physical properties related to the band structure. The development of such materials as standard components of mass-produced devices requires an elaborate optical characterization including the identification of dopant states or the revelation of band structure details.

This long list of highlights in recent years shows that the field of semiconductor optics is more vibrant than ever. The topics under investigation have gone far beyond the level that is described in standard textbooks like the one written by Claus Klingshirn [C.F. Klingshirn: *Semiconductor Optics*, 2nd corr. printing (Springer, Berlin Heidelberg New York 1997)], which gives one of the best introductions to the field. This present compilation of recent research activities appears on the occasion of C. Klingshirn's 60th birthday in honor of his outstanding contributions to the science and teaching in semiconductor optics. The contributors include some of his former students as well as close collaborators and longstanding scientific colleagues. We know that C. Klingshirn is very pleased by the fact that some of the topics and materials (like ZnO and GaN) which were "modern" at the time of his doctoral studies, have recently gained renewed attention in the quest for advanced concepts in semiconductor optics and its applications.

Karlsruhe,
March 2004

Heinz Kalt
Michael Hetterich

Contents

List of Contributors

Tommy W. Berg
Research Center COM
Technical University of Denmark
Ørsteds Plads 345V
2800 Lyngby, Denmark
twb@com.dtu.dk

Ravi D.R. Bhat
Department of Physics
University of Toronto
60 St. George Street
Toronto, Ontario, Canada M5S 1A7
rbhat@physics.utoronto.ca

Hans-Georg Breunig
Institut für Festkörperphysik
Universität Bremen
Postfach 330440
28334 Bremen, Germany
breunig@staff.uni-marburg.de

Hyatt M. Gibbs
Optical Sciences Center
The University of Arizona
Tucson, AZ 85721, USA
hyattgibbs@att.net

Jürgen Gutowski
Institut für Festkörperphysik
Universität Bremen
Postfach 330440
28334 Bremen, Germany
gutowski@ifp.uni-bremen.de

Hartmut Haug
J.W. Goethe-Universität Frankfurt
Institut für Theoretische Physik,
Robert-Mayer-Str. 8
60054 Frankfurt am Main, Germany
Haug@itp.uni-frankfurt.de

Michael Hetterich
Institut für Angewandte Physik
Universität Karlsruhe (TH)
Wolfgang-Gaede-Straße 1
76131 Karlsruhe, Germany
michael.hetterich@physik.uni-karlsruhe.de

Jørn M. Hvam
Research Center COM
Technical University of Denmark
Ørsteds Plads 345V
2800 Lyngby, Denmark
hvam@com.dtu.dk

Heinz Kalt
Institut für Angewandte Physik
Universität Karlsruhe (TH)
Wolfgang-Gaede-Straße 1
76131 Karlsruhe, Germany
heinz.kalt@physik.uni-karlsruhe.de

Mackillo Kira
Fachbereich Physik und
Wiss. Zentrum
für Materialwissenschaften
Philipps-Universität Marburg
Renthof 5
35032 Marburg, Germany
mackillo.kira@physik.
uni-marburg.de

Stephan W. Koch
Fachbereich Physik und
Wiss. Zentrum für
Materialwissenschaften
Philipps-Universität Marburg
Renthof 5
35032 Marburg, Germany
stephan.w.koch@physik.
uni-marburg.de

Gerrit Kocherscheidt
Experimentelle Physik IIb
Universität Dortmund
Otto-Hahn-Str. 4
44227 Dortmund, Germany
gerrit.kocherscheidt@gmx.de

Martin Kuball
H.H. Wills Physics Laboratory
University of Bristol
Tyndall Avenue
Bristol BS8 1TL, UK
martin.kuball@bristol.ac.uk

Wolfgang Langbein
Experimentelle Physik IIb
Universität Dortmund
Otto-Hahn-Str. 4
44227 Dortmund, Germany
wolfgang.langbein@
uni-dortmund.de

Ali Najmaie
Department of Physics
University of Toronto
60 St. George Street
Toronto, Ontario, Canada M5S 1A7
anajmaie@physics.utoronto.ca

Rolf Sauer
Abteilung Halbleiterphysik
Universität Ulm
Albert-Einstein-Allee 45
89069 Ulm, Germany
rolf.sauer@physik.uni-ulm.de

John E. Sipe
Department of Physics
University of Toronto
60 St. George Street
Toronto, Ontario, Canada M5S 1A7

Arthur L. Smirl
Laboratory for Photonics
& Quantum Electronics
138 IATL
University of Iowa
Iowa City, Iowa 52242, USA
art-smirl@uiowa.edu

Martin J. Stevens
Laboratory for Photonics
& Quantum Electronics
138 IATL
University of Iowa
Iowa City, Iowa 52242, USA
martin-stevens@uiowa.edu

Klaus Thonke
Abteilung Halbleiterphysik
Universität Ulm
Albert-Einstein-Allee 45
89069 Ulm, Germany
klaus.thonke@physik.uni-ulm.de

Henry M. van Driel
Department of Physics
University of Toronto
60 St. George Street
Toronto, Ontario, Canada M5S 1A7
vandriel@physics.utoronto.ca

Tobias Voss
Institut für Festkörperphysik
Universität Bremen
Postfach 330440
28334 Bremen, Germany
tvoss@ifp.uni-bremen.de

Martin Wegener
Institut für Angewandte Physik
Universität Karlsruhe (TH)
Wolfgang-Gaede-Straße 1
76131 Karlsruhe, Germany
martin.wegener@physik.uni-karlsruhe.de

Ulrike Woggon
Experimentelle Physik IIb
Universität Dortmund
Otto-Hahn-Str. 4
44227 Dortmund, Germany
ulrike.woggon@uni-dortmund.de

Peter Würfel
Institut für Angewandte Physik
Universität Karlsruhe (TH)
Wolfgang-Gaede-Straße 1
76131 Karlsruhe, Germany
peter.wuerfel@physik.uni-karlsruhe.de

Hui Zhao
Laboratory for Photonics
& Quantum Electronics
138 IATL
University of Iowa
Iowa City, Iowa 52242, USA
hui-zhao@uiowa.edu

Roland Zimmermann
Institut für Physik der
Humboldt-Universität zu Berlin
Newtonstr. 15
12489 Berlin, Germany
zim@physik.hu-berlin.de

Excitons in Semiconductors

Stephan W. Koch and Mackillo Kira

1 Introduction

The quantum mechanical problem of a single electron-hole pair in a homogeneous semiconductor leads to the Wannier equation, which, for a parabolic bandstructure is mathematically identical to the hydrogen atom Schrödinger equation [1]. This equation can be solved analytically in three and two dimensions, which is relevant for idealized bulk or quantum-well structures. Solutions are also available for quasi-one dimensional systems, however, their treatment requires the regularization of the Coulomb interaction potential [2, 3].

The bound state solutions of the Wannier equation are the excitons, which are characterized by a binding energy and a Bohr radius that, for systems with not too strong Coulomb attraction, significantly exceeds the characteristic length scale of the atomic unit cell. In particular the excitonic $1s$-state shows up as a pronounced resonance in the low-temperature linear optical absorption spectra of most good quality direct-gap semiconductors. Since the binding energy generally increases with the fundamental bandgap of the semiconductor material, II-VI bulk systems are well suited for studies of excitonic signatures [4]. In these materials, not only exciton absorption resonances, but also their saturation, the formation of an electron-hole plasma [2, 5], and even optical gain has been studied extensively [6].

An interesting topic in this context is the question under which conditions a truely incoherent excitonic population forms and how it can be identified experimentally. Excitonic features in linear absorption spectra are clearly not related to any population effect since only an induced optical polarization and no population exists in the linear regime. The presence of some form of an electron-hole-pair population is required to observe photoluminescence (PL) under incoherent conditions. However, the mere appearance of excitonic resonances in PL spectra is not sufficient to draw conclusions about the presence of excitons since also unbound electron-hole pairs can give rise to these features [7].

An unambiguous method to identify excitonic populations is to perform terahertz (THz) spectroscopy, i.e. to probe transitions between excitonic eigenstates [8, 9]. Under incoherent conditions, the observation of resonances due to these transitions is a clear signature of an exciton population. However, since these resonances correspond to differences between energy eigenvalues related to the relative electron-hole motion, they are independent of the exciton's center-of-mass energy. Therefore, the induced THz absorption is insensitive to the exciton distribution function.

In this article, we first summarize in Sect. 2 some of the well-known basic theoretical concepts to describe the semiconductor absorption. We supplement this discussion in Sect. 3 with aspects of the microscopic luminescence analysis. In Sect. 4 we then discuss the linear THz-response of a Coulomb interacting electron-hole system. In all cases, we pay attention to excitonic features and their microscopic origin.

2 Absorption Spectroscopy

To compute the semiclassical optical properties of a material system we have to solve Maxwell's equations. In order to have a simple geometry, we consider the configuration of an idealized quantum-well (QW) structure, where the light field is incident normal to the plane of the well. In its simplest form, the wave equation for one-dimensional propagation is given as

$$\left[\frac{\partial^2}{\partial z^2} + \frac{n^2(z)}{c^2}\frac{\partial^2}{\partial t^2}\right] E = -\mu_0 \frac{\partial^2}{\partial t^2} P, \tag{1}$$

where E is the electromagnetic field, z is the space coordinate normal to the surface of the QW, t denotes the time, and μ_0 is a constant prefactor depending on the system of units. The response of the material has been divided into a resonant part, treated dynamically in terms of the macroscopic optical polarization P, and the nonresonant part lumped into the (background) refractive index n.

For a system without disorder, it is convenient to expand the macroscopic polarization P into a Bloch basis [3]

$$P = \sum_{\mathbf{k}_\parallel} d_{cv}\, P_{\mathbf{k}_\parallel} + \text{c.c.}, \tag{2}$$

where d_{cv} is the electron-hole interband dipole matrix element between conduction and valence band. To account for the QW geometry, we split all three-dimensional momentum vectors $\mathbf{k}$ into their components in z direction and parallel to the plane of the QW,

$$\mathbf{k} = (\mathbf{k}_\parallel, k_z)\,. \tag{3}$$

The microscopic polarization $P_{\mathbf{k}_\parallel}$ and the carrier occupation probabilities $f^{e,h}_{\mathbf{k}_\parallel}$ constitute the diagonal and off-diagonal elements of the reduced single-particle density matrix ρ. Introducing creation and annihilation operators

$(a^\dagger_{\lambda,\mathbf{k}_\|}, a_{\lambda,\mathbf{k}_\|})$ for conduction $(\lambda = c)$ and valence $(\lambda = v)$ band electrons, we can write ρ as:

$$\rho = \begin{pmatrix} \langle a^\dagger_{c,\mathbf{k}_\|}\, a_{c,\mathbf{k}_\|}\rangle & \langle a^\dagger_{v,\mathbf{k}_\|}\, a_{c,\mathbf{k}_\|}\rangle \\ \langle a^\dagger_{c,\mathbf{k}_\|}\, a_{v,\mathbf{k}_\|}\rangle & \langle a_{v,\mathbf{k}_\|}\, a^\dagger_{v,\mathbf{k}_\|}\rangle \end{pmatrix} = \begin{pmatrix} f^e_{\mathbf{k}_\|} & P_{\mathbf{k}_\|} \\ P^\star_{\mathbf{k}_\|} & f^h_{\mathbf{k}_\|} \end{pmatrix} . \tag{4}$$

2.1 Semiconductor Bloch Equations

To obtain the equation of motion for ρ, we use the Heisenberg equation for the operators $\mathcal{O}$,

$$i\hbar \frac{\partial}{\partial t} \mathcal{O} = [\,\mathcal{O}, H\,] , \tag{5}$$

with the standard semiclassical many-body Hamiltonian

$$H = H_0 + H_C + H_I . \tag{6}$$

Here,

$$H_0 = \sum_k \varepsilon^c_{\mathbf{k}_\|} a^\dagger_{c,\mathbf{k}_\|}\, a_{c,\mathbf{k}_\|} + \sum_k \varepsilon^v_{\mathbf{k}_\|} a^\dagger_{v,\mathbf{k}_\|}\, a_{v,\mathbf{k}_\|} \tag{7}$$

is the single-particle part,

$$\begin{aligned} H_C = &\frac{1}{2} \sum_{\mathbf{k}_\|,\mathbf{k}'_\|,\mathbf{q}_\| \neq 0} V_{\mathbf{q}_\|} a^\dagger_{c,\mathbf{k}_\|+\mathbf{q}_\|} a^\dagger_{c,\mathbf{k}'_\|-\mathbf{q}_\|} a_{c,\mathbf{k}'_\|} a_{c,\mathbf{k}_\|} \\ &+\frac{1}{2} \sum_{\mathbf{k}_\|,\mathbf{k}'_\|,\mathbf{q}_\| \neq 0} V_{\mathbf{q}_\|} a^\dagger_{v,\mathbf{k}_\|+\mathbf{q}_\|} a^\dagger_{v,\mathbf{k}'_\|-\mathbf{q}_\|} a_{v,\mathbf{k}'_\|} a_{v,\mathbf{k}_\|} \\ &- \sum_{\mathbf{k}_\|,\mathbf{k}'_\|,\mathbf{q}_\| \neq 0} V_{\mathbf{q}_\|} a^\dagger_{c,\mathbf{k}_\|+\mathbf{q}_\|} a_{v,\mathbf{k}'_\|} a^\dagger_{v,\mathbf{k}'_\|-\mathbf{q}_\|} a_{c,\mathbf{k}_\|} \end{aligned} \tag{8}$$

is the Coulomb interaction Hamiltonian, and

$$H_I = -E(t) \sum_k (d^*_{cv} a^\dagger_{c,\mathbf{k}_\|} a_{v,\mathbf{k}_\|} + d_{cv} a^\dagger_{c,\mathbf{k}_\|} a_{v,\mathbf{k}_\|}) \tag{9}$$

describes the interaction with the classical electromagnetic field.

In Eq. (7), $\varepsilon^c_{\mathbf{k}_\|}$ ($\varepsilon^v_{\mathbf{k}_\|}$) denotes the single-particle energy of an electron in the conduction (valence) band with in-plane momentum $\mathbf{k}_\|$. In Eq. (8), the first two sums describe the repulsion among electrons in the same bands and the last term includes the interband attraction. Here, the Fourier transform of the Coulomb interaction potential is denoted by $V_{\mathbf{q}_\|}$ which incorporates the QW confinement. The system is coupled to the light field via the interband dipole matrix element d_{cv} in Eq. (9), showing that the light field either creates or destroys pairs of electrons and holes (missing valence-band electrons).

Calculating the equation of motion for the microscopic polarization $P_{\mathbf{k}_\|}$ by working out the commutators in Eq. (5), we obtain terms containing the single-particle density matrix (via $H_0 + H_I$) and a coupling to four-operator terms (via H_C). This is the beginning of the well known many-body hierarchy: If we derive equations of motion for the four operator expectation values, we obtain a coupling to six-operator terms, and so on. In order to close the coupled set of equations, we have to truncate that hierarchy at some stage in a self-consistent fashion [3, 10, 11].

At the level of a dynamical Hartree-Fock approximation, we use a decoupling scheme where all four-operator terms are factorized into products of two-operator terms. This leads to the Hartree-Fock Semiconductor Bloch Equations (HF-SBE). For a two-band system the full SBE can be written as [3, 10]

$$\begin{aligned}\left[i\hbar\frac{\partial}{\partial t} - \epsilon^e_{k_\|}(t) - \epsilon^h_{k_\|}(t)\right] P_{\mathbf{k}_\|}(t) &= -\left[1 - f^e_{\mathbf{k}_\|}(t) - f^h_{\mathbf{k}_\|}(t)\right] \Omega_{\mathbf{k}_\|}(t) \\ &\quad + \frac{\partial}{\partial t} P_{\mathbf{k}_\|}(t)|_{\mathrm{corr}} \\ \frac{\partial}{\partial t} f^a_{\mathbf{k}_\|}(t) &= -\frac{2}{\hbar}\mathrm{Im}\left[\Omega_{\mathbf{k}_\|}(t) P^\star_k(t)\right] + \frac{\partial}{\partial t} f^a_{\mathbf{k}_\|}(t)|_{\mathrm{corr}} \qquad (10)\end{aligned}$$

where the Hartree-Fock terms are shown explicitly and the contributions beyond Hartree-Fock are denoted by the subscript *corr* . In Eqs. (10)

$$\Omega_{\mathbf{k}_\|}(t) = d_{cv}\, E(t) + \sum_{k'_\| \neq k_\|} V_{|\mathbf{k}_\| - \mathbf{k}'_\||}\, P_{k'_\|}(t) \qquad (11)$$

is the renormalized field (Rabi energy) and

$$\epsilon^a_k(t) = \varepsilon^a_k - \sum_{k'_\| \neq k_\|} V_{|\mathbf{k}_\| - \mathbf{k}'_\||}\, f^a_{k'_\|}(t) \qquad (12)$$

is the renormalized transition energy and the index $a = e, h$.

At the Hartree-Fock level the SBE contain neither dephasing of the polarization nor screening of the interaction potential nor relaxation of the carrier distributions. To include these effects, we have to go beyond the Hartree-Fock approximation and specify the terms denoted with the subscript "corr".

2.2 Analytic Solution

Before we discuss correlation contributions to the SBE we first summarize the analytic solution of the linear problem. Starting from an unexcited semiconductor system, i.e., all polarizations and occupations vanish before the system is excited, we can linearize the polarization equation in the interaction with the external field. After a Fourier transform to real space, we obtain

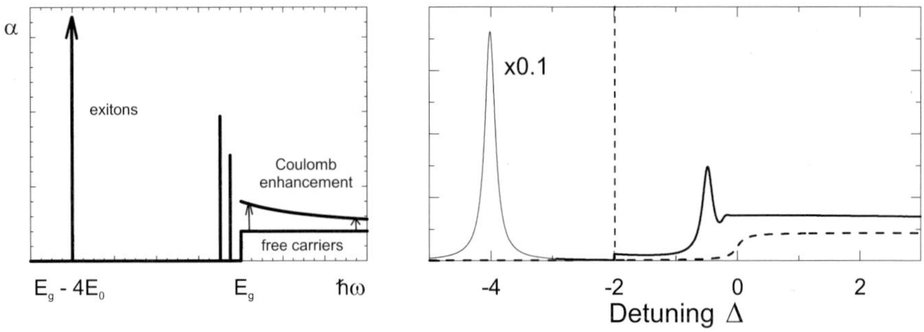

Fig. 1. Imaginary part of the linear susceptibility which is proportional to the linear optical absorption spectra in in the vicinity of the band gap for a two-dimensional direct semiconductor. The spectra in the left figure are calculated without broadening, whereas the results in the right figure have been obtained assuming a homogeneous broadening corresponding to 1/20th of the exciton binding energy.

$$\left[i\hbar \frac{\partial}{\partial t} + \frac{\hbar^2 \nabla^2_{\mathbf{r}_\parallel}}{2m_r} - V(r_\parallel) \right] P(\mathbf{r}_\parallel, t) = -d_{cv}\, E(t)\, \delta(\mathbf{r}_\parallel), \tag{13}$$

where m_r is the reduced electron-hole mass and $\mathbf{r}_\parallel$ is a position within the QW plane. The homogeneous part of Eq. (13) leads to the Wannier equation

$$\left[\frac{\hbar^2 \nabla^2_{\mathbf{r}_\parallel}}{2m_r} - V(r_\parallel) \right] \phi_\nu(\mathbf{r}_\parallel) = E_\nu \phi_\nu(\mathbf{r}_\parallel). \tag{14}$$

It is mathematically identical to the Schrödinger equation for the relative motion in the hydrogen problem. The bound state solutions of the Wannier equation are the exciton states and the unbound solutions represent the Coulomb interacting interband excitations. From the solution of the inhomogeneous Eq. (13) we obtain the electron-hole pair susceptibility, which yields the Elliott formula [1] for the linear semiconductor susceptibility

$$\chi(\omega) = 2|d_{cv}|^2 \sum_\lambda \frac{|\phi_\lambda(r=0)|^2}{E_\lambda - \hbar\omega - i\gamma}. \tag{15}$$

Here we wrote only the resonant contribution and introduced γ as phenomenological dephasing of the polarization. The Elliott formula shows that absorption resonances occur at the frequencies $\omega = E_\lambda/\hbar$. An example of such an excitonic spectrum is plotted in Fig. 1. For later reference we note at this point that in our theoretical formulation the resonances appear as a consequence of the Coulombic renormalization of the Rabi energy, Eq. (11).

2.3 Microscopic Scattering

Without phenomenological input, the HF-SBE do not include any dephasing or relaxation contributions. To describe such processes, we have to extend the theory beyond the Hartree-Fock level and derive equations for the correlation terms using systematic approximations. This can be done microscopically consistently, e.g. by using Green's function techniques, projection operators, or equations of motion methods [10–12].

Ignoring memory effects and restricting the analysis to two-particle collisions we obtain

$$\frac{\partial}{\partial t} P_{\mathbf{k}_\parallel}(t)|_{\mathrm{corr}} = -\Gamma_{\mathbf{k}_\parallel}(t) P_{\mathbf{k}_\parallel}(t) + \sum_{\mathbf{k}'_\parallel} \Gamma_{\mathbf{k}_\parallel,\mathbf{k}'_\parallel}(t) P_{\mathbf{k}'_\parallel}(t) \tag{16}$$

and

$$\begin{aligned} \frac{\partial}{\partial t} f^a_{\mathbf{k}_\parallel}(t)|_{\mathrm{corr}} &= -\Sigma^{\mathrm{out},a}_{\mathbf{k}_\parallel}(t) f^a_{\mathbf{k}_\parallel}(t) \\ &\quad + \Sigma^{\mathrm{in},a}_{\mathbf{k}_\parallel}(t)[1 - f^a_{\mathbf{k}_\parallel}(t)] + \Sigma^{\mathrm{pol},a}_{\mathbf{k}_\parallel}(t). \end{aligned} \tag{17}$$

The rates $\Sigma^{\mathrm{in,out}}_k$ describe carrier scattering into and out of the state k, and Σ^{pol} contains the nonlinear polarization scattering contributions. The explicit form of all these terms is given in Ref. [12]. In the incoherent limit, i.e. without polarization scattering, Eq. (17) reduces to the well-known Boltzmann collision term for carrier-carrier scattering. Also Eq. (16) has the structure of a Boltzmann scattering integral, however, the polarization and the rates Γ are complex quantities.

Figure 2 compares numerical results for the absorption of the exciton resonance with measured data obtained after optical pumping in the continuum with different intensities. In agreement with experiment, the calculation yields an intensity dependent resonance saturation and broadening [13]. This behavior is a signature of the excitation induced dephasing [14, 15] resulting from destructive interference among different microscopic polarization contributions [16].

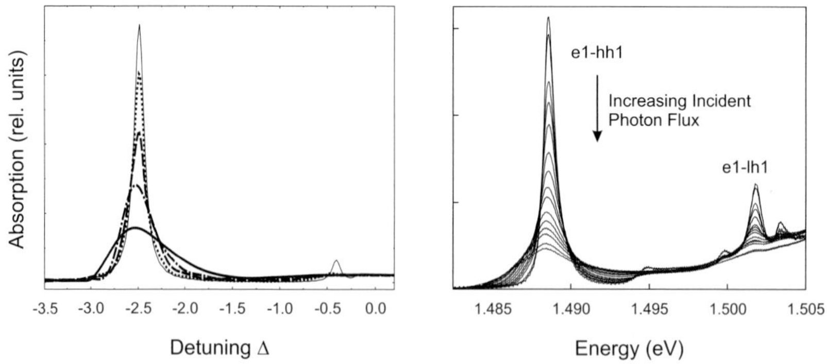

Fig. 2. Comparison of computed (left figure) and measured (right figure) excitonic absorption spectra for an InGaAs quantum well after optically pumping with a fs-pulse. With increasing pump intensity the exciton saturates and broadens, but shifts only weakly. (After Ref. [13].)

3 Photoluminescence

In order to gain more insight into excitonic features in semiconductors, it is illustrative to look at the spontaneous emission. Under incoherent conditions, the presence of an emission signal is a clear indication of radiatively decaying electron-hole-pair populations. However, without a detailed microscopic analysis it is not obvious how much one can deduce about the state of the population, i.e. about the possible presence of incoherent excitons and their distribution.

Since the process of light emission via spontaneous electron-hole recombination is an intrisically quantum mechanical effect, we have to quantize the light field. For this purpose, we introduce bosonic creation and annihilation operators $B^\dagger_{\mathbf{q}}$ and $B_{\mathbf{q}}$ where $\mathbf{q}$ refers to a specific light mode with momentum $\mathbf{q}$ and energy $\hbar\omega_{\mathbf{q}} = \hbar c|\mathbf{q}|$. We split the three-dimensional momentum $\mathbf{q}$ into in-plane and perpendicular components according to Eq. (3).

We replace the semiclassical light-matter interaction Hamiltonian (9) by the fully quantized form

$$H_{\mathrm{QED}} = \sum_{\mathbf{k}_\parallel,\mathbf{q},q_z} \left(i\mathcal{F}_{\mathbf{q}} B_{q_z,\mathbf{q}_\parallel} a^\dagger_{\mathbf{k}_\parallel+\mathbf{q}_\parallel,c} a_{\mathbf{k}_\parallel,v} + i\mathcal{F}_q B_{q_z,\mathbf{q}_\parallel} a^\dagger_{\mathbf{k}_\parallel+\mathbf{q}_\parallel,v} a_{\mathbf{k}_\parallel,c} \right) + \mathrm{h.c.}, \tag{18}$$

where $\mathcal{F}_{\mathbf{q}} = d_{cv}\mathcal{E}_{\mathbf{q}}u_{\mathbf{q}}$ is the strength of the interaction which can expressed in terms of the dipole-matrix element, vacuum-field amplitude $\mathcal{E}_{\mathbf{q}}$, and the mode strength $u_{\mathbf{q}}$ at the QW position. Microscopically, H_{QED} includes all possible processes where electron-hole pairs are created (annihilated) by the simultaneous absorption (emission) of a photon. Only the momentum in the quantum-well plane is conserved for the planar structures studied here, the momentum in z-direction is not conserved.

3.1 Semiconductor Luminescence Equations

In our quantum treatment, we recover the classical properties of the light from the expectation value $\langle B_{\mathbf{q}}\rangle$ which uniquely determines the electric field $E(t) \equiv \langle \hat{E}\rangle = \sum_{\mathbf{q}} \mathcal{E}_{\mathbf{q}} u_{\mathbf{q}} \langle B_{\mathbf{q}}\rangle$+c.c.. The intensity of the emitted light is obtained from $\langle \hat{E}\hat{E}\rangle$ which contains both the classical $\langle B^\dagger\rangle\langle B\rangle$ and the quantum-optical parts

$$\Delta\langle B^\dagger B\rangle \equiv \langle B^\dagger B\rangle - \langle B^\dagger\rangle\langle B\rangle \quad . \tag{19}$$

For pulsed excitation conditions there is a time at which the coherent polarization and the electric field have decayed. Then the system enters into the incoherent regime where $P_{\mathbf{k}_\parallel}$ as well as the classical fields $\langle B_{\mathbf{q}}\rangle$ vanish. Thus, the emitted light intensity follows entirely from the quantum-optical intensity correlations $\Delta\langle B^\dagger B\rangle$. For the direction perpendicular to the QW, this photon-number-like dynamics can be computed from

$$i\hbar\frac{\partial}{\partial t}\Delta\langle B_{q_z}^{\dagger}B_{q'_z}\rangle = \hbar\left(\omega_{q'_z}-\omega_{q_z}\right)\Delta\langle B_{q_z}^{\dagger}B_{q'_z}\rangle + i\sum_{\mathbf{k}_{\|}}\left(\mathcal{F}_{q_z}\Delta\langle B_{q'_z}a_{c,\mathbf{k}_{\|}}^{\dagger}a_{v,\mathbf{k}_{\|}}\rangle + \mathcal{F}_{q'_z}^{\star}\Delta\langle B_{q_z}^{\dagger}a_{v,\mathbf{k}_{\|}}^{\dagger}a_{c,\mathbf{k}_{\|}}\rangle\right). \quad (20)$$

This equation shows that the dynamics of $\Delta\langle B^{\dagger}B\rangle$ is coupled to the photon assisted polarization $\Delta\langle B^{\dagger}a_v^{\dagger}a_c\rangle$ which is the amplitude for a process where an electron is lowered from the conduction to the valence band under simultaneous emission of a photon.

The dynamics of the photon assisted polarization is given by

$$\begin{aligned} i\hbar\frac{\partial}{\partial t}\Delta\langle B_{q_z}^{\dagger}a_{v,\mathbf{k}_{\|}}^{\dagger}a_{c,\mathbf{k}_{\|}}\rangle &= \left(\varepsilon_k^c-\varepsilon_k^v-\hbar\omega_{q_z}\right)\Delta\langle B_{q_z}^{\dagger}a_{v,\mathbf{k}_{\|}}^{\dagger}a_{c,\mathbf{k}_{\|}}\rangle \\ &- \left(1-f_k^e-f_k^h\right)\Omega_{\mathbf{k}_{\|},q_z}^{\mathrm{ST}} \\ &+ i\mathcal{F}_{q_z}\left(f_k^e f_k^h + \sum_{k'}\Delta\langle a_{c,\mathbf{k}'_{\|}}^{\dagger}a_{v,\mathbf{k}_{\|}}^{\dagger}a_{c,\mathbf{k}_{\|}}a_{k',v}\rangle\right) \\ &+ i\hbar\frac{\partial}{\partial t}\Delta\langle B_{q_z}^{\dagger}a_{v,\mathbf{k}_{\|}}^{\dagger}a_{c,\mathbf{k}_{\|}}\rangle|_{\mathrm{scatt}}, \end{aligned} \quad (21)$$

where we have defined the renormalized stimulated contribution

$$\Omega_{k,q_z}^{ST} \equiv \sum_{q'_z} i\mathcal{F}_{q'_z}\Delta\langle B_{q_z}^{\dagger}B_{q'_z}\rangle + \sum_{\mathbf{k}'_{\|}} V_{\mathbf{k}_{\|}-\mathbf{k}'_{\|}}\Delta\langle B_{q_z}^{\dagger}a_{v,\mathbf{k}'_{\|}}^{\dagger}a_{c,\mathbf{k}'_{\|}}\rangle, \quad (22)$$

in analog to the renormalized field, Eq. (11). The last term in Eq. (21) contains the coupling to higher order scattering terms resulting from the Coulomb and phonon interactions.

Equations (20) and (21) constitute the *semiconductor luminescence equations* (SLE) [7, 17]. The first and second lines of Eq. (21) resemble the SBE; renormalized kinetic energies ε and phase-space filling, $1-f^e-f^h$, can be identified directly and Ω^{ST} is the generalization of Ω, Eq. (11). In contrast to the SBE, however, the SLE have source terms, which are listed in the third line of Eq. (21). As long as carriers are excited in the system, $f^e f^h$ drives the photon-assisted polarization, which in turn is the source in Eq. (20) describing incoherent light emission. The factor $f_k^e f_k^h$ shows that spontaneous emission recombination takes place if an electron and hole are present simultaneously. An additional source for the spontaneous emission is given by the correlated part, $\Delta\langle a_c^{\dagger}a_v^{\dagger}a_c a_v\rangle$, describing the possibility to emit light directly from exciton populations.

3.2 Analytic Solution

In order to gain some insights into the spectral features and the population dependence of the semiconductor luminescence, we look for an analytic solution.

In many experimentally relevant situations, the carrier system changes slowly such that the populations f^λ and $\Delta\langle a_c^\dagger a_v^\dagger a_c a_v\rangle$ can be taken as constant. For such conditions, the incoherent Eqs. (20) and (21) are closed. We furthermore assume weak excitation where the carrier densities are sufficiently low such that we may use the Wannier equation and ignore the specific form of the microscopic scattering.

Within this approximation scheme, we obtain a simple expression for the photoluminescence spectrum, which is given as the steady-state photon flux,

$$I_{\mathrm{PL}}(\omega_q) \equiv \frac{\partial}{\partial t}\Delta\langle B_q^\dagger B_q\rangle = \frac{2|\mathcal{F}_q|^2}{\hbar}\mathrm{Im}\left[\sum_\lambda \frac{|\phi_\lambda(r=0)|^2 N_\lambda}{E_\lambda - \hbar\omega_q - i\gamma}\right]. \qquad (23)$$

As mentioned above, this result is valid only for low densities and negligible broadening, however, the extension to higher densities is possible and follows in a form similar to the nonlinear absorption calculations [17].

Comparing Eqs. (23) and (15) we note strong similarities. In particular, the frequency dependence of both equations is governed by the same denominator giving rise to excitonic resonances. Hence, we can already conclude at this point, that the appearance of these resonances is independent of the detailed structure of the population factor appearing in the numerator of Eq. (23). This factor determines the strength of the PL. It can be written as

$$N_\lambda = \sum_k |\phi_\lambda(k)|^2 f_k^e f_k^h + \sum \phi_\lambda^\star(k)\phi_\lambda(k')\Delta\langle a^\dagger_{c,\mathbf{k}_\parallel} a^\dagger_{v,\mathbf{k}'_\parallel} a_{c,\mathbf{k}'_\parallel} a_{v,\mathbf{k}_\parallel}\rangle. \qquad (24)$$

The first term originates from carriers in the single particle distributions f_k and the second contribution is a correlated two-particle quantity that defines the number of excitons with vanishing center-of-mass momentum, $\mathbf{q}_\parallel = \mathbf{0}$. Since the electron-hole plasma and exciton population terms appear additively in N_λ, both contributions can lead to luminescence and therefore also to emission at the excitonic resonances. Thus, the mere appearance of luminescence at, e.g., at the $1s$-resonance cannot be taken as a unique signature of exciton populations.

To illustrate these features, we show in Fig. 3 normalized PL spectra that have been computed with (right column) and without (left column) an incoherent excitonic population. First of all, we note that the spectra are dominated by a strong $1s$-resonance in both cases. Whereas it is basically impossible to identify exciton population effects in the normalized spectra on a linear scale (top figures), we see that the logarithmic scale (bottom figures) reveals a characteristic difference in the ratio of the $1s$ and $2s$ or bandedge luminescence. A quantitative analysis of this ratio in a series of experimental measurements for different excitation conditions makes it possible to extract informations about the excitonic population of the optically active states [18]. However, the total exciton population is not directly accessible via PL experiments.

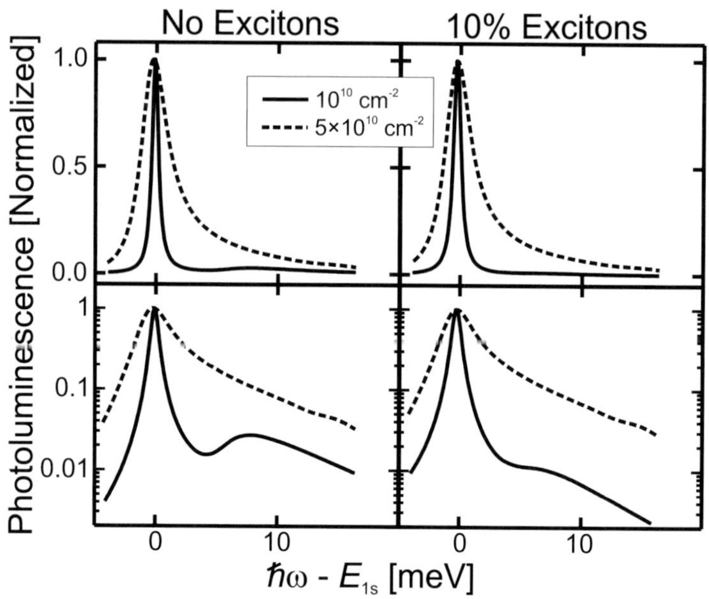

Fig. 3. Comparison of computed luminescence spectrum on a linear (top figures) and a logarithmic scale (bottom figures). The figures on the LHS show the results for the quasi-equilibrium electron-hole luminescence at a temperature of 77 K without any excitonic populations and the figures on the RHS include 10% excitons in a thermal distribution. The solid (dashed) lines have been computed for carrier densities of 10^{10} cm^{-2} (5×10^{10} cm^{-2}).

4 Terahertz Spectroscopy

One way to detect small concentrations of a particular species of atoms or molecules is to use an optical probe that is sensitive to transitions between the eigenstates of the respective species. If the characteristic absorption resonances are observed in the probe spectrum, the atoms or molecules must be present, and through proper normalization of the respective oscillator strength one can deduce their relative concentration.

In order to apply this concept to the identification of excitonic populations in III-V or II-VI semiconductors one needs terahertz (THz) fields, since these can be resonant with transitions from the exciton's $1s$ to $2p$ state [8, 9]. The energy corresponding to the THz regime is way below the fundamental bandgap of most direct-gap semiconductors. Hence, THz fields do not generate electron-hole interband excitations in those systems. Instead, they lead to intraband transitions, i.e. to transitions between different many-body states.

For the theoretical description of THz processes, the light-matter interaction discussed in this article has to be extended to include the coupling to intraband quantities. Microscopically, this interaction follows from

$$\hat{H}_{\mathrm{THz}} = -A(t) \sum_{k} \left[j_c(\mathbf{k}_\parallel) a^\dagger_{c,\mathbf{k}_\parallel} a_{c,\mathbf{k}_\parallel} + j_v(\mathbf{k}_\parallel) a^\dagger_{v,\mathbf{k}_\parallel} a_{v,\mathbf{k}_\parallel} \right] + \frac{Q^2 A^2(t)}{2m_0} \sum_{k} \left[a^\dagger_{c,\mathbf{k}_\parallel} a_{c,\mathbf{k}_\parallel} + a^\dagger_{v,\mathbf{k}_\parallel} a_{v,\mathbf{k}_\parallel} \right], \quad (25)$$

which includes the current-matrix element

$$j_\lambda(\mathbf{k}_\parallel) \equiv \frac{Q\hbar \mathbf{k}_\parallel \cdot \mathbf{e}_\sigma}{m_\lambda}, \quad (26)$$

with the effective mass m_λ and polarization direction of the field $\mathbf{e}_\sigma$ which lies in the QW plane.

4.1 Microscopic Terahertz Equations

As for the optical pulse, the propagation of THz fields is described by the one-dimensional wave equation

$$\left[\frac{\partial^2}{\partial z^2} - \frac{n^2(z)}{c^2} \frac{\partial^2}{\partial t^2} \right] A(z,t) = \mu_0 J, \quad (27)$$

where J is the macroscopic current of the system. For Coulomb interacting systems, the effect of the A^2 term is largely suppressed, such that the dominant contributions to the current follow from

$$J \equiv \sum_{k} \left[j_e(\mathbf{k}_\parallel) f^e_{\mathbf{k}_\parallel} + j_h(\mathbf{k}_\parallel) f^h_{\mathbf{k}_\parallel} \right]. \quad (28)$$

In order to compute THz effects self-consistently, we have to determine the carrier dynamics

$$\frac{\partial}{\partial t} f^{e(h)}_{\mathbf{k}_\parallel} = -\frac{2}{\hbar} \mathrm{Im} \Big[\Omega^\star_k P_{\mathbf{k}_\parallel} + \sum_{\mathbf{k}_\parallel, \mathbf{k}'_\parallel, \mathbf{q}_\parallel, \lambda} V_{\mathbf{k}'_\parallel + \mathbf{q}_\parallel - \mathbf{k}_\parallel} \Delta \langle a^\dagger_{c(v),\mathbf{k}_\parallel} a_{\lambda,\mathbf{k}'_\parallel} a_{c(v),\mathbf{k}'_\parallel + \mathbf{q}_\parallel} a_{\lambda, \mathbf{k}_\parallel - \mathbf{q}_\parallel} \rangle \Big] - \frac{2}{\hbar} \mathrm{Im} \left[\sum_{q_z} i \mathcal{F}^\star_{q_z} \Delta \langle \hat{B}^\dagger_{q_z} \hat{a}^\dagger_{v,\mathbf{k}_\parallel} \hat{a}_{c,\mathbf{k}_\parallel} \rangle \right]. \quad (29)$$

Even with the THz field included, the carrier dynamics has the same structure as in the semiconductor Bloch equations (10), however, now we have written the Coulomb scattering contributions more explicitly. The last term in Eq. (29) is of quantum-optical origin describing photon assisted electron-hole recombination. We see from Eq. (29) that the carrier density and thus

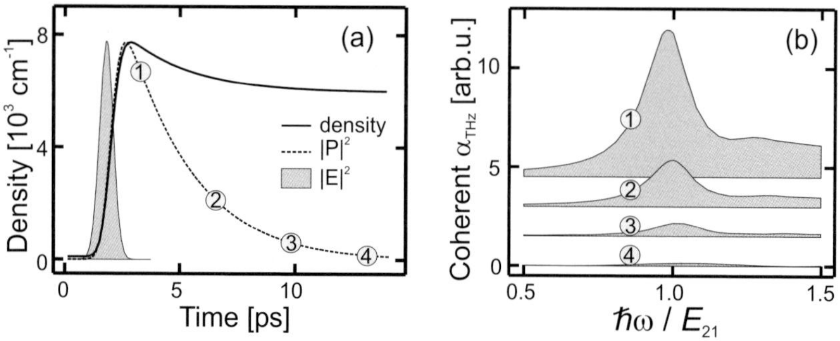

Fig. 4. Figure (a) shows the temporal dynamics of the excitation pulse (shaded area), of the optically excited excitonic polarization (absolute square of the polarization, dashed line), and of the electron-hole density (solid line). Figure (b) displays snapshots of those parts of the THz absorption spectrum that are induced by the coherent polarization. E_{21} is the energy difference between the $2p$ and $1s$ exciton resonances. The times are marked in (a). The area under the spectra decreases with increasing time due to the decay of the polarization.

the current J does not enter directly. Hence, the coupling to the THz field has to follow indirectly, either via the polarization and/or via the four-point correlations.

Under simultaneous optical and THz excitation the polarization dynamics obeys the equation

$$\left[i\hbar\frac{\partial}{\partial t} - \epsilon^e_{\mathbf{k}_\parallel}(t) + \epsilon^h_{\mathbf{k}_\parallel}(t)\right] P_{\mathbf{k}_\parallel}(t) = j(\mathbf{k}_\parallel)A(t)P_{\mathbf{k}_\parallel} \tag{30}$$

$$- \left[1 - f^e_{\mathbf{k}_\parallel}(t) - f^h_{\mathbf{k}_\parallel}(t)\right] \Omega_{\mathbf{k}_\parallel}(t) + \frac{\partial}{\partial t} f^a_{\mathbf{k}_\parallel}(t)|_{\mathrm{corr}} \quad . \tag{31}$$

In comparison to the polarization equation of the SBE, Eq. (31) has an additional driving term $j(\mathbf{k}_\parallel)A(t)P_{\mathbf{k}_\parallel} \equiv \left[j_e(\mathbf{k}_\parallel) + j_h(\mathbf{k}_\parallel)\right] A(t)P_{\mathbf{k}_\parallel}$ that provides a coupling between $A(t)$ and $f^{e(h)}$. This term leads to coherently induced currents and thus to a THz response.

To illustrate the signatures of an optical polarization in the THz spectrum we numerically solved the coupled equations without the contribution of $\frac{\partial}{\partial t} f^a_{\mathbf{k}_\parallel}|_{\mathrm{corr}}$. Assuming resonant excitation at the $1s$-exciton peak of the absorption spectrum, we obtain the results shown in Fig. 4. In the THz absorption spectrum we clearly see a resonance corresponding to the $1s$ to $2p$ transition. The transitions to higher excitonic states are spectrally merged providing a smooth background. Since we only included the coherent contributions in this calculation, the induced THz absorption signal vanishes as the interband polarization decays.

For completely incoherent situations, i.e. if neither an optical field nor a polarization is present, a THz current can only be induced via the four-point correlation

$$c_X^{\mathbf{q}_\parallel,\mathbf{k}'_\parallel,\mathbf{k}_\parallel} \equiv \langle a^\dagger_{c,\mathbf{k}_\parallel} a^\dagger_{v,\mathbf{k}'_\parallel} a_{c,\mathbf{k}'_\parallel+\mathbf{q}_\parallel} a_{v,\mathbf{k}_\parallel-\mathbf{q}_\parallel} \rangle. \tag{32}$$

In the incoherent regime, this quantity obeys the equation

$$\begin{aligned} i\hbar\frac{\partial}{\partial t} c_X^{\mathbf{q}_\parallel,\mathbf{k}'_\parallel,\mathbf{k}_\parallel} &= \left(\varepsilon^v_{k-q} + \varepsilon^c_{k'+q} - \varepsilon^v_{k'} - \varepsilon^c_k\right) c_X^{\mathbf{q}_\parallel,\mathbf{k}'_\parallel,\mathbf{k}_\parallel} \\ &+ j(k'+q-k)A(t)c_X^{\mathbf{q}_\parallel,\mathbf{k}'_\parallel,\mathbf{k}_\parallel} \\ &+ V_{\mathbf{k}'_\parallel+\mathbf{q}_\parallel-\mathbf{k}_\parallel}\left[\bar{f}^e_{\mathbf{k}_\parallel}\bar{f}^h_{\mathbf{k}_\parallel-\mathbf{q}_\parallel} f^e_{\mathbf{k}'_\parallel+\mathbf{q}_\parallel} f^h_{\mathbf{k}'_\parallel} - f^e_{\mathbf{k}_\parallel} f^h_{\mathbf{k}_\parallel-\mathbf{q}_\parallel}\bar{f}^e_{\mathbf{k}'_\parallel+\mathbf{q}_\parallel}\bar{f}^h_{\mathbf{k}'_\parallel}\right] \\ &+ \left[1 - f^e_{\mathbf{k}_\parallel} - f^h_{\mathbf{k}_\parallel-\mathbf{q}_\parallel}\right]\sum_{\mathbf{l}_\parallel} V_{\mathbf{k}_\parallel-\mathbf{l}_\parallel} c_X^{\mathbf{q}_\parallel,\mathbf{k}'_\parallel,\mathbf{l}_\parallel} \\ &- \left[1 - f^e_{\mathbf{k}'_\parallel+\mathbf{q}_\parallel} - f^e_{\mathbf{k}'_\parallel}\right]\sum_{l} V_{\mathbf{k}'_\parallel-\mathbf{l}_\parallel} c_X^{\mathbf{q}_\parallel,\mathbf{l}_\parallel,\mathbf{k}_\parallel} \\ &+ D^{\mathbf{q}_\parallel,\mathbf{k}'_\parallel,\mathbf{k}_\parallel}_{\mathrm{X\ rest}} + T^{\mathbf{q}_\parallel,\mathbf{k}'_\parallel,\mathbf{k}_\parallel}_{\mathrm{X}}, \end{aligned} \tag{33}$$

where we have introduced the abbreviation $\bar{f}^\lambda_{\mathbf{k}_{||}} \equiv 1 - f^\lambda_{\mathbf{k}_{||}}$. Explicitly shown in Eq. (33) are the terms providing the coupling to $A(t)$ as well as the dominant Coulomb sums. The additional terms, $D_{\mathrm{X,rest}}$ contain the remaining Coulomb sums as well as the contributions from spontaneous emission while T includes the higher order scattering effects [19].

Equation (33) shows that c_{X} is directly coupled to the THz field in a similar way as the polarization, Eq. (31). Thus, an applied THz field induces changes to c_{X}, which generates a THz current via Eq. (29) such that one obtains a finite THz response also in the incoherent regime. In Eq. (33), the dominant Coulomb sums represent the electron-hole attraction. The phase-space factors multiplying these terms implement the fundamental Pauli-blocking effects between these two fermions. The solution of the c_{X} dynamics can describe exciton formation under the appropriate conditions [9, 19, 20]. In this article, however, we analyze how exciton populations can be monitored via the THz signal.

As an example, we show in Fig. 5 the build-up of the incoherently induced THz absorption after resonant excitation. For exactly the same conditions as in Fig. 4 we evaluated the incoherent contributions to study the rise of the THz signal while the coherent polarization is gradually converted into incoherent excitonic population. We note that the spectral shape of the coherently and the incoherently induced THz absorption is basically indistinguishable. This is understandable since both result from transitions between the same excitonic states. In a real THz experiment, one always sees the sum of the results shown in Figs. 4b and 5b.

4.2 Analytic Solution

In order to obtain some analytic insights we now make similar adiabatic approximations as in our analysis of the semiconductor luminescence equations,

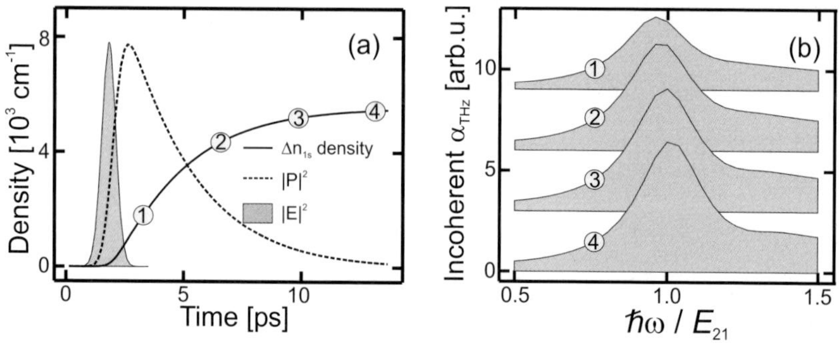

Fig. 5. Figure (a) shows the temporal dynamics of the excitation pulse (shaded area), of the optically induced polarization (absolute square of the polarization, dashed line), and of the density of incoherent $1s$ excitons (solid line). Figure (b) displays snapshots of those parts of the THz absorption spectrum that are induced by the incoherent excitonic population. E_{21} is the energy difference between the $2p$ and $1s$ exciton resonances. The times are marked in (a). The area under the spectra increases with increasing time due to the build-up of excitonic population.

Sect. 3.2. We consider the completely incoherent regime and assume a slow temporal variation of those parts of $f^{e(h)}$ and c_X that are independent of the THz field. For such a situation, it is convenient to represent the c_X in an exciton basis

$$\Delta\langle \hat{X}^\dagger_{\nu,\mathbf{q}_{||}} \hat{X}_{\nu',\mathbf{q}_{||}}\rangle = \sum_{\mathbf{k}_{||},\mathbf{k}'_{||}} \phi_\nu(\mathbf{k}_{||} - \tfrac{m_e}{M}\mathbf{q}_{||})\phi^\star_{\nu'}(\mathbf{k}'_{||} + \tfrac{m_h}{M}\mathbf{q}_{||})c_X^{\mathbf{q}_{||},\mathbf{k}'_{||},\mathbf{k}_{||}} , \quad (34)$$

with $M = m_e + m_h$. Assuming that a weak THz field is applied to the incoherent system we can write the analytic solution for the linear THz susceptibility

$$\chi_{\mathrm{THz}}(\omega) = -\frac{1}{\epsilon_0 \hbar\omega^2(\omega + i\Gamma)} \sum_{\nu,\lambda} \left(S^{\nu,\lambda}_{\omega} \Delta n_{\nu,\lambda} - \left[S^{\nu,\lambda}_{-\omega} \Delta n_{\nu,\lambda} \right]^\star \right) \quad , \quad (35)$$

with decay constants Γ for the THz currents and γ for the four-point correlations, respectively. To simplify the notation, we introduce the quantities

$$J_{\nu,\lambda} \equiv \sum_{\mathbf{k}_{||}} \phi^\star_\nu(\mathbf{k}_{||}) j(\mathbf{k}_{||}) \phi_\lambda(\mathbf{k}_{||}) \quad , \quad (36)$$

$$\Delta n_{\nu,\lambda} \equiv= \sum_{\mathbf{q}_{||}} \Delta\langle \hat{X}^\dagger_{\nu,\mathbf{q}_{||}} \hat{X}_{\lambda,\mathbf{q}_{||}}\rangle \quad . \quad (37)$$

The expression for $J_{\nu,\lambda}$ implies the typical dipole selection rules, i.e. J is vanishing for $\nu = \lambda$, whereas e.g. $\nu = 1s$ and $\lambda = 2p$ leads to non-zero contributions. It is interesting to note that the dipole transitions between the different excitonic bound states emerge only after considering the dynamics of the Coulomb correlated four-point quantities $\Delta n_{\nu,\lambda}$.

The THz response can be obtained from the response function

$$S_{\omega}^{\nu,\lambda} = \sum_{\beta} \frac{(E_{\nu} - E_{\beta}) J_{\nu,\beta} J_{\beta,\lambda}}{E_{\beta} - E_{\nu} - \hbar\omega - i\gamma} \quad . \tag{38}$$

Equations (35)–(38) represent a general formula to calculate linear THz response from an arbitrary configuration of electron-hole correlations. We notice that the THz signal follows entirely from the correlated two-particle quantities in the incoherent regime.

In the case that only diagonal correlations exist,

$$\Delta n_{\nu,\lambda} = \delta_{\nu,\lambda} \Delta n_{\nu,\nu} \equiv \delta_{\nu,\lambda} \Delta n_{\nu} \quad , \tag{39}$$

Eq. (35) reduces to

$$\chi_{\text{atom}}(\omega) = -\frac{1}{\epsilon_0} \sum_{\nu} \left(S_{\text{atom}}^{\nu}(\omega) - [S_{\text{atom}}^{\nu}(-\omega)]^{\star} \right) \Delta n_{\nu} \quad , \tag{40}$$

$$S_{\text{atom}}^{\nu}(\omega) = \sum_{\beta} \frac{|D_{\nu,\beta}|^2}{E_{\beta} - E_{\nu} - \hbar\omega - i\gamma} \quad , \tag{41}$$

where

$$D_{\nu,\lambda} = \int d^2\mathbf{r}_{\|} \, \phi_{\nu}^{\star}(\mathbf{r}_{\|}) \, \left[Q\mathbf{r}_{\|} \cdot \mathbf{e}_{\sigma} \right] \, \phi_{\lambda}(\mathbf{r}_{\|}) \tag{42}$$

defines the excitonic dipole-matrix element.

Equation (41) has the form that is typical for an atomic absorption spectrum when different atomic levels are populated according to Δn_{ν} [21]. Consequently, the correlation Δn_{ν} has a one-to-one correspondence to the atom number in state ν. Using this equivalence, we can define

$$\Delta N_{\nu,\mathbf{q}_{\|}} \equiv \Delta \langle \hat{X}_{\nu,\mathbf{q}_{\|}}^{\dagger} \hat{X}_{\nu,\mathbf{q}_{\|}} \rangle \quad , \tag{43}$$

which establishes a physically meaningful concept to introduce a center-of-mass distribution for excitons in the state ν.

Our theory shows that the THz response follows from the entire distributions since Eq. (35) contains a sum over all center-of-mass momenta $\mathbf{q}_{\|}$. If we compare Eq. (35) to (40), we notice that it contains also off-diagonal correlations $\Delta n_{\nu,\lambda}$. These terms include the contributions resulting from the underlying fermionic character of electrons and holes even when they are bound into excitons. Thus, Eq. (35) can be viewed as a many-body generalization of the multi atom absorption.

The microscopic analysis shows that the incoherent THz response results only form the correlated parts of four-point quantities in contrast to the PL, where also the factorized two-point term $f^e f^h$ acts as a source. The THz absorption can therfore positively identify truely incoherent exciton populations in all center-of-mass momentum states, i.e. both bright and dark excitons.

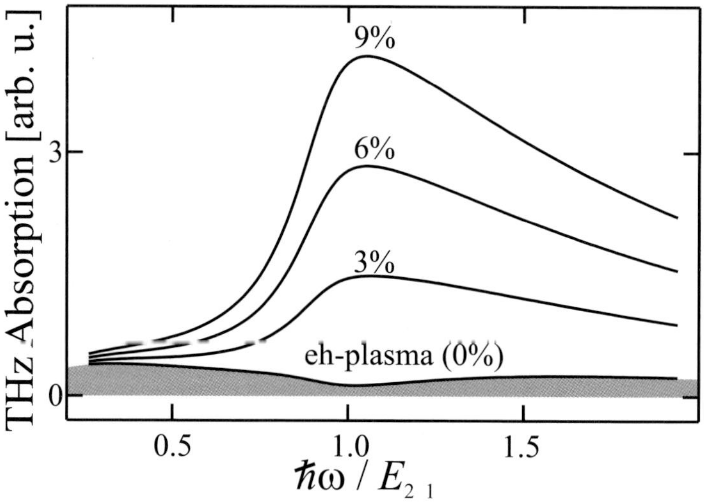

Fig. 6. Computed THz absorption spectra with different percentages of incoherent excitons. Without any excitons one sees the typical plamsa like THz response. With increasing exciton concentration the characteristic $1s$ to $2p$ transition resonance emerges and becomes gradually more pronounced. E_{21} is the energy difference between the $2p$ and $1s$ exciton resonances. The area under the spectra is proportional to the total exciton concentration. (From Ref. [20].)

To illustrate these features, we compare in Fig. 6 computed THz absorption spectra for different concentrations of excitons. In the absence of excitons we obseve the characteristic spectrum af a correlated electron-hole plasma which exhibits an increase toward low frequencies. On top of this plasma spectrum we note the gradual development of the $1s$ to $2p$ transition resonance already seen in Figs. 4 and 5. The strength of this resoance increases with increasing exciton concentration. In experimentally observed spectra one could thus use the area under the resonance to deduce the total exciton concentration.

5 Conclusions

In summary, excitonic effects play an important role in many aspects of semiconductor spectroscopy. Characteristic resonances and enhancement signatures can be observed in absorption and emission spectra. Microscopic theory attributes these features to the electron-hole Coulomb attraction that leads to renormalizations in the semiclassical and photon-assisted polarization.

The mere presence of excitonic peaks in optical spectra is not sufficient to draw conclusions about exciton populations in the semiconductor system.

Such informations can be gained via detailed quantitative analysis of luminescence spectra, by the THz spectrosopy of intra-exciton transitions, or by other techniques sensitive to the detailed nature of the incoherent electronic excitations.

Despite the advanced level of understanding of light-matter interaction effects in semiconductors, there are still many open questions and challenges. These include problems related to the dynamics of many-body correlations, quantum optical effects in semiconductors, quantum statistical and condensation effects, and many more. Hence, one can expect many interesting and stimulating results from ongoing and future semiconductor research.

Acknowledgments

This work has benefited greatly from many discussions and collaborations with H.M. Gibbs, W. Hoyer, F. Jahnke, G. Khitrova, T. Meier, P. Thomas, and members of their research groups. We acknowledge financial support through the Max-Planck Research prize of the Max-Planck and Humboldt Societies, the Deutsche Forschungsgemeinschaft, and the Optodynamics Center at the Philipps-University Marburg.

References

1. R.J. Elliott in: *Polarons and Excitons*, ed by C.G. Kuper, G.D. Whitefield (Oliver and Boyd 1963) pp 269
2. C. Klingshirn, H. Haug: Phys. Rep. **70**, 315 (1981) and references therein
3. For a textbook discussion see H. Haug, S. W. Koch: *Quantum Theory of the Optical and Electronic Properties of Semiconductors*, 4th ed. (World Scientific, Singapore 2004)
4. For a textbook discussion see, e.g., C.F. Klingshirn: *Semiconductor Optics*, 2nd corr. printing (Springer, Berlin Heidelberg New York 1997)
5. K. Bohnert, M. Anselment, G. Kobbe, C. Klingshirn, H. Haug, S.W. Koch, S. Schmitt-Rink, F.F. Abraham: Z. Physik B **42**, 1 (1981)
6. S.W. Koch, H. Haug, G. Schmieder, K. Bohnert, C. Klingshirn: Phys. Status Solidi (b) **89**, 431 (1978); C. Klingshirn, W. Maier, B. Hönerlage, H. Haug, S.W. Koch: Solid State Electronics **21**, 1357 (1978)
7. M. Kira, F. Jahnke, S. W. Koch: Phys. Rev. Lett. **81**, 3263 (1998)
8. R.H.M. Groeneveld, D. Grischkowsky: J. Opt. Soc. Am. B **11**, 2502 (1994); J. Cerne et al.: Phys. Rev. Lett. **77**, 1131 (1996)
9. M. Kira, W. Hoyer, T. Stroucken, S.W. Koch: Phys. Rev. Lett. **87**, 176401 (2001)
10. M. Lindberg, S.W. Koch: Phys. Rev. B **38**, 3342 (1988)
11. W. Schäfer: Journ. Opt. Soc. Am. B **13**, 1291 (1996)
12. F. Jahnke, M. Kira, S.W. Koch: Z. Physik B **104**, 559 (1997)

13. F. Jahnke, M. Kira, S.W. Koch, G. Khitrova, E.K. Lindmark, T.R. Nelson, D.V. Wick, J.D. Berger, O. Lyngnes, H.M. Gibbs, K. Tai: Phys. Rev. Lett. **77**, 5257 (1996)
14. G. Khitrova, H.M. Gibbs, F. Jahnke, M. Kira, S.W. Koch: Rev. Mod. Phys. **71**, 1591 (1999)
15. H. Wang, K. Ferrio, D.G. Steel, Y.Z. Hu, R. Binder, S.W. Koch: Phys. Rev. Lett. **71**, 1261 (1993)
16. S.W. Koch, T. Meier, F. Jahnke, P. Thomas: Appl. Phys. A **71**, 511 (2000)
17. M. Kira, W. Hoyer, F. Jahnke, S.W. Koch: Prog. Quantum Electron. **23**, 189 (1999)
18. S. Chatterjee et al.: Phys. Rev. Lett. **92**, 067402 (2004)
19. W. Hoyer, M. Kira, S.W. Koch: Phys. Rev. B **67**, 155113 (2003)
20. M. Kira, W. Hoyer, S.W. Koch: Solid State Commun. (2004)
21. L. Allen, J.H. Eberly: *Optical Resonance and Two-Level Atoms* (Wiley, New York 1975)

Hot Excitons in ZnSe Quantum Wells

Hui Zhao and Heinz Kalt

1 Introduction

In a semiconductor, an incident photon with suitable energy can excite an electron from the valence band to the conduction band, leaving a hole in the valence band. The Coulomb attraction between the electron and the hole may lead to the formation of a hydrogen-like bound state, called exciton. Beside the relative motion of the electron and the hole within the exciton, the exciton as a whole has its center-of-mass momentum and kinetic energy. In the case that the electron-hole pair is excited with some excess energy, the exciton can be formed with a significant kinetic energy and momentum. The hot exciton then travels in the semiconductor according to its momentum, and relaxes to its band minimum by giving its excess energy to the lattice through phonon scattering processes. Relaxed excitons, also called cold excitons, have an average excess energy comparable to the thermal energy of the lattice. At this stage, a quasi-equilibrium state of exciton and phonon system is reached. Eventually, the relaxed exciton recombines radiatively, resulting in photoluminescence (PL), or nonradiatively, converting the energy into heat. At high temperatures, thermal dissociation processes can break the excitons into electron-hole pairs. When the temperature is low enough that the phonon population is not sufficient for these processes, the exciton is very stable and dominates many optical processes of semiconductors.

This picture of the fate of excitons in semiconductors is a well accepted concept since many decades and has been nicely reviewed e.g. in [1, 2]. Only recently this concept has been challenged and the traditional interpretation of PL experiments has been questioned (see e.g. the contribution of S.W. Koch and M. Kira in this book). In particular, theory and experiments indicate that a PL signal at the exciton resonance in III-V semiconductor structures might not necessarily be a signature for a real exciton population [3, 4]. The situation turns out to be quite different in semiconductors with high excitonic binding energy and strong polar coupling allowing for an efficient exciton formation after nonresonant excitation. We will show in this contribution that quantum structures with binary ZnSe wells are an outstanding model system to study the dynamics of hot excitons. In particular, we will describe relaxation and transport properties of excitons on the time scale of inelastic scattering and on the length scale of the light wavelength.

We will discuss in this chapter hot-exciton processes starting from hot-exciton formation followed by relaxation, real-space transport, and radiative recombination. We restrict our discussion to two-dimensional excitons in semiconductor quantum wells. In the next section, we will begin with the discussion of the very first stage of the hot-exciton dynamics – the exciton formation process. After a brief review of several optical methods used in hot-exciton studies, we will outline the principle and implementation of phonon-sideband spectroscopy. We will emphasize the advantage of this method being that excitonic states with non-zero momentum can be accessed directly. The application of phonon-sideband spectroscopy to studies of hot-exciton dynamics in energy and real space will be discussed in Sects. 3 and 4, respectively. In Sect. 3 we monitor directly the energy relaxation of the hot-exciton distribution. In Sect. 4 the different transport regimes of excitons and various optical techniques for transport studies are reviewed. This is followed by a presentation of experimental results obtained by spatially resolved zero-phonon-line spectroscopy and phonon-sideband spectroscopy in the individual transport regimes. The investigations also include the regime of coherent or quasi-ballistic transport of excitons. The chapter is closed by some brief conclusions.

2 Hot-Exciton Formation and Spectroscopy

2.1 Hot-Exciton Formation

Excitons in quantum wells can be excited with different initial properties depending on the photon energy of the excitation laser. In the case of resonant excitation, the laser photon energy is tuned right into the exciton resonance. In this case one starts with coherent optical excitations which turn into real excitons after a dephasing process typically involving a scattering with acoustic phonons or with other (in)coherent excitons [5, 6]. Thus, the excitons are generated with very small momentum and kinetic energy. In the polar semiconductor ZnSe such 'cold' excitons can also be created quasi-resonantly, i.e. after the emission of one or more longitudinal optical (LO) phonons, as will be shown below.

In the following we want to focus on 'hot' excitons which have a significant momentum and kinetic energy. Hot-exciton formation is only possible by non-resonant excitation. But since the photon's momentum is almost negligible, a direct hot-exciton generation by the laser photon is nearly impossible. This is obvious when one compares the very different dispersion curves of exciton and photon (see Fig. 1). (There is a possibility to create hot excitons by nearly resonant excitation within the inhomogeneous linewidth. We will return to this special excitation mode in Sect. 4). Instead one can tune the laser photon energy to the continuum states. Starting from the interband polarization related to coherent electron-hole pairs two relaxation scenarios are possible:

Firstly, the electron-hole pairs may immediately form hot excitons assisted by the ultrafast interaction with phonons . In this process, the change in energy and momentum is provided by the phonons (unless disorder relaxes the conservation laws). Secondly, the electrons and holes may relax toward their band minima individually and form excitons during or after this relaxation process. Generally speaking, these two exciton formation channels are both possible in most situations, and the dominant one is determined by their relative rates.

Currently there is a debate in the literature on which formation process is actually the dominant one in III-V semiconductors. Several PL experiments [7–9] and theoretical studies [10–12] on GaAs have indicated efficient hot-exciton formation assisted by the emission of longitudinal-optical LO phonons. However, recent theoretical studies proposed that the PL signal is not necessarily related to a real exciton population [3]. This questioned the previous understanding of exciton formation processes obtained mainly by PL experiments. Latest investigations using terahertz (THz) absorption between internal exciton states indicated that the exciton population is built up on the time scale of several hundred picoseconds [4]. It is obvious that the exciton formation processes in GaAs are not well understood yet.

The exciton formation processes in the more polar II-VI semiconductors can be quite different from the case of GaAs. The excitonic binding energy of bulk ZnSe is five times larger than in GaAs (20 versus 4.2 meV), and the Fröhlich coupling between electrons and LO phonons is seven times stronger (0.43 versus 0.06). These differences suggest that in II-VI systems, the LO-phonon-assisted exciton formation processes can be very efficient, thus dominating over the individual relaxation of electrons and holes. And indeed this is reflected by a variety of experiments. Hot-exciton photoluminescence is generally observed in II-VI systems, indicating the efficient hot-exciton formation process [13–15]. Efficient relaxation of excitons via LO phonon emission is evidenced by the LO phonon cascade found in the photoluminescence excitation (PLE) spectra of numerous II-VI systems [16], while this effect is not seen in GaAs. Our spatially resolved transport experiments as described below reveal a similar cascade in the transport length which can only be explained by a relaxation of hot excitons [17]. More importantly, several experiments do not analyze the PL at the excitonic resonance and thus avoid the problems of interpretation described above. In particular, temporally resolved phonon-sideband spectroscopy of ZnSe quantum wells monitors directly the temporal evolution of exciton population on its dispersion curve. It has been found that the hot excitons are formed on a picosecond time scale [18]. Ultrafast, quantum-kinetics experiments show LO phonon-mediated beats in the four-wave mixing signal revealing the coupling of solely the electrons to the LO phonons in GaAs, while the electron-hole pairs as entities have an efficient coupling to one or even two LO phonons in ZnSe [19]. All these experiments support the efficient hot-exciton formation assisted by LO phonon emission.

2.2 Optical Detection of Hot Excitons

There are several possibilities to study exciton dynamics in semiconductors by optical methods [5]. Most of these techniques like photoluminescence or pump-probe experiments exploit the direct coupling between excitons and photons, without involving any other quasi-particles, like phonons. The observed resonance in PL and absorption spectra is therefore called zero-phonon line (ZPL). The underlying law of momentum conservation, however, imposes strong limitations on the interpretation of the optical experiments. Since its momentum is negligibly small, the photon can only couple to excitonic states with near-zero momentum. Hot excitons with a large center-of-mass kinetic energy and momentum are 'dark' states, and cannot be directly studied by ZPL spectroscopy. This invisibility of hot excitons in ZPL spectroscopy makes the study of hot-exciton processes a hard task. Interpretation of the ZPL data is difficult since they involve several successive processes including exciton formation, relaxation, and recombination. But only the final step of the entire process can be monitored.

Recently, THz absorption has been developed to be a powerful tool in detecting hot excitons [4, 20]. Beside the center-of-mass motion, the relative motion of electron and hole inside the exciton forms the inner structure of excitonic energy states. The separation in energy between the ground state and the first excited state is typically on the order of meV, corresponding to emission and absorption in the THz spectral regime. Thus, the absorption of a carefully tuned THz beam monitors the population of excitons inside the sample. Since this resonance is insensitive to the center-of-mass kinetic energy, both cold and hot excitons can be detected simultaneously. Although still in its very first stage of development, this technique has been successfully applied in getting important information like exciton formation in GaAs quantum wells [4].

On the positive side, the insensitivity of internal energy structure of excitons on its center-of-mass kinetic energy enables the THz spectroscopy to probe the whole exciton ensemble including hot excitons. On the negative side, however, the same fact implies that this technique cannot distinguish excitons with different center-of-mass energies. So in this kind of experiment, one can get the overall exciton population, but without any information on its energy distribution. Generally, such an energy resolution is highly desired for studying hot-exciton dynamics, especially the energy relaxation and related processes. In the next subsection, we will see that phonon-sideband spectroscopy can provide this information, thus being an ideal tool to study hot-exciton processes.

2.3 Phonon-Sideband Spectroscopy

Phonon-Assisted Recombination of Hot Excitons

Hot excitons cannot couple to photons directly due to their much larger momentum. But this limitation can be overcome if other quasi-particles like phonons are involved in the recombination and take away the excess momentum of hot excitons. Indeed, phonon-assisted recombination and absorption of excitons have been observed in many materials. The related peaks in PL or absorption spectra are called phonon-sidebands (PSB) or phonon replica. Due to the multi-particle feature of these processes, the PSB is typically much weaker than the ZPL. But still, due to the strong coupling of excitons to LO phonons in polar semiconductors, the PSB has been frequently exploited for the study of hot excitons. For a summary of early results see e.g. [21], very recent results on II-VI compounds and group-III nitrides are found e.g. in [13–15] and [22].

In the following discussion, we will focus on LO-phonon-assisted recombination in ZnSe quantum wells. Figure 1 shows schematically the hot-exciton dynamics in ZnSe quantum wells and the resulting PL spectrum composed of ZPL and PSB's. The excitons are formed with high center-of-mass kinetic energy E_k and momentum. In the zero-phonon process, they need to relax to the band minimum by emitting acoustic phonons before recombination. However, in the phonon-assisted recombination process, the hot exciton can recombine by emitting one photon together with one or more LO phonons, to fulfill the conservation of energy and momentum:

$$E_0 + E_\mathrm{k} = E_\mathrm{photon} + nE_\mathrm{LO} \tag{1}$$

and

$$\hbar \boldsymbol{k} = n\hbar \boldsymbol{k}_\mathbf{LO}. \tag{2}$$

Here, E_0 is the energy of excitonic resonance, which is assumed to be equal to the energy of the ZPL, E_photon and E_LO are the energies of the emitted photon and LO phonon, respectively. The latter is measured to be 31.8 meV in ZnSe quantum wells. The photon momentum is neglected since it is much smaller than those of exciton and phonon. The number of LO phonons involved is given by n.

The spectral shape of the nth order PSB, $I_\mathrm{PSB}^{(n)}(E_\mathrm{photon}) = I_\mathrm{PSB}^{(n)}(E_0 + E_\mathrm{k} - nE_\mathrm{LO})$, is determined by the energy distribution of the excitons, $f(E_\mathrm{k})$, and the probability of nth order phonon-assisted recombination process, $W_n(E_\mathrm{k})$:

$$I_\mathrm{PSB}^{(n)}(E_0 + E_\mathrm{k} - nE_\mathrm{LO}) \propto f(E_\mathrm{k}) \times W_n(E_\mathrm{k}). \tag{3}$$

The probability of the phonon-assisted recombination of hot excitons has been analyzed theoretically by Permogorov [21]. In the case of first order LO-phonon-assisted recombination, the probability is

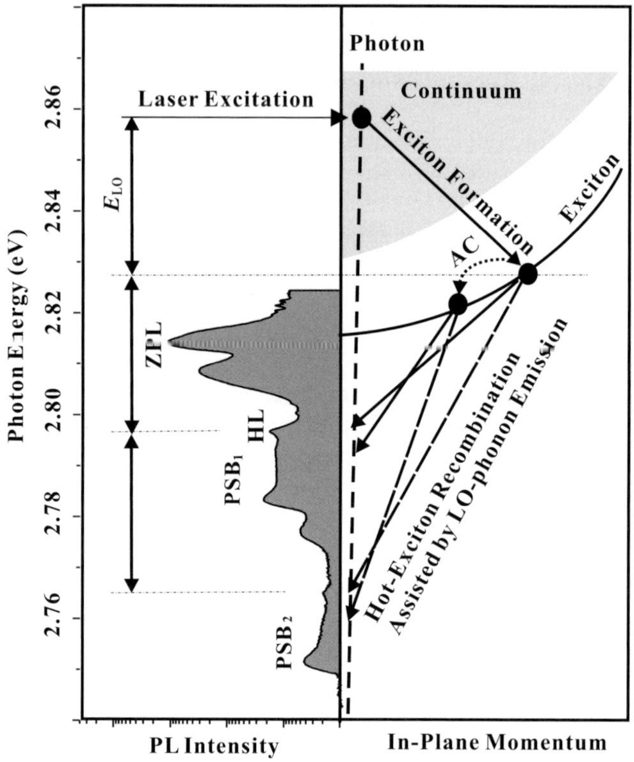

Fig. 1. Schematic drawing of the excitonic dynamics (right panel) and the resulting PL spectrum (left panel) measured from a 7.3 nm ZnSe/ZnSSe quantum well sample at a temperature of 7 K. The shown excitonic dynamics include the laser excitation, the hot-exciton formation achieved by LO phonon emission, the energy relaxation of hot excitons by acoustic phonon (AC) emission, and the LO-phonon-assisted recombination of hot excitons. The PL spectrum is composed of a ZPL as well as the first and second order PSB's.

$$W_1(E_\mathrm{k}) \propto \left| \sum_i \frac{\psi_i(0)H}{E_\mathrm{k} - E_\mathrm{LO}} \right|^2 , \tag{4}$$

where $|\psi_i(0)|^2$ is the oscillator strength for the optical transition from the state i in dipole approximation. The summation is over all possible intermediate states. For $1s$ excitons, the main intermediate state is $1s(k \approx 0)$. H is the Fröhlich scattering matrix element:

$$H(q) \propto F \frac{E_\mathrm{LO}}{\hbar a_0 q} \left\{ [1 + (a_0 q_\mathrm{c}/2)^2]^{-2} - [1 + (a_0 q_\mathrm{v}/2)^2]^{-2} \right\} . \tag{5}$$

Here a_0 is the exciton Bohr radius, q is the wavevector of the LO phonon, $q_\mathrm{c,v} = q(\mu/m_\mathrm{c,v})$, where $\mu = m_\mathrm{c}m_\mathrm{v}/(m_\mathrm{c} + m_\mathrm{v})$ with the electron and hole effective band masses m_c and m_v, respectively, and F is the Fröhlich exciton–phonon coupling constant. In order to deduce $f(E_\mathrm{k})$ from I_PSB, we only need the *energy dependence* of the recombination probability. Thus, we omit the energy-independent parameters, and get

$$W_1(E_\mathrm{k}) \propto \left| \frac{[1+(a_0 q_\mathrm{c}/2)^2]^{-2} - [1+(a_0 q_\mathrm{v}/2)^2]^{-2}}{q(E_\mathrm{k}-E_\mathrm{LO})} \right|^2 . \quad (6)$$

In the second order phonon-assisted process, the recombination probability has been shown to be independent of the exciton kinetic energy [21]. This is induced by the fact that the involved two intermediate states in this process have different parities. As a consequence the second order PSB is equivalent to the energy distribution of excitons.

Energy-Dependent Huang–Rhys Factor

The Huang–Rhys factor S [23] has been generally used to describe the relative intensities of the ZPL and PSB's. In this context, the intensity of the nth order PSB is related to that of the ZPL by

$$I_\mathrm{PSB}^{(n)} = \frac{S^n}{n!} I_\mathrm{ZPL} . \quad (7)$$

This factor describes the coupling strength of the radiative transition to the LO phonon polarization field. From (7) we get the Huang–Rhys factor as the ratio of the relative intensities of the first and second order PSB's,

$$S(E_\mathrm{k}) = 2\frac{I_\mathrm{PSB}^{(2)}(E_\mathrm{k}+E_0-2E_\mathrm{LO})}{I_\mathrm{PSB}^{(1)}(E_\mathrm{k}+E_0-E_\mathrm{LO})} . \quad (8)$$

By using (3), and recalling that W_2 is independent of E_k, we have,

$$S(E_\mathrm{k}) \propto \frac{1}{W_1(E_\mathrm{k})} . \quad (9)$$

Here, we note that the Huang–Rhys factor is not a constant number, but a function of the kinetic energy of the excitons. From the measured first and second order PSB's, we obtain the Huang–Rhys factor by using (8). The result is shown in Fig. 2 as the solid curve [24]. We also calculate this factor from (6) and (9), as shown as the dashed curve in Fig. 2. Except for an unknown scaling factor, the theoretical curve coincides well with the experimental result for excitons with $E_\mathrm{k} > 3\,\mathrm{meV}$. For cold excitons, the discrepancy between theory and experiment can be attributed to the influence of disorder that may localize the excitons.

In contrast to the energy-dependent Huang–Rhys factor discussed here, early experiments usually measured the Huang–Rhys factor by using the

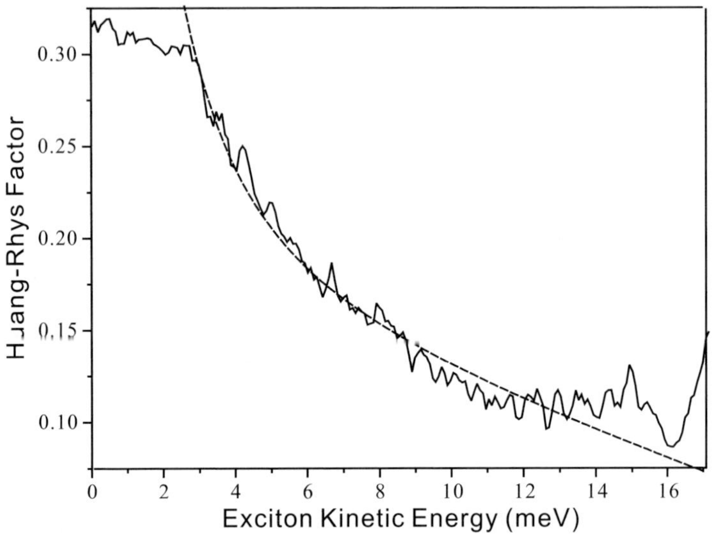

Fig. 2. Energy-dependent Huang–Rhys factor of excitons in a 7.3 nm ZnSe/ZnSSe quantum well measured by evaluating the PSB spectra (solid curve) and calculated from (6) and (9) (dashed curve).

spectrally integrated intensity of PSB. This yielded of course a spectrally integrated Huang–Rhys factor. The values of S measured in this way show a strong dependence on experimental conditions like sample temperature and excitation wavelength [25, 26]. These artificial effects can be explained by the change of exciton distribution function when one varies the experimental conditions. This change in turn alters the integration weight for the deduction of the spectrally integrated Huang–Rhys factor. We confirm experimentally that the energy-dependent Huang–Rhys factor shown in Fig. 2 is independent of sample temperature and excitation wavelength [24]. This parameter now directly describes the intrinsic properties of exciton–phonon interaction.

Based on the fact that W_2 is independent of E_k, it is reasonable to assume that the best way to determine the exciton distribution $f(E_\mathrm{k})$ is to measure directly the second order PSB. However, the second order PSB is too weak to provide a reasonable signal-to-noise ratio in temporally or spatially resolved experiments. To overcome this problem in practice, we preform temporally and spatially integrated measurement to get the PL spectrum of a particular sample with high signal-to-noise ratio. From this spectrum we deduce the energy-dependent Huang–Rhys factor like the one shown in Fig. 2. Then the temporally or spatially resolved spectroscopy can be performed focusing on the first order PSB, and the $f(E_\mathrm{k})$ can be easily and accurately obtained by using the experimentally determined Huang–Rhys factor.

Phonon-Sideband Spectroscopy of Coherent Excitons

We have shown how PSB spectroscopy can be used to deduce the energy distribution of excitons. Now we want to argue that PSB spectroscopy can even be exploited to monitor the presence of coherent excitons.

In the PSB spectrum, we observe a sharp peak at its upper energy limit (noted as HL in Fig. 1). The peak locates at $2E_{\mathrm{LO}}$ below the excitation photon energy. When we tune the laser photon energy, the peak shifts to keep this energy difference unchanged. Generally, the peak observed at an integer times E_{LO} below the excitation photon energy can be related to LO-phonon-assisted hot-exciton photoluminescence (HL) as we discussed above and/or resonant Raman scattering (RRS) [27]. The difference between these two processes is whether a real excitonic population is involved or not [28]. The HL is composed of two distinct one-photon processes, i.e., absorption followed by emission, while RRS is a single two-photon process with an intermediate state of infinitively short lifetime [13]. Recently, it has been proven that HL can be the dominant process at low temperatures in several systems, e.g., II-VI [13, 14, 29, 30], III-V [31] and group IV [32] semiconductors.

In the ZnSe quantum well samples discussed in this chapter, we confirm that the sharp peak is dominated by HL by the following experimental facts [33]. Firstly, when raising the temperature we observe a pronounced thermal broadening and quenching of this peak. This temperature behavior clearly distinguishes HL from RRS [27, 32]. Secondly, increasing the excitation photon energy by E_{LO}, the linewidth of the peak becomes ~2–3 times wider. If the peak were dominated by the RRS process, such an increase of the excitation energy corresponds to a change from second order RRS to third order RRS. In this case the linewidth should not change significantly [14]. On the other hand, since the peak is induced by the HL process, the above tuning of the excitation energy introduces an additional relaxation step by LO phonon emission before the LO-phonon-assisted recombination. The result is a broadening of the linewidth as has been described already for ZnTe quantum wells [14].

Having excluded a significant Raman contribution it is straightforward to attribute the HL peak to LO-phonon-assisted recombination of hot excitons right after their formation and before the first step of the energy relaxation, i.e., before the first acoustic phonon scattering event. Hot excitons that have already emitted or absorbed one or more acoustic phonons cannot contribute to this peak due to the involved change in energy. In this sense, the HL peak monitors specific excitonic states with a kinetic energy well defined by the excitation photon energy.

During its lifetime an exciton experiences numerous elastic and inelastic scattering events. The former ones only change the direction of the exciton momentum leaving the energy of the exciton unchanged. During an elastic scattering event, the exciton wavefunction retains its coherence. An inelastic scattering event, however, changes the exciton energy and thus the phase of its wavefunction. It destroys the coherence of the individual exciton. One can view at excitonic relaxation in a sense that the wavefunction of an individual exciton is in a coherent state between each successive inelastic scattering events. Since the HL peak monitors the excitons before the very first inelastic scattering event, i.e. acoustic phonon scattering, it reflects the presence of individually coherent excitons in that energy state. We will return to this fact in more detail in Sect. 4 where we will exploit the HL peak to deduce the spatial coherence of excitons during non-classical transport.

3 Energy Relaxation of Hot Excitons

In order to understand the temporal and spatial dynamics of excitons one needs to monitor the actual energy distribution given by the statistical distribution function and the density of states. In quasi-equilibrium the steady-state energy distribution of excitons obeys Bose–Einstein statistics

$$f_{\mathrm{BE}}(E_{\mathrm{k}}) = \{\exp[(E_{\mathrm{k}} - \mu)/k_{\mathrm{B}}T] - 1\}^{-1}, \quad (10)$$

with μ being the chemical potential of the excitons and T the temperature of the exciton system [2]. When the exciton gas is in quasi-equilibrium with the lattice, the temperature of the exciton distribution equals the lattice temperature. In quantum wells the density of states is given by a step function in energy. Within each subband the density of states is constant.

In the case of monochromatic laser excitation followed by a fast hot-exciton formation process as we discuss in the previous section, the exciton distribution in ZnSe quantum wells formed right after the excitation is very far from the quasi-equilibrium distribution f_{BE}. Instead, the distribution is a delta function (ignoring homogeneous broadening) at the energy of

$$E_{\mathrm{k}} = E_{\mathrm{exc}} - nE_{\mathrm{LO}}, \quad (11)$$

where n is an integer. Such an initial energy distribution evolves towards quasi-equilibrium via interactions among the excitons and with the lattice.

Here, we want to distinguish two processes during this evolution, thermalization and relaxation. By thermalization, we describe the process in which the excitons exchange energy within the exciton subsystem through exciton–exciton interaction. In this process, the average energy of the exciton subsystem keeps constant, but the energy distribution evolves from the delta function towards the f_{BE}. The time scale of the thermalization is determined by the rate of exciton–exciton scattering and thus by the density of the excitons. Only for exciton populations with high density, the excitons exchange their energy rapidly, and the thermalized distribution f_{BE} can be established within picosecond. However, the exciton temperature of this distribution is still different from the lattice temperature. When the sample is kept at liquid-He temperature, the exciton temperature will typically be higher. The evolution of the exciton temperature towards lattice temperature is achieved by the other process – relaxation.

In the energy-relaxation process, the excitons lose their excess energy to the lattice. This is done by phonon scattering processes, including optical phonon and acoustic phonon emissions. Thus, the time scale of the relaxation is determined by the scattering rates of these processes. In polar materials, the rate of optical phonon emission is usually orders of magnitudes higher than that of acoustic phonon emission. For example in ZnSe quantum wells, calculations proved that the LO phonon emission time is about 100 fs, while the acoustic phonon emission time is in the range of 30 ps [18]. Therefore, if the hot exciton has enough energy to emit optical phonons, it follows this efficient path of relaxation rapidly before any acoustic phonon emission event can take place. The number of optical phonons that a hot exciton emits depends on its excess energy. This sequence of relaxation is confirmed e.g. by the modelling of details of the LO phonon cascades in photoluminescence excitation spectra of ZnSe quantum wells [16]. The rest of the relaxation is then achieved by the much slower acoustic phonon emission. Considering that the average energy of the acoustic phonon emitted in each scattering event is only a few meV, one anticipates a relaxation time of several hundred picoseconds.

Experimental investigation of these thermalization and relaxation processes in conventional ZPL spectroscopy is difficult because the hot excitons reside mainly in the dark high-momentum states. One typically measures the rise time and decay time of the ZPL after a pulsed excitation [7, 13, 14, 34–40]. The interpretation of the data is not straightforward since the contributions of exciton formation, thermalization, relaxation, and recombination processes are hard to separate. In contrast to this, the time-resolved PSB spectroscopy provides an ideal way to study the energy relaxation process since the energy distribution of excitons can be directly monitored in this technique, as we discussed in Sect. 2.

In the time-resolved PSB spectroscopy discussed here, the samples of ZnSe multiple quantum wells are excited by picosecond laser pulses from a mode-locked Ti:Sapphire laser. The pulse energy is reduced to 4 pJ, corresponding to a carrier density of $10^9/\,cm^2$, to exclude the influence of exciton–exciton scattering. The sample is installed in a cryostat and cooled down to 4 K by liquid helium. The PL from the sample is spectrally dispersed by a spectrometer, then temporally resolved by a streak camera.

On the left panel of Fig. 3 we show a series of the first order PSB spectra (corresponding to the spectral portion of PSB_1 in the time-integrated spectrum shown in Fig. 1) measured at different time delays after the pulsed excitation, as indicated beside each curve [18]. By using the measured Huang–Rhys factor shown in Fig. 2, we deduce from these PSB spectra the exciton distribution function $f(E_{\mathrm{k}})$ at each time delay, as shown in the right panel of Fig. 3. In this experiment, the excitation laser photon energy is 2.8313 eV.

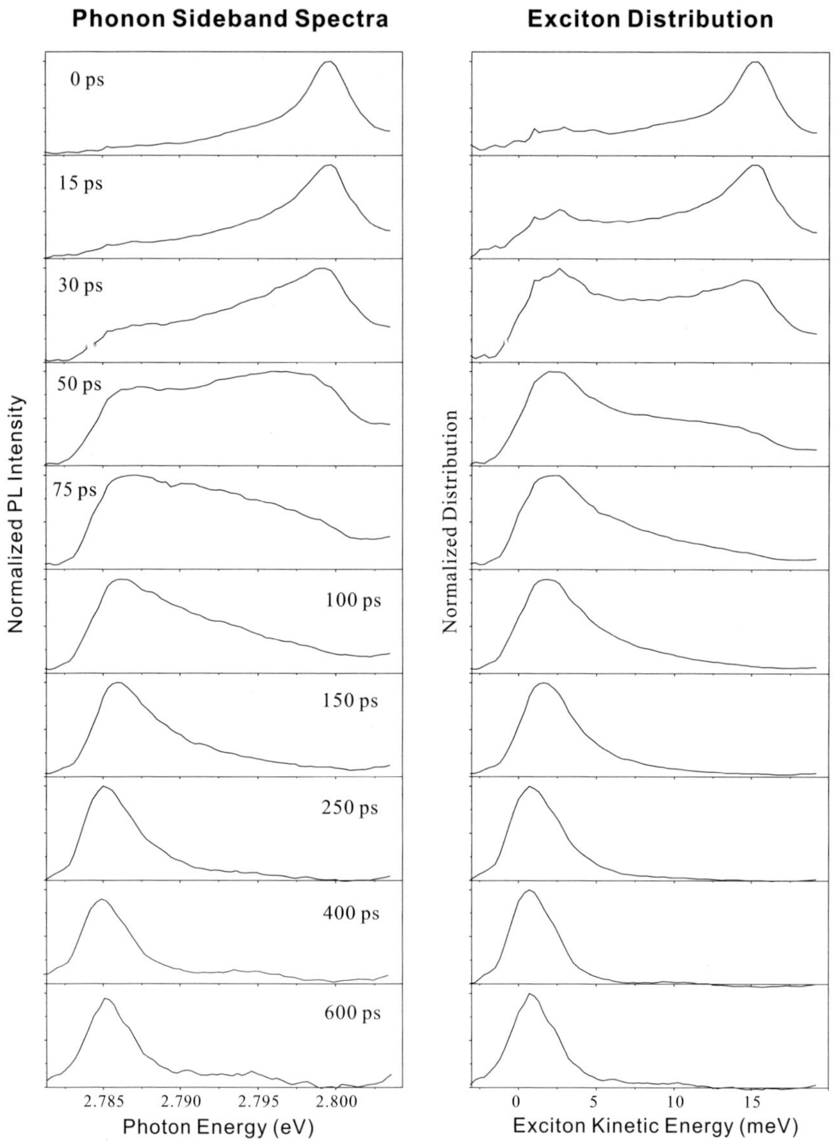

Fig. 3. Temporally resolved PSB spectroscopy on excitonic relaxation in a 7.3 nm ZnSe/ZnSSe quantum well. The left panel shows the measured PSB spectra at various time delays after pulsed laser excitation. The corresponding energy distribution functions deduced from these spectra are shown in the right panel.

This defines the initial kinetic energy of excitons to be 15 meV. In Fig. 3 we see that the initial exciton distribution is narrow and centered at 15 meV. The width of the initial distribution can be attributed to the spectral resolution of the system. The peak at 15 meV gradually disappears, showing that the excitons are driven out of the initial state on a time scale of few 10 ps. After 400 ps, the distribution function keeps unchanged, indicating the end of the relaxation process.

These experiments nicely reveal the relaxation dynamics of hot excitons. The observed timescales of the relaxation processes are well reproduced by detailed theoretical analysis of the exciton–phonon interaction [18].

4 Hot-Exciton Transport

4.1 Transport Regimes of Quantum Well Excitons

In a semiconductor quantum well, the vertical motion of excitons along the growth direction is confined by the potential barriers. Within the quantum well plane, the exciton can move freely unless it is localized by potential fluctuations [41]. This lateral transport of quantum well excitons is an important part of the overall excitonic dynamics since the hot excitons have rather high kinetic energies. But only recently this coupling of energy relaxation and transport has been studied on the relevant scales namely the timescale of exciton–phonon interaction and the length scale of quasi-ballistic transport [17, 33, 42, 43].

Generally, the excitonic transport process is determined by the interplay between the free propagation of the exciton wave packet according to its group velocity and various scattering processes. Since the former can be easily modelled, the knowledge of the influence of scattering processes on the transport is essential in understanding the transport behavior. We have already mentioned the scattering processes that a quantum well exciton encounters during its lifetime. In particular, elastic scattering processes conserve the energy of the exciton and thus do not destroy the coherence of the exciton wavefunction [44]. For excitons in a realistic semiconductor quantum well, the elastic scattering mechanisms are the scattering by interface roughness, lattice defects and impurities. The scattering by acoustic phonons, optical phonons and the inter-exciton scattering are the main inelastic scattering mechanisms.

Excitonic transport displays quite different features when one observes it on different length and time scales. The average distance the exciton travels between two successive scattering events is called mean free path. Transport processes observed on a length scale shorter than the mean free path are called ballistic transport since there is no disturbance by scattering. On a length scale larger than this ballistic transport length, the exciton undergoes scattering. However, if the scattering events are all of elastic nature, the exciton energy is conserved. This transport regime is called quasi-ballistic.

More importantly, since the elastic scattering does not change the phase of its wavefunction, the exciton stays in the regime of coherent transport. Enlarging the observation length scale further, inelastic scattering occurs, and the exciton loses the coherence during the transport. The details of the transport on such a length scale depend on the energy of the exciton. In the case that the exciton is still hot with an kinetic energy larger than the thermal energy, it relaxes the energy and decreases the velocity during the transport. This phase is called hot-exciton transport. After the completion of the energy relaxation, the exciton system reaches quasi-equilibrium in terms of energy. But the spatial inhomogeneity of the exciton distribution leads to further transport in terms of a classical diffusion that can be described by the diffusion equation with a single constant diffusivity.

To summarize, the ballistic transport occurs between two successive scattering events, no matter whether they are inelastic or elastic. Between two successive inelastic scattering events, the exciton shows quasi-ballistic transport. Both of these processes are coherent transport processes. Transport of hot excitons over a large scale is coupled with energy relaxation, while cold excitons diffuse in real space. The length and time scales of these processes depend on the rate of the involved scattering mechanisms.

4.2 Experimental Methods to Study Excitonic Transport

Excitonic transport in semiconductor quantum wells has been extensively studied during the past two decades. The main optical techniques used in this kind of studies include transient gratings, spatially resolved pump-probe and spatially resolved PL spectroscopy. In transient-grating experiments [45–51], one creates a spatially periodic density distribution of excitons by the interference of two laser pulses impinging on the sample surface with some angle between them. The decay of this density grating is then detected by the diffraction of a time-delayed pulse. Analysis shows that the decay rate of the grating can be written as

$$\tau_{\mathrm{G}} = \frac{4\pi^2 D}{\Lambda^2} + \frac{1}{\tau_{\mathrm{R}}}, \tag{12}$$

where D is the diffusivity, Λ is the grating period, and τ_{R} is the recombination rate. By measuring τ_{G}, one can deduce D since τ_{R} can be measured independently. In pump-probe experiments, an intense laser pulse (pump) is used to generate excitons. The evolution of the exciton density is then detected by measuring the absorption of a time-delayed and much weaker probe beam. In spatially resolved pump-probe experiments, the pump and probe beams are focused tightly and spatially separated. In this way the spatial distribution of excitons is monitored [52–57].

In contrast to these nonlinear techniques, experiments based on PL can be performed in the low-density regime, where the interactions between excitons are negligible. Two types of techniques have been exploited during the past decade for PL spectroscopy of exciton transport, namely time-of-flight experiments and spatially resolved PL.

In time-of-flight experiments, one excites excitons within a small spot by short laser pulses. Then one detects the decay of the PL signal from this spot which is related to the transport of the excitons out of the spot. One way to achieve the local excitation and detection is to cover the sample surface by a mask with an etched hole [58–63]. Alternatively, one can use a fiber tip with near-field coupling to the sample surface to guide the excitation laser beam and the PL signal, i.e. near-field microscopy [64–66]. By using a coated fiber tip or by reducing the etched hole to a nano-aperture, one can get high spatial resolution of 100 nm or less. However, the disadvantage of the time-of-flight method is that, the analysis of the data is rather indirect. Rather than measuring the overall spatial PL distribution, only the PL dynamics from a single spot is detected. Typically the diffusion equation is used to model the data and deduce the diffusivity and mobility. However, as we discussed in the previous subsection, the dynamics of the excitons is expected to be non-diffusive for small excitation spots.

One step towards the resolution of these difficulties is to measure directly the spatial distribution of the PL signal that reflects (to some extend) the spatial distribution of the excitons. This is done e.g. in a spatially resolved PL technique based on confocal microscopy (typically called μ-PL) [67–72]. Figure 4 shows a typical configuration of this kind of setup [73]. The laser beam is focused to a tiny spot on the sample through a microscope objective. The PL is collected by the same objective. A shiftable pinhole is installed in the image plane of the microscope to choose the detection spot, which is independent of the excitation spot. The signal is then dispersed by a spectrometer and detected by a CCD camera. By scanning the pinhole one can obtain the PL spectrum at different locations, even far outside the excitation spot. Alternatively, the pinhole and the spectrometer can be removed so that the PL image is directly imaged on a CCD camera. For time-resolved

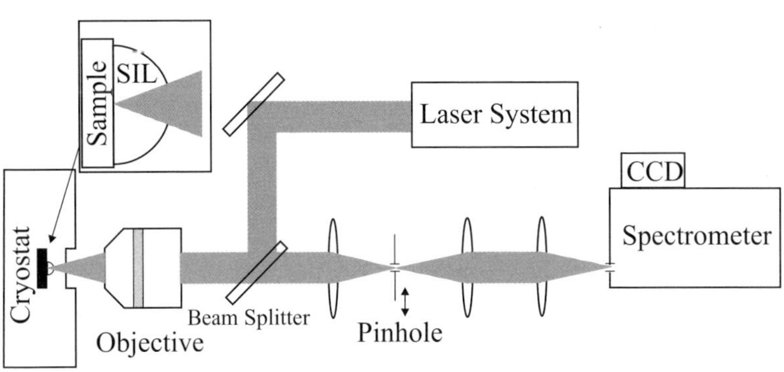

Fig. 4. Schematic drawing of a confocal μ-PL setup enhanced by a solid immersion lens.

experiments, a streak camera can be attached to the spectrometer to provide temporal resolution.

With this kind of system, the spatial distribution of PL after a spatially narrow excitation can be measured directly. However, the spatial resolution is restricted to the diffraction limit and is in the order of μm. This prevents the observation of ballistic and coherent transport processes occurring on a sub-μm length scale. To enhance the spatial resolution, we put a hemispherical solid immersion lens (SIL) on the sample surface. It modifies the numerical aperture of the microscope and thus enhances the resolution to about 200 nm in terms of the half width at half maxima (HWHM) of the laser spot [73]. Additionally, the SIL also improves the collection efficiency by about five times [73]. This feature is very important for the case of weak signals like in PSB spectroscopy. In the following, we will call this experimental arrangement nano-PL .

4.3 The Role of Initial Kinetic Energy on Hot-Exciton Transport

Using the SIL-enhanced nano-PL we study the transport properties of excitons in ZnSe quantum wells at a temperature of 7 K. We start with experimental results obtained from zero-phonon-line (ZPL) spectroscopy under continuous-wave (cw) laser excitation. In the left panel of Fig. 5 we show the profile of the PL spot measured by scanning the pinhole in the image plane of

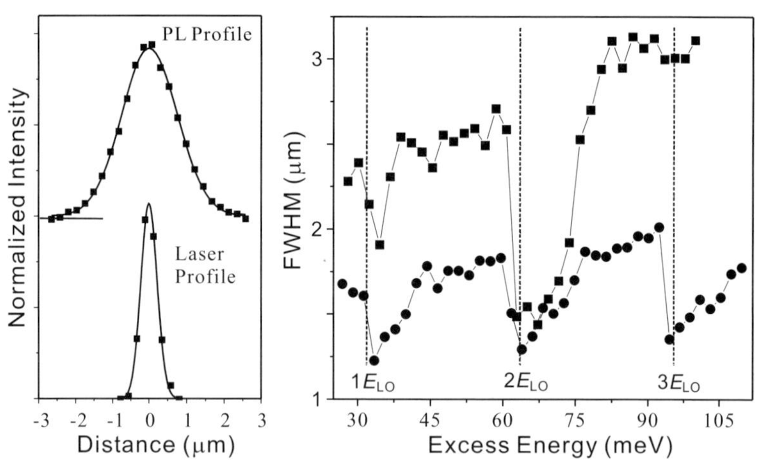

Fig. 5. Time-integrated ZPL spectroscopy of excitonic transport in ZnSe quantum wells. Left panel: Spatial profiles of the PL intensity (upper) and the laser spot (lower) measured by using the pinhole-scanning technique. Right panel: The PL-spot size (full width at half maximum, FWHM) as a function of the excitation excess energy measured for two different ZnSe quantum wells: A 5 nm ZnSe single quantum well with ZnMgSSe barriers (upper) and a 7.3 nm ZnSe/ZnSSe multiple quantum well (lower).

the microscope. The profile of the cw laser spot is also measured by the same method. The PL spot is significantly larger than the laser spot, reflecting the occurrence of excitonic transport.

To characterize the excitonic transport, we measure the average transport length, which can be deconvoluted from the half width at half maximum (HWHM) of the PL spot, as a function of the excitation excess energy. The right panel of Fig. 5 shows the results measured from two different ZnSe quantum well samples. We observe a general trend of increasing transport length with excess energy. But more importantly, we find a pronounced oscillating feature with a period equal to the LO phonon energy. Recalling the fast relaxation assisted by LO phonon emission and the slow relaxation assisted by acoustic phonon emission, we can understand this feature as follows: Increasing the excitation excess energy, the velocity of hot excitons formed by the first LO phonon emission (see Fig. 1) becomes larger. This results in the increase of the transport length. However, when the excess energy is further increased such that the hot exciton has enough energy to emit one additional LO phonon, it does so rapidly to relax the kinetic energy. Consequently, the initial velocity of excitons for the transport process is quenched.

These observations show the importance of the initial velocity on the transport process, implying that the transport even in this cw experiment is far from being dominated by classical diffusion of relaxed excitons [17]. Also, the difference between the transport lengths measured from two samples implies that the sample imperfection plays an important role in controlling the transport process.

4.4 Spatio-Temporal Exciton Dynamics

In order to reveal the dynamics of the excitons simultaneously in space and time, we perform time-resolved nano-PL (nano-TRPL) experiments. A streak camera with a CCD camera in photon-counting mode is attached to the spectrometer, providing 5 ps temporal resolution. The sample is locally excited by ps laser pulses. We have chosen the laser photon energy such that the hot excitons are formed with a kinetic energy of 9 meV. The evolution of the PL signal from various locations on the sample with different distances from the laser spot are measured by scanning the pinhole. From these time-resolved PL data we obtain the spatial distribution of the PL intensity at various delay times, shown as symbols in the left panel of Fig. 6. These profiles allow us to obtain the PL spot size (FWHM) as a function of time. If one assumes that the transport was a classical diffusion process, the diffusion equations predict that the square of the FWHM should increase linearly with time, with the slope related to the diffusivity [53]. On the right panel of Fig. 6, we plot the square of the FWHM that we obtained from the left panel as function of time (symbols). A nonlinear expansion of the spot size is clearly observed up to 300 ps which is consistent with the transport of hot excitons.

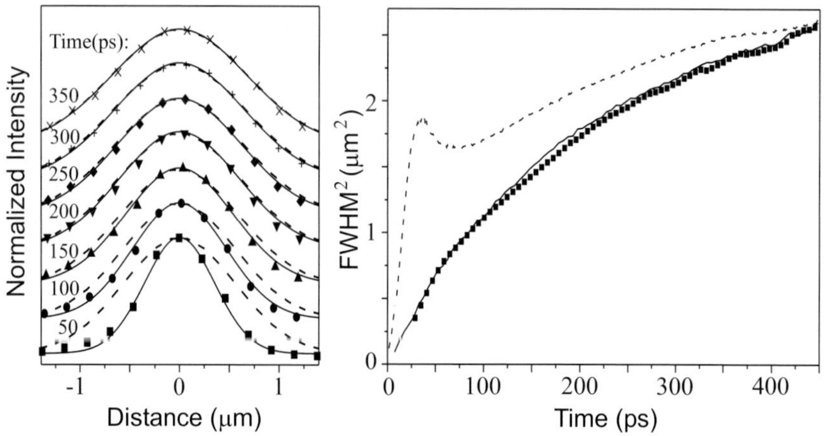

Fig. 6. Temporally resolved ZPL spectroscopy on excitonic transport in a 8 nm ZnSe/ZnMgSSe multiple quantum well. Left panel: Symbols: Spatial profiles of PL spot measured at different time delays. Solid curve: The simulated PL profiles at corresponding time delays. Dashed curve: The simulated spatial profile of the exciton density at corresponding time delays. Right panel: Symbols: The measured time-evolution of the PL spot size. Solid curve: The simulated time-evolution of the PL spot size. Dashed curve: The simulated time-evolution of the size of the spatial exciton density profile.

To reveal the details of the spatio-temporal dynamics of the hot excitons, we reproduce this dynamics including exciton formation, relaxation, real-space motion, and recombination by Monte Carlo simulation [42]. Microscopic processes including acoustic phonon absorption, acoustic phonon emission, and interface-roughness scattering are calculated explicitly. By adjusting the sole fit parameter (the correlation length of the interface roughness), the experimental data are nicely reproduced (solid curves in Fig. 6). The simulation also yields the temporal evolution of the average exciton energy relaxing from the initial 9 meV to the thermal energy. This relaxation process lasts about 300–400 ps, being consistent with the time scale of the nonlinear expansion of the PL spot [42].

The simulation also reveals that there is a pronounced difference between the spatial distribution of the PL intensity and that of the exciton population. The latter is plotted as the dashed lines in Fig. 6. For a relaxed exciton distribution, the whole ensemble is within the recombination window at the band minimum. Then the PL intensity is proportional to the exciton population. But this state is reached only after several 100 ps. However, if the exciton system is still in the relaxation process, a significant portion of the ensemble can still be in high-energy states which are dark for ZPL spectroscopy. In this situation, the PL intensity reflects only the population of the cold excitons with $E_{\mathrm{k}} \sim 0$. It definitely does not reflect the population of the whole exciton ensemble. This fact has been used to explain the delayed radiative recombination of excitons leading to PL decay times much longer that the intrinsic exciton lifetime. But unfortunately, it is often overlooked when analyzing excitonic transport. For example, the PL intensity distribution is typically used as an input to the diffusion equation rather than the exciton density. Figure 6 reveals that these two profiles can be quite different when the exciton distribution is hot which is unavoidable after non-resonant excitation.

4.5 Spatial Density Oscillations and Coherent Transport of Excitons

The most striking feature in Fig. 6 is actually the peak in the exciton density distribution at about 30 ps. It indicates a breathing-like behavior of the hot-exciton transport at early times. After a rapid expansion of the exciton density profile due to the high initial kinetic energy of the excitons, the spot size reduces to finally increase again. The Monte Carlo simulations reveal the origin of this spectacular behavior. While interface roughness scattering, which is much faster than the inelastic scattering, is dominantly directed towards small scattering angles, the emission of acoustic phonons has the largest matrix element for a backward scattering of the excitons. Starting their quasi-ballistic movement out of the excitation spot after their LO-phonon-assisted generation, a significant number of excitons reverse their direction of propagation when the first acoustic phonon is emitted after an average time of about 30 ps.

Up to now, this spatial oscillation of the exciton profile is only indirectly deduced from the simulations. But tuning the excitation laser such, that the excitons are formed quasi-resonantly (i.e. within the inhomogeneous linewidth of the ZPL) after the emission of two LO phonons, we can directly observe this oscillation in the PL spot diameter (Fig. 7) [74]. The latter now reflects directly the exciton density distribution. Here the excitons still have enough kinetic energy (about 1 meV) to be able to emit an acoustic phonon. The experimental data are nicely reproduced by the simulations.

Since the first inelastic scattering event marks the end of the quasi-ballistic or coherent transport phase, we can determine from our maximum in Fig. 7 simultaneously the coherence length and time of the excitons. The deduced values of 800 nm (deconvoluted from the half width at half maximum of the PL spot) and 29 ps are consistent with independent measurements using cw nano-PL (see next subsection) and four-wave mixing [6], respectively. These quantities are of crucial importance for the assessment of excitons in semiconductors for possible applications like coherent information processing exploiting the spatial part of excitonic wavefunction.

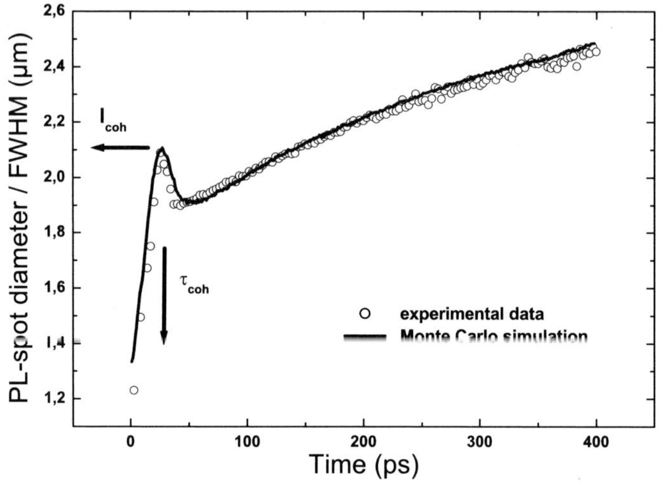

Fig. 7. Simultaneous determination of coherence length l_{coh} and coherence time τ_{coh} from the oscillation in the PL profile under quasi-resonant excitation. The symbols are the experimental data points from a 8 nm ZnSe/ZnMgSSe multiple quantum well, the solid line is calculated from Monte Carlo simulations.

4.6 Phonon-Sideband Spectroscopy of Excitonic Transport

Except for quasi-resonant excitation, the experiments of ZPL spectroscopy do not directly reflect the excitonic density distribution. But this is overcome by combining the nano-PL technique with the PSB spectroscopy which we discussed in Sect. 2.3.

In these experiments, the ZnSe quantum wells are excited by a cw laser beam that is focused to a spot of about half a µm FWHM. The size of the detection spot is 460 nm, defined by the size of the pinhole that is installed in the image plane of the microscope. By scanning the pinhole, we detect the PSB spectra at various distances from the excitation spot. On the left panel of Fig. 8, we show a series of the measured PSB spectra. Increasing the distance, we see from the top to the bottom a systematic change of the spectral shape of the PSB's. This indicates the occurrence of energy relaxation during the excitonic transport. From these PSB spectra, we obtain the corresponding exciton distribution functions by the same method as we used in the temporally resolved PSB spectroscopy discussed in Sect. 3. These are plotted on the right panel of Fig. 8. The spatial evolution of the energy distribution of excitons are clearly resolved. We note that over a transport length of about 1.5 µm, there is still a significant population of excitons on high-energy states [43].

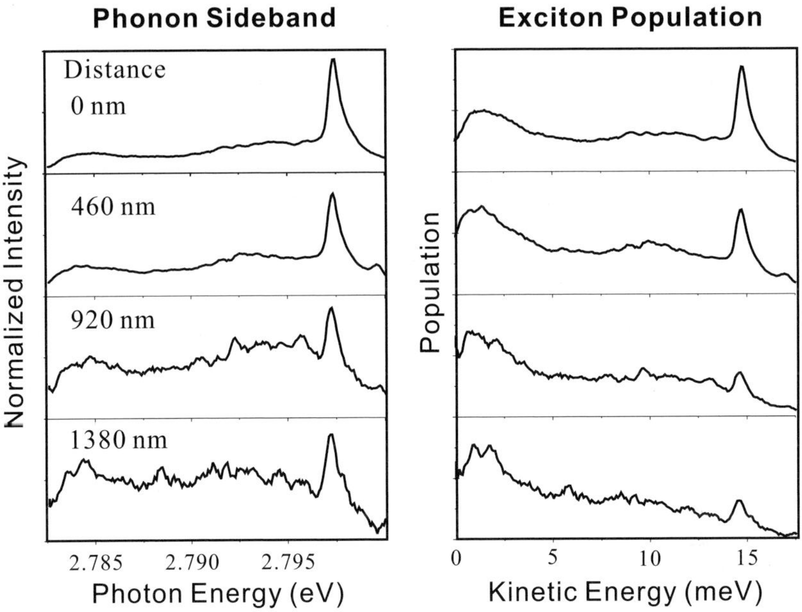

Fig. 8. Spatially resolved PSB spectroscopy on excitonic transport in a 7.3 nm ZnSe/ZnSSe multiple quantum well. Left panel: The spectra of PSB measured at different positions on the sample. The distances from each location to the excitation laser spot are indicated by the numbers in the figure. Right panel: The corresponding energy distribution functions deduced from these PSB spectra by using the Huang–Rhys factor.

In order to get a quantitative description of this coupled relaxation and transport process, we calculate the average energy of excitons from the deduced exciton distribution function, and plot it as a function of the excitation–detection distance in Fig. 9. The data-sets correspond to different values of the excitation excess energy, thus of the initial kinetic energy of excitons. We find that when the excitons travel out of the excitation spot, their average energy is reduced. However, over a distance of about 1.5 μm, the reduction is only 20%. This implies that there is still a long way for the excitons to go before they finish the relaxation so that the classical diffusion can take place.

By comparing the slope of different curves, we note that the spatial energy relaxation rate is slower for excitons with less kinetic energy. This is consistent with the fact that the rate of acoustic phonon emission increases with kinetic energy. From these two features, we can conclude that on the length scale of at least several micrometers, the transport is coupled to the energy relaxation. Actually the excitons never reach the diffusive transport regime during their lifetime when the initial excess energy is sufficiently high.

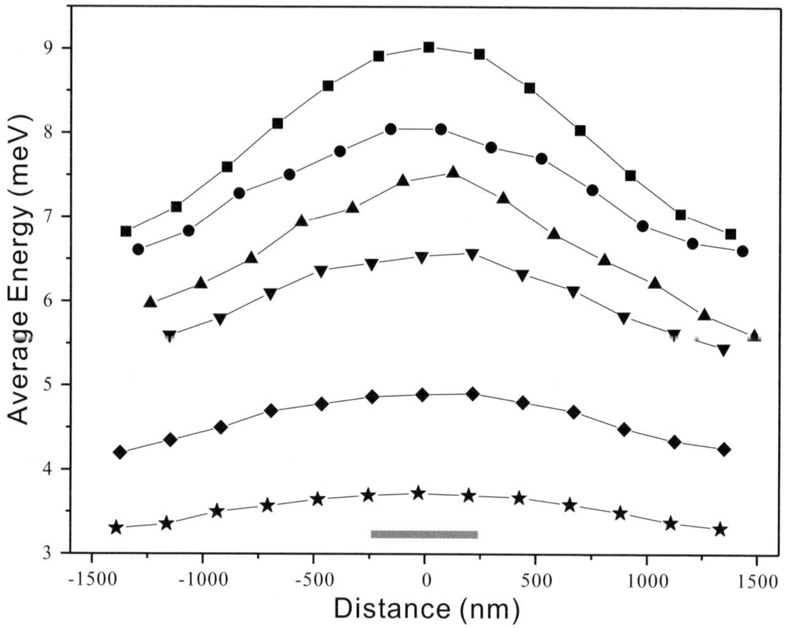

Fig. 9. The average kinetic energy of excitons as a function of the distance between the excitation and the detection spots. The initial kinetic energies of the excitons are (from top to bottom) 20.7, 17.6, 14.3, 11.0, 7.9, and 4.8 meV, respectively. The bar above the horizontal axis indicates the FWHM of the excitation laser spot.

We now want to zoom in again to a smaller length scale, where quasi-ballistic transport has been observed in the time-resolved nano-PL. We can deduce the coherence length of the excitonic wavefunction during transport in real space in an alternative way.

When we discussed the PSB spectroscopy in Sect. 2, we have clarified that the sharp peak at the upper limit of the PSB (HL in Fig. 1) monitors the presence of coherent excitons. Correspondingly, the spatial distribution of that sharp peak monitors the spatial propagation of the coherent exciton wavefunction before decoherence. By PSB nano-PL, we measure the spatial profile of the sharp peak, and obtain the coherence length of excitons defined as the HWHM of that profile after de-convolution of the finite excitation and detection spot sizes in the experiments.

Figure 10 shows the deduced coherence length as a function of exciton kinetic energy [33]. The coherence length is in the range of 300 to 400 nm, decreasing slowly with the kinetic energy. At first sight, one would anticipate an increase of the coherence length with kinetic energy due to the increased group velocity. But in fact, the scattering time of acoustic phonon emission, which is here identical to the coherence time, decreases with exciton kinetic energy. The observed slow decrease of coherence length can be interpreted

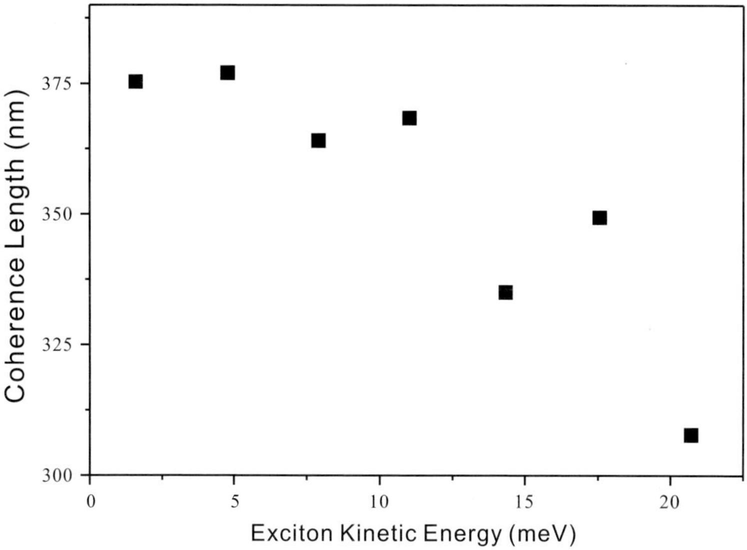

Fig. 10. Coherence length of excitons as a function of kinetic energy in a 7.3 nm ZnSe/ZnSSe multiple quantum well.

as the compensation of these two opposite trends of velocity and scattering time.

Since elastic scattering has no influence on the HL peak, the present technique is not sensitive to elastic scattering events. Therefore, we are not able to experimentally distinguish the ballistic and quasi-ballistic transport regimes. Nevertheless, we can interpret the measured coherence length as a quasi-ballistic rather than a ballistic transport length, since the scattering rate of interface roughness is typically much larger than that of the acoustic phonon emission in real quantum wells. Indeed, Monte Carlo simulations show a ballistic transport length of about 200 nm [42], being smaller than the measured coherence length. Further experimental separation of the ballistic and quasi-ballistic processes is quite challenging since it will require not only a better spatial resolution but also a way to monitor elastic scattering events.

5 Conclusions

The model system of ZnSe-based quantum wells has been used to study the spatio-temporal dynamics of hot excitons. The photoluminescence-type experiments exploit the strong polar coupling of excitons to LO phonons which ensures that excitons form on a subpicosecond timescale with a well defined initial kinetic energy. The coupling also enables a direct observation of the excitonic population within its parabolic dispersion. The thermalization and relaxation dynamics of the hot excitons can be monitored as a function of time. Non-classical transport processes of excitons are identified by nano-PL on the length scale of the light wavelength and the time scale of inelastic scattering with phonons. In particular, a periodic quenching of the exciton transport length related to the LO phonon cascade in excitation spectroscopy is found. A breathing-like spatial oscillation of the exciton density is observed and modelled by Monte Carlo simulations. The latter features prove that, unlike in GaAs or GaInAs, excitonic properties are observed in the photoluminescence experiments on ZnSe. We further illustrated the mutual coupling of energy relaxation and spatial propagation and we determined for the first time the coherence length and time of excitons in one single experiment.

Acknowledgments

The results reviewed in this chapter have been achieved by a fruitful collaboration with numerous students and colleagues in Karlsruhe. Without trying to be complete, we acknowledge valuable contributions by B. Dal Don, S. Moehl, G. Schwartz, M. Umlauff, D. Lüerßen, W. Langbein, and A. Hepting. We also acknowledge the groups of M. Heuken (Aachen) and of D. Hommel (Bremen) for providing the high-quality ZnSe quantum well samples. Financial support came from the Deutsche Forschungsgemeinschaft, partly within the Center for Functional Nanostructures in Karlsruhe. Finally, we want to thank Claus Klingshirn for his continuous support of our research activities, for countless inspiring discussions, and last but not least for introducing one of us (H.K.) into the exciting field of semiconductor optics.

References

1. C. Klingshirn, H. Haug: Phys. Rev. **70**, 315 (1981)
2. C.F. Klingshirn: *Semiconductor Optics*, 2nd corr. printing (Springer, Berlin Heidelberg New York 1997)
3. M. Kira, W. Hoyer, T. Stroucken, S.W. Koch: Phys. Rev. Lett. **87**, 176401 (2001)
4. R.A. Kaindl, M.A. Carnaban, D. Hägele, R. Lövenich, D.S. Chemla: Nature **423**, 734 (2003)
5. See e.g. J. Gutowski, H.G. Breunig, T. Voss: *Dynamics of Excitons and Exciton Complexes in Wide-Gap Semiconductors*, this book
6. S. Wachter, M. Maute, H. Kalt, I. Galbraith: Phys. Rev. B **65**, 205314 (2002)
7. T.C. Damen, J. Shah, D.Y. Oberli, D.S. Chemla, J.E. Cunningham, J.M. Kuo: Phys. Rev. B **42**, 7434 (1990)
8. P.W.M. Blom, P.J. van Hall, C. Smit, J.P. Cuypers, J.H. Wolter: Phys. Rev. Lett. **71**, 3878 (1993)
9. M. Hurioli, P. Borri, M. Colocci, M. Gulia, F. Rossi, E. Molinari, P. Selbmann, P. Lugli: Phys. Rev. B. **58**, 13403 (1998)
10. M. Gulia, F. Rossi, E. Molinari, P. Selbmann, P. Lugli: Phys. Rev. B. **55**, 16049 (1997)

11. I.-K. Oh, J. Singh, A. Thilagam, A.S. Vengurlekar: Phys. Rev. B. **62**, 2045 (2001)
12. K. Siantidis, V.M. Axt, T. Kuhn: Phys. Rev. B. **65**, 035303 (2001)
13. R.P. Stanley, J. Hegarty, R. Fischer, J. Feldmann, E.O. Göbel, R.D. Feldman, R.F. Austin: Phys. Rev. Lett. **67**, 128 (1991)
14. N. Pelekanos, J. Ding, Q. Fu, A.V. Nurmikko, S.M. Durbin, M. Kobayashi, R.L. Hunshor: Phys. Rev. B **43**, 9354 (1991)
15. J.H. Collet, H. Kalt, L.S. Dang, J. Cibert, K. Saminadayar, S. Tatarenko: Phys. Rev. B **43**, 6843 (1991)
16. H. Kalt, M. Umlauff, J. Hoffmann, W. Langbein, J.M. Hvam, M. Scholl, J. Söllner, M. Heuken, B. Jobst, D. Hommel: J. Cryst. Growth. **184/185**, 795 (1998)
17. H. Zhao, S. Moehl, S. Wachter, H. Kalt: Appl. Phys. Lett. **80**, 1391 (2002)
18. M. Umlauff, J. Hoffmann, H. Kalt, W. Langbein, J.M. Hvam, M. Scholl, J. Söllner, M. Heuken, B. Jobst, D. Hommel: Phys. Rev. B **57**, 1390 (1998)
19. D. Steinbach, G. Kocherscheidt, M.U. Wehner, H. Kalt, M. Wegener, K. Ohkawa, D. Hommel, V.M. Axt: Phys. Rev. B **60**, 12079 (1999)
20. J. Cerne, J. Kono, M.S. Sherwin, M. Sundaram, A.C. Gossard, G.E.W. Bauer: Phys. Rev. Lett. **77**, 1131 (1996)
21. S. Permogorov: Optical emission due to exciton scattering by LO phonons in semiconductors. In: *Excitons*, ed by E.I. Rashba, M.D. Sturge (North-Holland, Amsterdam New York Oxford 1982) pp 177–204
22. X.B. Zhang, T. Taliercio, S. Kolliakos, P. Lefebvre: J. Phys.: Condens. Matter **13**, 7053 (2001)
23. K. Huang, A. Rhys: Proc. R. Soc. London, Ser. A **204**, 406 (1950)
24. H. Zhao, H. Kalt: Phys. Rev. B **68**, 125309 (2003)
25. M. Wojdak, A. Wysmolek, K. Pakula, J.M. Baranowski: Phys. Status Solidi B **216**, 95 (1999)
26. W. Liu, M.F. Li, S.J. Xu, K. Uchida, K. Matsumoto: Semicond. Sci. Technol. **13**, 769 (1998)
27. S. Permogorov: Phys. Status Solidi B **48**, 9 (1975)
28. Y.R. Shen: Phys. Rev. B **9**, 622 (1974)
29. D. Some, A.V. Nurmikko: Phys. Rev. B **48**, 4418 (1993)
30. S.S. Prabhu, A.S. Vengurlekar, J. Shah: Phys. Rev. B **53**, 10465 (1996)
31. R. Nötzel, L. Däweritz, N.N. Ledentsov, K. Ploog: Appl. Phys. Lett. **60**, 1615 (1992)
32. I.G. Ivanov, T. Egilsson, A. Henry, B. Monemar, E. Janzén: Phys. Rev. B **64**, 085203 (2001)
33. H. Zhao, S. Moehl, H. Kalt: Phys. Rev. Lett. **87**, 097401 (2002)
34. H. Kalt, J. Collet, S.D. Baranovskii, R. Saleh, P. Thomas, L.S. Dang, J. Cibert: Phys. Rev. B **45**, 4253 (1992)
35. F. Yang, G.R. Hayes, R.T. Phillips, K.P. O'Donnell: Phys. Rev. B **53**, 1697 (1996)
36. Z.L. Yuan, Z.Y. Xu, W. Ge, J.Z. Xu, B.Z. Zheng: J. Appl. Phys. **79**, 424 (1996)
37. J. Kusano, Y. Segawa, Y. Aoyagi, S. Namba, H. Okamoto: Phys. Rev. B **40**, 1685 (1989)
38. Y. Masumoto, S. Shionoya, H. Kawaguchi: Phys. Rev. B **29**, 2324 (1984)
39. A. Schülzgen, F. Kreller, F. Henneberger, M. Lowisch, J. Puls: J. Cryst. Growth **138**, 575 (1994)

40. U. Neukirch, D. Weckendrup, J. Gutowski, D. Hommel, G. Landwehr: J. Cryst. Growth **138**, 861 (1994)
41. See e.g. W. Langbein, G. Kocherscheidt, R. Zimmermann: *Probing Localized Excitons by Speckle Analysis of Resonant Light Scattering*, this book
42. H. Zhao, B. Dal Don, S. Moehl, H. Kalt: Phys. Rev. B **67**, 035306 (2003)
43. H. Zhao, S. Moehl, H. Kalt: Appl. Phys. Lett. **81**, 2794 (2002)
44. V.V. Mitin, V.A. Kochelap, M.A. Stroscio: *Quantum heterostructures* (Cambridge University Press, UK 1999) pp 18–20
45. J. Hegarty, L. Goldner, M.D. Sturge: Phys. Rev. B **30**, 7346 (1984)
46. K. Hattori, T. Mori, H. Okamoto, Y. Hamakawa: Appl. Phys. Lett. **51**, 1259 (1987)
47. H. Schwab, K.-H. Pantke, J.M. Hvam, C. Klingshirn: Phys. Rev. B **46**, 7528 (1992)
48. J. Erland, B.S. Razbirin, K.-H. Pantke, V.G. Lyssenko, J.M. Hvam: Phys. Rev. B **47**, 3582 (1993)
49. D. Oberhauser, K.-H. Pantke, J.M. Hvam, G. Weimann, C. Klingshirn: Phys. Rev. B **47**, 6827 (1993)
50. V. Mizeikis, V.G. Lyssenko, J. Erland, J.M. Hvam: Phys. Rev. B **51**, 16651 (1995)
51. A.C. Schaefer, J. Erland, D.G. Steel: Phys. Rev. B **54**, 11046 (1996)
52. F.A. Majumder, H.-E. Swoboda, K. Kempf, C. Klingshirn: Phys. Rev. B **32**, 2407 (1985)
53. L.M. Smith, D.R. Wake, J.P. Wolfe, D. Levi, M.V. Klein, J. Klem, T. Henderson, H. Morkoç: Phys. Rev. B **38**, 5788 (1988)
54. L.M. Smith, J.S. Preston, J.P. Wolfe, D.R. Wake, J. Klem, T. Henderson, H. Morkoç: Phys. Rev. B **39**, 1862 (1989)
55. H.W. Yoon, D.R. Wake, J.P. Wolfe, H. Morkoç: Phys. Rev. B **46**, 13461 (1992)
56. M. Achermann, B.A. Nechay, F. Morier-Genoud, A. Schertel, U. Siegner, U. Keller: Phys. Rev. B **60**, 2101 (1999)
57. B.A. Nechay, U. Siegner, F. Morier-Genoud, A. Schertel, U. Keller: Appl. Phys. Lett. **74**, 61 (1999)
58. H. Hillmer, S. Hansmann, A. Forchel, M. Morohashi, E. Lopez: Appl. Phys. Lett. **53**, 1973 (1988)
59. H. Hillmer, A. Forchel, S. Hansmann, M. Morohashi, E. Lopez, H.P. Meier, K. Ploog: Phys. Rev. B **39**, 10901 (1989)
60. H. Hillmer, A. Forchel, R. Sauer, C.W. Tu: Phys. Rev. B **42**, 3220 (1990)
61. H. Hillmer, A. Forchel, C.W. Tu: Phys. Rev. B **45**, 1240 (1992)
62. Y. Takahashi, K. Muraki, S. Fukatsu, S.S. Kano, Y. Shiraki, R. Ito: Jpn. J. Appl. Phys. **32**, 5586 (1993)
63. Y. Takahashi, K. Muraki, S. Fukatsu, S.S. Kano, Y. Shiraki, R. Ito: Solid State Comm. **88**, 677 (1993)
64. G. von Freymann, D. Lüerßen, C. Rabenstein et al: Appl. Phys. Lett. **76**, 203 (2000)
65. H. Akiyama, T. Matsusue, H. Sakaki: Phys. Rev. B **49**, 14523 (1994)
66. A. Vertikov, I. Ozden, A. Nurmikko: Appl. Phys. Lett. **74**, 850 (1999)
67. A.F.G. Monte, S.W. da Silva, J.M.R. Cruz, P.C. Morais, H.M. Cox: J. Appl. Phys. **85**, 2866 (1999)
68. A.F.G. Monte, S.W. da Silva, J.M.R. Cruz, P.C. Morais, A.S. Chaves: Phys. Rev. B **62**, 6924 (2000)

69. L.-L. Chao, G.S. Gargill–III, E. Snoeks, T. Marshall, J. Petruzzello, M. Pashle: Appl. Phys. Lett. **74**, 741 (1999)
70. F.P. Logue, D.T. Fewer, S.J. Hewlett et al: J. Appl. Phys. **81**, 536 (1997)
71. A. Vertikov, I. Ozden, A.V. Nurmikko: J. Appl. Phys. **86**, 4697 (1999)
72. V. Malyarchuk, J.W. Tomm, V. Talalaev, C. Lienau, F. Rinner, M. Baeumler: Appl. Phys. Lett. **81**, 346 (2002)
73. S. Moehl, H. Zhao, B. Dal Don, S. Wachter, H. Kalt: J. Appl. Phys. **93**, 6265 (2003)
74. B. Dal Don, H. Zhao, G. Schwartz, H.Kalt: Phys. Stat. Sol. (c) **1**, 462 (2004)

Probing Localized Excitons by Speckle Analysis of Resonant Light Scattering

Wolfgang Langbein, Gerrit Kocherscheidt, and Roland Zimmermann

1 Introduction

The determination of microscopic optical properties of an ensemble of localized excitons in a solid-state environment using speckle analysis of their resonant light scattering is reviewed. Time-resolved and spectrally resolved variants of the speckle analysis technique are discussed, and the properties of localized excitons in disordered semiconductor quantum wells are investigated. The results cover spin relaxation, radiative coupling, acoustic phonon interaction, and exciton level statistics. The theoretical description is based on density-matrix theory in terms of the disorder eigenstates for the exciton center-of-mass motion. Kinetic equations are derived and solved numerically for time-resolved as well as frequency-resolved spectra with speckles. A key role is played by intensity-intensity correlation functions which give the average speckle shape and allow to extract information on the underlying microscopic properties.

2 The Concepts of Speckle Analysis of Resonant Emission

The speckle analysis is a well established optical technique to investigate static and dynamic disorder of media. In the far field, the light scattering shows an irregular pattern of bright and dark spots which are called speckle. In the transparent (off-resonant) regime, spatiotemporal inhomogeneities of the refractive index are probed, while incoherent scattering involving real populations of intermediate electronic states is negligible. In contrast, for resonant emission, the contribution of incoherent scattering is important, and can be even dominate the total emission of the system. Since only the coherent part of the emission creates speckle, the speckle contrast is a convenient measure of the coherence degree in the emission. Then, the ultrafast dynamics of the speckle contrast gives a direct measure of the conversion from coherent to incoherent emission, involving in the present case coherent or incoherent exciton densities. Furthermore, using temporal and spectral speckle correlations, more information on the underlying microscopic properties can be extracted.

In order to describe the concept of speckle analysis we consider first a simplified model of the solid in terms of local optical resonances with running index j. The local polarization in response to an impulsive optical excitation of unit area at $t = 0$ is written as

$$P_j(t) = f_j \, \mathrm{e}^{(\mathrm{i}\omega_j - \gamma_j)t} \tag{1}$$

and contains the transition frequency ω_j and the corresponding oscillator strength f_j. In assuming an exponentially damped oscillator with damping rate γ_j we have already accounted for the experimental situation of averaging over a large number of individual pulses, each finding a different phase pattern of the environment (thermal bath). The polarizations of all the individual oscillators add up to a total polarization in the far field,

$$P(\boldsymbol{k}, t) = \sum_j \mathrm{e}^{-\mathrm{i}\boldsymbol{k}\boldsymbol{R}_j} P_j(t) \,, \tag{2}$$

where $\boldsymbol{R}_j$ denotes the spatial position of oscillator j. Here, the spatial extension of the oscillator is assumed to be small compared to the wavelength of light. Measuring on a macroscopic sample means that the directional dependence on $\boldsymbol{k}$ cannot be resolved, and we are left with an ensemble average of the individual dynamics,

$$P_e(t) = \langle P(\boldsymbol{k}, t) \rangle_e = \langle f_j \, \mathrm{e}^{-\mathrm{i}\boldsymbol{k}\boldsymbol{R}_j + \mathrm{i}\omega_j t - \gamma_j t} \rangle_e \,. \tag{3}$$

Now, the dynamics can show a quite different behavior compared to the individual (microscopic) one. For instance, the macroscopic ensemble is characterized by a probability distribution (optical density)

$$D(\omega) = \langle f_j \, \delta(\omega - \omega_j) \rangle_e \tag{4}$$

whose width gives the inhomogeneous broadening. The macroscopic polarization decays just on a timescale inversely proportional to this inhomogeneous width, which can completely obscure the often slower dephasing decay (scale $1/\gamma_j$). Therefore, in the presence of inhomogeneous broadening, properties of $P_j(t)$ cannot be extracted from a linear optical experiment. To overcome this limitation, nonlinear optical experiments measuring the third-order susceptibility have been employed in the past, such as spectral hole burning in frequency domain or photon echo experiments in time-domain [1, 2]. Under some assumptions about the non-linearity present in the system, the dephasing dynamics can be extracted. However, to get a sizable non-linear signal, the excitation intensity has to be rather high, so that interactions between the local oscillators can have an unwanted influence on the measured dynamics.

The resonant speckle analysis is a linear optical technique to measure the microscopic intensity and polarization dynamics. It takes advantage of the random but fixed spatial distribution of the individual oscillators in the solid, allowing the coherently emitted light of the oscillators to create an interference pattern, the speckles. To perform time-resolved speckle analysis, one measures the directionally resolved intensity $I(\boldsymbol{k}, t)$,

$$I(\boldsymbol{k}, t) = \left| \sum_j P_j(t) \mathrm{e}^{-\mathrm{i}\boldsymbol{k}\boldsymbol{R}_j} \right|^2 = \sum_{j,l} \langle P_j(t) P_l^*(t) \rangle_q \, \mathrm{e}^{-\mathrm{i}\boldsymbol{k}(\boldsymbol{R}_j - \boldsymbol{R}_l)} \,. \tag{5}$$

Here, it is important to introduce the average over subsequent pulses explicitly, which is equivalent to the quantum mechanical average $\langle .. \rangle_q$ over the bath degrees of freedom. As shown in more detail below for the excitonic oscillators interacting with thermal acoustic phonons, different oscillators behave still independently, $\langle P_j(t) P_l^*(t) \rangle_q = \langle P_j(t) \rangle_q \langle P_l^*(t) \rangle_q$ for $j \neq l$, and we get

$$I(\boldsymbol{k}, t) = \underbrace{\sum_{j,l} \langle P_j(t) \rangle_q \langle P_l^*(t) \rangle_q \mathrm{e}^{\mathrm{i}\boldsymbol{k}(\boldsymbol{R}_j - \boldsymbol{R}_l)}}_{I_{\mathrm{coh}}(\boldsymbol{k},t)} + \underbrace{\sum_j \left[\langle P_j(t) P_j^*(t) \rangle_q - |\langle P_j(t) \rangle_q|^2 \right]}_{I_{\mathrm{inc}}(t)} \,. \tag{6}$$

The first part is given directly by the square of the q-averaged polarization $P(\boldsymbol{k}, t)$ and is therefore called coherent intensity I_{coh}, it shows the directional dependence (speckling). The second part is independent of the emission direction, and represents the difference between the total emitted intensity and the coherent one. It is therefore called incoherent intensity $I_{\mathrm{inc}}(t)$.

The coherent polarization $P(\boldsymbol{k}, t)$ as given in Eq. (2) is a sum of many complex numbers with random phases, which results in independent Gauss distributions for both the real and imaginary part. The coherent intensity, $I_{\mathrm{coh}}(\boldsymbol{k}, t) \propto (\Re P)^2 + (\Im P)^2$, exhibits therefore an exponential distribution of intensity values I,

$$\mathcal{P}(I) = \frac{1}{\langle I_{\mathrm{coh}} \rangle_{\boldsymbol{k}}} \mathrm{e}^{-I/\langle I_{\mathrm{coh}} \rangle_{\boldsymbol{k}}} \,. \tag{7}$$

The speckle average is performed by a directional average over the speckle pattern, denoted by $\langle .. \rangle_{\boldsymbol{k}}$. To quantify the coherence in the total emission, the *coherence degree* c is introduced as the ratio of the speckle-averaged coherent to total emitted intensity,

$$c = \frac{\langle I_{\mathrm{coh}} \rangle_{\boldsymbol{k}}}{\langle I \rangle_{\boldsymbol{k}}} \,, \tag{8}$$

dropping for the moment the time variable. While I_{coh} has a contrast c of unity, the incoherent light shows no random intensity fluctuations over the speckle ensemble, and thus constitutes only an intensity offset. A unique feature of the exponential distribution Eq. (7) is that its second moment is twice as large as the first one (i.e. the average value). Thus, the coherence degree can be directly extracted from the fluctuations of the total emission intensity,

$$c = \frac{\sqrt{\langle (I_{\rm coh} - \langle I_{\rm coh}\rangle_{\boldsymbol{k}})^2\rangle_{\boldsymbol{k}}}}{\langle I\rangle_{\boldsymbol{k}}} = \sqrt{\frac{\langle I^2\rangle_{\boldsymbol{k}}}{\langle I\rangle^2_{\boldsymbol{k}}} - 1}\,. \tag{9}$$

This expression has to be modified due to the finite directional and/or temporal resolution in the measurement. A simple argument considers the resolution as an averaging over $N_{\rm eff}$ individual speckles. Then, the contrast of the measured speckle pattern is reduced to $1/\sqrt{N_{\rm eff}}$, and the coherence degree of the emission is determined by

$$c = \sqrt{N_{\rm eff}\left(\frac{\langle I^2\rangle_{\boldsymbol{k}}}{\langle I\rangle^2_{\boldsymbol{k}}} - 1\right)}\,. \tag{10}$$

Knowing $N_{\rm eff}$, one can still determine the coherence degree of the emission using the directional variance of the measured intensity. The displaced exponential distribution changes into a more rounded one, approaching a Gauss distribution at poor resolution. Looking more careful, the experimental resolution does not refer to a certain number of independent speckles but rather to a given averaging window (in direction or time). A corresponding functional analysis [3] revealed that the distribution shape for given $N_{\rm eff}$ changes with increasing dimensionality of averaging.

An example for a measured and calculated speckle intensity distribution is given in Fig. 1. The displacement due to the incoherent emission, and the rounded exponential decay due to $N_{\rm eff}$ are obvious. Still, the evaluation of the variance is not complicated, and gives e.g. for the case of Gauss-type

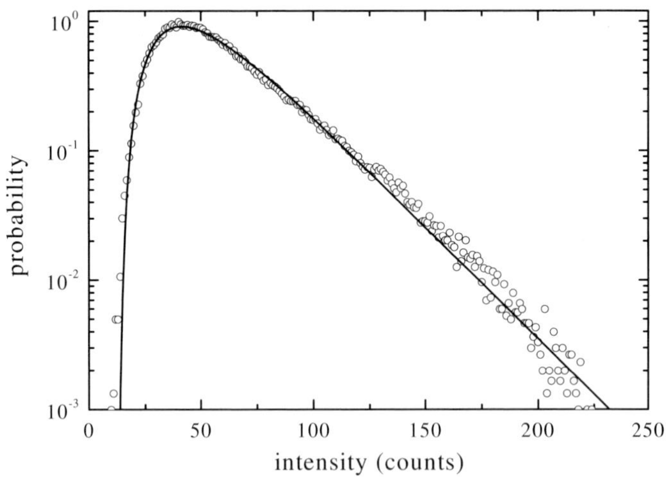

Fig. 1. Intensity probability distribution of the emission from a 9 nm GaAs SQW 8 ps after excitation at T=5 K. The calculated distribution (line) corresponds to a one-dimensional Gaussian averaging [3] with $N_{\rm eff} = 2.9$ and a coherence degree of $c = 0.83$.

temporal resolution with variance t_0 a correction in the coherence degree Eq. (10) with $N_{\text{eff}} = \sqrt{1 + 4t_0^2\sigma_0^2}$. Here, σ_0 is the variance of the transition frequency distribution ω_j.

Note that the formation of the directional speckle pattern only implies the mutual temporal coherence of the emission within the excited sample region, and not necessarily the temporal coherence with respect to the excitation. However, if one assumes independent dephasing mechanisms at different positions within the excited region, which is typically a good approximation in the study of semiconductor nanostructures, the mutual coherence implies the coherence with respect to the excitation, too.

The emission coherence, either time or frequency resolved, is a result of the speckle analysis relying only on one speckle pattern. More insight can be gained if the correlation of speckle patterns occurring for different values of experimental parameters is considered. These parameters can be the emission time, the emission frequency, the emission polarization, or additional parameters like external magnetic or electric fields applied to the sample. The general expression for the speckle correlation between patterns A and B reads

$$C_{A,B} = \frac{\langle I_A I_B \rangle_{\boldsymbol{k}}}{\langle I_A \rangle_{\boldsymbol{k}} \langle I_B \rangle_{\boldsymbol{k}}} - 1 \,. \tag{11}$$

Obviously, the coherence degree of a given pattern is $c^2 = C_{A,A}$.

3 Theoretical Description of Localized Excitons

For the investigation of speckles in the resonant emission from semiconductor quantum wells (QW), the excitons play the role of local oscillators. The disorder which results in localization comes from interface roughness and/or alloy mixing in ternary compounds. We write the exciton Schrödinger equation for the semiconductor two-band model in effective mass approximation,

$$\left(-\frac{\hbar^2}{2m_e}\Delta_{\boldsymbol{r}_e} - \frac{\hbar^2}{2m_h}\Delta_{\boldsymbol{r}_h} - \frac{e^2/\epsilon_0}{|\boldsymbol{r}_e - \boldsymbol{r}_h|} + W_e(\boldsymbol{r}_e) + W_h(\boldsymbol{r}_h) - \epsilon_j\right)\Psi_j(\boldsymbol{r}_e, \boldsymbol{r}_h) = 0 \,. \tag{12}$$

Apart from the standard kinetic term and the Coulomb attraction, we have introduced the confinement potentials $W_a(\boldsymbol{r}_a)$ ($a = e, h$) which describe the spatial variation of the local band edges. The z-axis is taken along the growth direction. If interface roughness dominates the disorder, the band edge difference between barrier and well material Δ_a and the local well width $L_z(\boldsymbol{\rho}) = L_z + \Delta L_z(\boldsymbol{\rho})$ enter ($\boldsymbol{\rho}$ is the in-plane vector).

If the exciton binding energy is small compared to the (energetic) sublevel distance, but also well above the disorder-induced broadening, only the lowest exciton bound state $1s$ at the fundamental sublevel transition has to be considered. Consequently, the total wave function can be factorized into

$$\Psi_j(\boldsymbol{r}_e, \boldsymbol{r}_h) = u_e(z_e)\, u_h(z_h)\, \phi_{1s}(\boldsymbol{\rho}_e - \boldsymbol{\rho}_h)\, \psi_j(\boldsymbol{R}) \,, \tag{13}$$

introducing the 2D center-of-mass (COM) coordinate $\boldsymbol{R} = (m_e\boldsymbol{\rho}_e + m_h\boldsymbol{\rho}_h)/M$ with the exciton kinetic mass $M = m_e + m_h$. Both the confinement wave functions $u_a(z_a)$ and the relative wave function $\phi_{1s}(\boldsymbol{\rho})$ obey Schrödinger equations of the QW structure with (average) thickness L_z. Finally one is left with the COM equation

$$\left(-\frac{\hbar^2}{2M}\Delta_{\boldsymbol{R}} + V(\boldsymbol{R}) + \hbar\omega_x \right) \psi_j(\boldsymbol{R}) = \epsilon_j\, \psi_j(\boldsymbol{R}) \,, \tag{14}$$

where $\hbar\omega_x$ is the exciton transition energy of the averaged QW. The random COM potential resulting from well-width fluctuations has been derived elsewhere [4] and reads

$$V(\boldsymbol{R}) = \int \mathrm{d}\boldsymbol{R}' \sum_{a=e,h} \eta_a^2\, \phi_{1s}^2(\eta_a(\boldsymbol{R} - \boldsymbol{R}'))\, E_a'\, \Delta L_z(\boldsymbol{R}') \,. \tag{15}$$

with the band edge change $E_a' = \mathrm{d}E_a/\mathrm{d}L_z$. Due to the mass factors $\eta_e = M/m_h$, $\eta_h = M/m_e$, different weight is given to the electron and the hole contribution. Even if the band offset for the hole is less than the electron one – as it is the case for the GaAs/AlGaAs system – the hole may dominate the potential fluctuations: Due to its larger mass, the hole part in the exciton visits a smaller volume, and averages not as efficient as the electron.

A reasonable assumption for the well width fluctuations in the ensemble is

$$\langle \Delta L_z(\boldsymbol{R})\, \Delta L_z(\boldsymbol{R}') \rangle_e = h^2 \exp(-|\boldsymbol{R} - \boldsymbol{R}'|^2/2\zeta^2) \tag{16}$$

introducing a typical thickness fluctuation h and a correlation length ζ which may be considered as a typical island size. As before, $\langle .. \rangle_e$ denotes the average over different disorder realizations within the statistical ensemble. The convolution with the $1s$ wave function gives a final potential correlation length ξ which depends on the (mass-weighted) exciton Bohr radius a_B as well. If $a_\mathrm{B} > \zeta$ holds, the potential Eq. (15) is constructed as a large sum of random independent contributions, and the central limit theorem holds: The potential values are Gauss distributed with variance $\sigma_V^2 = \langle V^2(\boldsymbol{R}) \rangle_e$. Evaluated for an exciton $1s$ wave function of exponential type with Bohr radius $a_\mathrm{B} > \zeta$, Eq. (15) gives

$$\sigma_V^2 = \frac{2h^2\zeta^2}{a_\mathrm{B}^2} \left((E_e'\eta_e)^2 + 8E_e'E_h' + (E_h'\eta_h)^2 \right) \,. \tag{17}$$

The energetic fluctuations of the COM potential are reduced by the ratio between the statistically independent areas of size ζ^2 and the exciton averaging area a_B^2. Therefore, a direct assignment of σ or of the exciton linewidth to the energy fluctuations on an atomic scale is not possible. In general, the confinement energies fluctuate on a much larger energy scale than the resulting inhomogeneous exciton linewidth seems to predict.

In the opposite limit $a_{\mathrm{B}} < \zeta$, the potential values attain discrete values, and the exciton line splits into a multiplet related to different monolayer energies [5]. This may be achieved in high-quality samples with growth interruption, which will not be considered here. If a specific roughness model is not of interest, it is enough to construct a random potential via the potential correlation function alone,

$$g_V(\boldsymbol{R}-\boldsymbol{R}') = \langle V(\boldsymbol{R})\, V(\boldsymbol{R}')\rangle_e \,. \tag{18}$$

In the numerics, either an exponential type is used which models the 2D exciton wave function, or a Gaussian one which emphasizes more the random island structure. In the latter case, $g_V(\boldsymbol{R}) = \sigma_V^2 \exp(-R^2/(2\xi^2))$ holds. In both cases, only two parameters characterize the potential: variance σ and correlation length ξ, where ξ cannot be smaller than the (mass-weighted) exciton Bohr radius. More details on the numerical procedures are given in [6–9].

4 Density Matrix Formulation of Exciton Scattering and Light Emission

For the description of exciton scattering processes we work in the basis of localized eigenstates from the preceding section. Therefore, all disorder effects are already included on an exact level by defining exciton creation (destruction) operators $X_j^\dagger$ (X_j). The interaction with the light field gives rise to both excitation and emission processes, while the coupling to acoustic phonons $a_{\boldsymbol{q}}^\dagger(a_{\boldsymbol{q}})$ governs the relaxation of excitons between different states. The corresponding deformation-potential coupling function $t_{jl}^{\boldsymbol{q}}$ ensures that only spatially overlapping exciton states can scatter via phonons. The resulting Hamiltonian reads

$$\mathcal{H} = \sum_j \epsilon_j X_j^\dagger X_j + \sum_{j,\boldsymbol{k}} \left(\mathcal{E}(\boldsymbol{k},t)\psi_j^*(\boldsymbol{k}) X_j^\dagger + \text{h.c.}\right) + \sum_{j,l,\boldsymbol{q}} t_{jl}^{\boldsymbol{q}} X_j^\dagger X_l \left(a_{\boldsymbol{q}}^\dagger + a_{-\boldsymbol{q}}\right) . \tag{19}$$

The free phonon and light field parts have to be added as usual. In the coupling to the light field, dipole matrix element as well as the relative exciton wave function at zero electron-hole distance are included into the definition of the transverse field operator $\mathcal{E}(\boldsymbol{k},t)$. Only the state-dependent factor $\psi_j(\boldsymbol{k})$ is made explicit, being just the Fourier-transformed wave function

$$\psi_j(\boldsymbol{k}) = \int \mathrm{d}\boldsymbol{R}\, \mathrm{e}^{\mathrm{i}\boldsymbol{k}\cdot\boldsymbol{R}}\, \psi_j(\boldsymbol{R}) \,. \tag{20}$$

The basic quantities are excitonic polarization $P_j(t) = \langle X_j^\dagger(t)\rangle_q$ and excitonic density matrix $N_{jl}(t) = \langle X_j^\dagger(t) X_l(t)\rangle_q$ which obey equations of motion. We do not go into the details of their derivation which can be found elsewhere [4]. A crucial step is the neglect of memory effects in the phonon scattering (Markoff limit) which allows to derive equations being local in time. The polarization obeys

$$\mathrm{d}P_j(t)/\mathrm{d}t = (\mathrm{i}\omega_j - \gamma_j)P_j(t) + (\mathrm{i}/\hbar)\widetilde{\psi}_j(\boldsymbol{k}_{\mathrm{exc}})\, E_0^*(t) \tag{21}$$

with the classical driving field $E(\boldsymbol{R},t) = E_0(t)\exp(\mathrm{i}\boldsymbol{k}_{\mathrm{exc}}\boldsymbol{R})$. To be more precise, we have included a finite focus size in the excitation which is represented by a focus function $g(\boldsymbol{R})$. Then, the focus-limited Fourier transform

$$\widetilde{\psi}_j(\boldsymbol{k}_{\mathrm{exc}}) = \int \mathrm{d}\boldsymbol{R}\, \mathrm{e}^{\mathrm{i}\boldsymbol{k}_{\mathrm{exc}}\cdot\boldsymbol{R}}\, g(\boldsymbol{R})\, \psi_j(\boldsymbol{R}) \tag{22}$$

denoted by a tilde enters the source term. The dephasing rate is composed from the phonon-assisted transitions $\gamma_{l\leftarrow j}$ and the radiative rate γ_j^{rad} as

$$\gamma_j = \sum_l \gamma_{l\leftarrow j} + \gamma_j^{\mathrm{rad}} . \tag{23}$$

For the density matrix, the state-diagonal contribution follows from the kinetic equation

$$\mathrm{d}N_{jj}(t)/\mathrm{d}t = \sum_l 2\gamma_{j\leftarrow l}N_{ll} - 2\gamma_j N_{jj} + 2\Re\left((\mathrm{i}/\hbar)\widetilde{\psi}_j(\boldsymbol{k}_{\mathrm{exc}})P_j(t)E_0(t)\right), \tag{24}$$

while the non-diagonal ones factorize, $N_{jl}(t) = P_j(t)P_l^*(t)(j \neq l)$. Both polarization and diagonal density can be calculated from their respective differential equations, once the disorder states and the corresponding phonon rates have been generated before.

The excitonic polarizations driven by a short pulse $E_0(t) = A_0\delta(t)$ sum up to give as induced field in direction $\boldsymbol{k}$

$$P(\boldsymbol{k},t) = \Theta(t)(\mathrm{i}/\hbar)\sum_j \psi_j^*(\boldsymbol{k})\, \widetilde{\psi}_j(\boldsymbol{k}_{\mathrm{exc}})\, \mathrm{e}^{(\mathrm{i}\omega_j - \gamma_j)t}\, A_0 . \tag{25}$$

Comparing with the local oscillator model of Sect. 1, Eq. (3), we have $f_j \exp(-\mathrm{i}\boldsymbol{k}\boldsymbol{R}_j) = \psi_j^*(\boldsymbol{k})\, \psi_j(\boldsymbol{k}_{\mathrm{exc}})$. Upon Fourier transforming into frequency space, we can extract the optical density or absorption lineshape of 1s excitons

$$D(\omega) = \frac{1}{A}\sum_j |\psi_j(\boldsymbol{k}=0)|^2\, \frac{\gamma_j}{(\omega-\omega_j)^2 + \gamma_j^2} . \tag{26}$$

A is the wave function normalization area, and normal incidence for excitation and detection is assumed. In a given potential realization, Eq. (26) consists of a series of narrow Lorentzian peaks. Only after performing a large number of simulation runs with different potential realizations, a smooth

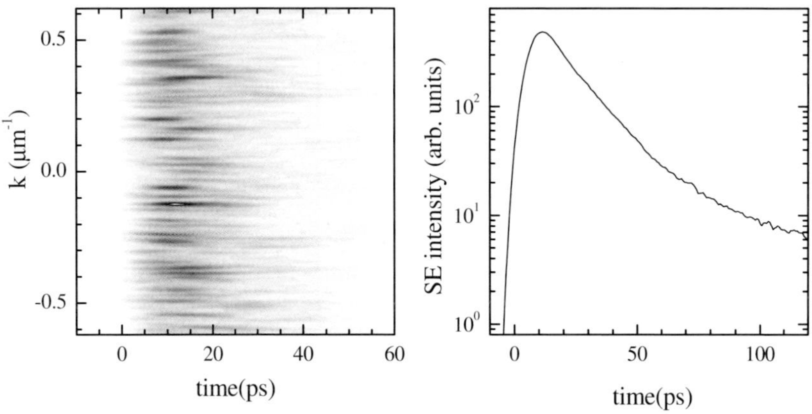

Fig. 2. Time-resolved emission intensity from a 15 nm wide GaAs quantum well at T=5 K, resonantly excited at the 1s heavy-hole exciton by a 1 ps short pulse at $t = 0$. Left: Directionally resolved, on a linear grey scale. Right: Directionally averaged.

curve results. For the absorption, this average corresponds to a transmission experiment with large focus and a finite aperture. If we are interested in the angle-resolved scattered intensity, this kind of ensemble average is not appropriate, and the irregular spiky behavior has to be preserved. It is responsible for the speckles both in energy and time.

The coherent time-resolved intensity after short-pulse excitation reads according to Eq. (25)

$$I_{\mathrm{coh}}(\boldsymbol{k},t) = \frac{A_0^2}{\hbar^2}\left|\sum_j \psi_j^*(\boldsymbol{k})\widetilde{\psi}_j(\boldsymbol{k}_{\mathrm{exc}})\mathrm{e}^{(\mathrm{i}\omega_j-\gamma_j)t}\right|^2 . \tag{27}$$

As discussed in Sec. 2 (Eq. (6)), the time-resolved incoherent intensity is determined by the state-diagonal difference

$$I_{\mathrm{inc}}(t) = \sum_j \psi_j^*(\boldsymbol{k})\widetilde{\psi}_j(\boldsymbol{k}_{\mathrm{exc}})N_j^{\mathrm{inc}}(t) \quad \text{with} \quad N_j^{\mathrm{inc}}(t) = N_{jj}(t) - |P_j(t)|^2 . \tag{28}$$

The speckle formation both in direction and in time is clearly visible in the experimental example of Fig. 2, indicating a strong contribution of $I_{\mathrm{coh}}(\boldsymbol{k},t)$. While the directional speckle relates to the factors $\psi_j^*(\boldsymbol{k})$ in Eq. (27), the temporal speckle originates from the factors $\exp((\mathrm{i}\omega_j - \gamma_j)t)$, with a phase dynamics given by the distribution of eigenfrequencies ω_j. The $\boldsymbol{k}$ averaged time dynamics shows a non-exponential decay for times larger than 50 ps. This is mainly due to the influence of $I_{\mathrm{inc}}(t)$, which has a slower dynamics than $I_{\mathrm{coh}}(t)$ and dominates for long times. For the spectrally resolved intensities one gets

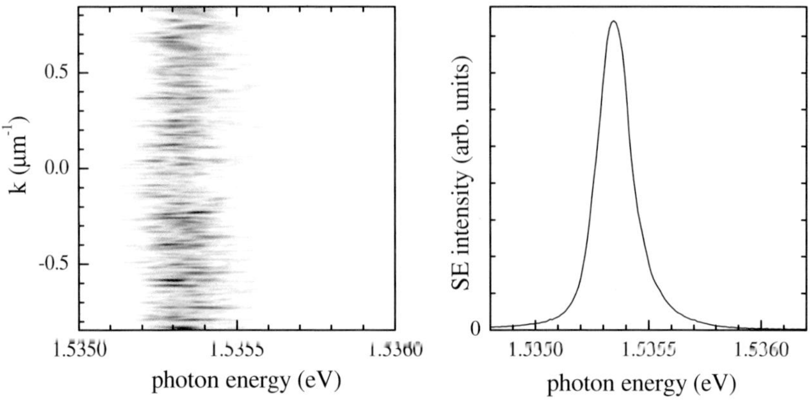

Fig. 3. Spectrally-resolved emission intensity, otherwise as Fig. 2.

$$I_{\rm coh}(\boldsymbol{k},\omega) = \frac{A_0^2}{\hbar^2}\left|\sum_j \frac{\psi_j^*(\boldsymbol{k})\widetilde{\psi}_j(\boldsymbol{k}_{\rm exc})}{\omega-\omega_j-{\rm i}\gamma_j}\right|^2 , \tag{29}$$

$$I_{\rm inc}(\boldsymbol{k},\omega) = \sum_j |\psi_j(\boldsymbol{k})|^2 \frac{\gamma_j}{(\omega-\omega_j)^2+\gamma_j^2}\int {\rm d}t N_j^{\rm inc}(t)\,. \tag{30}$$

The coherent and incoherent parts are spectrally overlapping. They can be distinguished using the speckle formation of the coherent part. An experimental example is given in Fig. 3 for pulsed excitation of a 15 nm GaAs SQW. The speckle formation both in direction and in frequency is visible, with a strong contribution of $I_{\rm coh}(\boldsymbol{k},\omega)$. The spectral speckle relate to the combination $\psi^*(\boldsymbol{k})/(\omega-\omega_j-{\rm i}\gamma_j)$ in Eq. (29).

For a time-independent excitation, $E_0(t) = E_0{\rm e}^{{\rm i}\omega_0 t}$, the polarizations have the same frequency as the excitation, and follow from Eq. (21) as

$$P_j(t) = E_0{\rm e}^{{\rm i}\omega_0 t}\frac{\widetilde{\psi}_j(\boldsymbol{k}_{\rm exc})}{\hbar(\omega_0-\omega_j-{\rm i}\gamma_j)}\,. \tag{31}$$

Here, the spectrally resolved intensity $I(\boldsymbol{k},\omega)$ contains all information and is given by

$$I(\boldsymbol{k},\omega) = \int {\rm d}t_1\,{\rm d}t_2\,{\rm e}^{{\rm i}\omega(t_1-t_2)}\sum_{jl}\psi_j^*(\boldsymbol{k})\psi_l(\boldsymbol{k})\langle X_j^\dagger(t_1)\,X_l(t_2)\rangle_q\,. \tag{32}$$

A careful inspection of the two-time density matrix within the Markoff approximation leads to the decomposition [4]

$$\langle X_j^\dagger(t_1)X_l(t_2)\rangle_q = P_j(t_1)\,P_l^*(t_2) + \delta_{jl}\,{\rm e}^{{\rm i}\omega_j(t_1-t_2)}\,{\rm e}^{{\rm i}\gamma_j|t_1-t_2|}\,N_j^{\rm inc}(\min(t_1,t_2)) \tag{33}$$

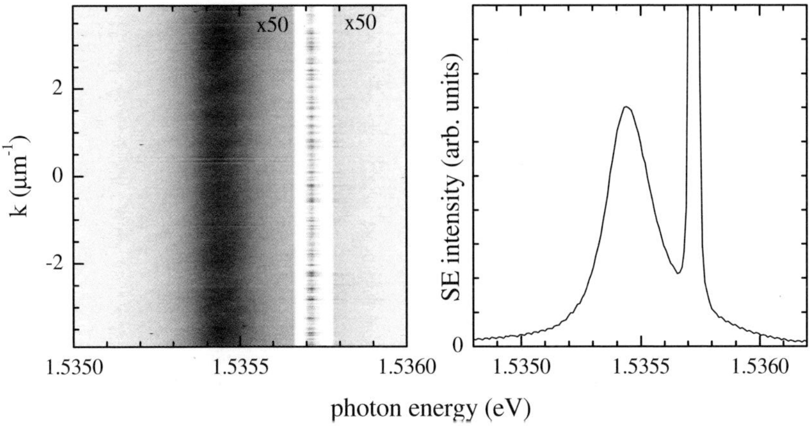

Fig. 4. Spectrally resolved emission intensity from the sample of Fig. 2, resonantly excited at $\hbar\omega_0 = 1.53572\,\mathrm{eV}$. Left: Directionally resolved, on a linear grey scale. The non-resonant emission is scaled by a factor 50. Right: Directionally averaged.

into a coherent (polarization product) and an incoherent term. Using Eq. (31), the coherent intensity can be evaluated as

$$I_{\mathrm{coh}}(\boldsymbol{k},\omega) = \delta(\omega-\omega_0)\frac{E_0^2}{\hbar^2}\left|\sum_j \frac{\psi_j^*(\boldsymbol{k})\widetilde{\psi}_j(\boldsymbol{k}_{\mathrm{exc}})}{\omega_0-\omega_j-\mathrm{i}\gamma_j}\right|^2 \tag{34}$$

and is found in full resonance with the driving field. Eq. (34) represents a speckle pattern due to the $\boldsymbol{k}$ dependence of the phases of $\psi_j^*(\boldsymbol{k})$, where all states in a γ window around ω_0 contribute.

The second (incoherent) part of Eq. (33) reduces in Eq. (32) to the incoherent intensity

$$I_{\mathrm{inc}}(\boldsymbol{k},\omega) = \sum_j N_j^{\mathrm{inc}}|\psi_j(\boldsymbol{k})|^2\frac{\gamma_j/\pi}{(\omega-\omega_j)^2+\gamma_j^2}\,. \tag{35}$$

Here, the phase information of $\psi_j(\boldsymbol{k})$ is skipped due to the absolute square, and no speckles are found in this part. An example for the spectrally and directionally resolved secondary emission $I(\boldsymbol{k},\omega)$ of a 15 nm GaAs SQW at low temperatures is shown in Fig. 4. The coherent part is fully resonant to the excitation and shows speckle formation. In contrast, the incoherent part is much broader due to phonon assisted relaxation, but has no dependence on the direction up to systematic experimental errors. The typical dephasing rates in this emission are $\hbar\gamma \approx 20\,\mu\mathrm{eV}$.

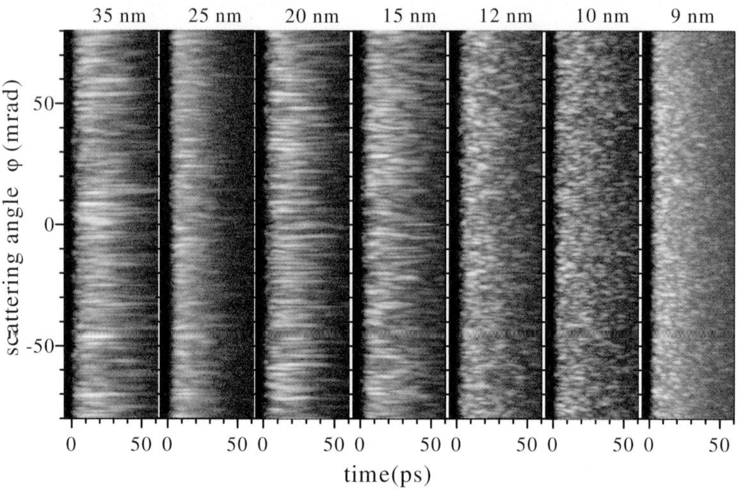

Fig. 5. Time- and directionally resolved emission intensity from GaAs quantum wells of different widths at T=5 K. Logarithmic grey scale over 2 decades.

5 Time-Resolved Speckle Analysis

In this section we discuss the experimental details of time-resolved speckle analysis and corresponding results on GaAs quantum wells.

5.1 Experimental Details

In the time-resolved speckle analysis, the intensity of secondary emission (SE) is resolved in direction and in time, and the time-resolved coherence degree $c(t)$ and emission intensity $I(t) = \langle I(\boldsymbol{k}, t)\rangle_{\boldsymbol{k}}$ is determined, as discussed in Sect. 2. To perform time-resolved speckle analysis, one has to use a detector with sufficient time resolution to resolve the temporal speckle dynamics, which is given by the inverse line width of the investigated resonance. In semiconductor nanostructures, this requires typically a time-resolution of better than 10 ps, which can be achieved by two techniques: (i) a streak-camera, which is limited to 1 ps time-resolution up to now, (ii) by optical gating techniques, which have a time-resolution of typically 100 fs given by the optical gating pulse, but a small sensitivity due to the sequential measurement of the delay-time dependence, and a gating efficiency of typically less than 10%.

We discuss here experimental results concerning the SE dynamics of the 1s heavy-hole exciton in GaAs quantum wells (QWs) with varying thicknesses. The samples were kept at a temperature of 5 K, and the exciton transition was resonantly excited by optical laser pulses of $\sim$ 1 ps duration. The SE was imaged into a monochromator and its intensity was detected time, energy, and directionally resolved by a synchroscan streak camera. The energy

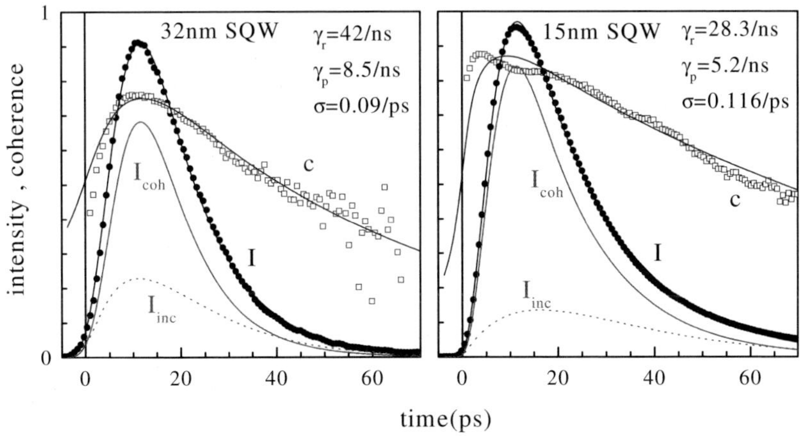

Fig. 6. Time-resolved speckle analysis for GaAs SQWs of 32 nm and 15 nm width at T=5 K. The symbols are the experimental data, while the curves for $I(t)$, $I_{\mathrm{coh}}(t)$, $I_{\mathrm{inc}}(t)$, and $c(t)$ are fits according to the simple oscillator model with the parameters given in the plots.

resolution of the Czerny-Turner type monochromator was adjusted by an aperture on the collimation mirror to about 2 meV. This was small enough to isolate the signal of the resonantly excited QW from the one of the GaAs substrate, and large enough to not reduce the temporal resolution of the streak camera of 3 ps. Using the second dimension of the streak-camera, the RRS dynamics is directionally resolved in one dimension, resulting in the simultaneous detection of about 100 independent RRS intensities (speckles). Repeated measurements of displaced cuts of the two-dimensional directional RRS allowed for the detection of several thousand speckles.

The time- and directionally resolved SE intensity as function of time after excitation is shown in Fig. 5 for single QWs (SQWs) of thicknesses between 32 nm and 9 nm. All SQWs exhibit a finite emission rise time, and a subsequent decay on the time-scale of several tens of picoseconds. The SE contains a large fraction of coherent emission, as is evident from the strong directional intensity fluctuations, the speckles. The directional size of the speckles is determined by the inverse spot size [10]. The temporal speckle correlation, on the other hand, changes with decreasing well width due to the increasing inhomogeneous broadening of the exciton resonances [11].

The measured directionally resolved SE intensity is used to determine the time-resolved emission coherence degree with (10), and is shown for a 32 nm and 15 nm GaAs SQW in Fig. 6 together with the directionally averaged coherent (I_{coh}) and incoherent (I_{inc}) emission intensity. The number of effectively averaged speckles N_{eff} for these measurements was 1.9, as determined from the speckle intensity distribution. The finite rise time of the coherent emission intensity, decreasing with increasing inhomogeneous broadening, is evident. For long times, the emission intensity decays, and a transfer from coherent to incoherent intensity takes place. It is also interesting to note that the coherence degree is initially increasing, a direct evidence for a faster initial dynamics of phonon scattering as compared to the disorder scattering [12]. The predicted dynamics in the simple oscillator model [13] is in good agreement with the experiments, giving values for the density decay rate $2\gamma_r$, the pure polarization decay γ_p, and the inhomogeneous broadening σ as displayed in Fig. 6.

Further results on the time-resolved speckle analysis of secondary emission from the heavy-hole exciton have been published on the temperature dependence [11, 14] and on the incoherence of non-resonant emission [15]. Also the instantaneous resonant Rayleigh scattering (RRS) from sub-ensembles of the exciton resonance was demonstrated [16].

5.2 Spin Coherence and Emission Polarization

The coupling of the exciton states to the light provides not only a decay mechanism for the polarization, but also influences the energy spectrum. Even for a single exciton state, the long-range exchange interaction results for spin-degenerate states with broken circular symmetry, which is the case for exciton states in a disorder potential, in a lifting of the spin degeneracy into two linearly polarized eigenstates [17, 18]. In principle, not only the coupling of a state with itself (diagonal coupling) should be taken into account, but also the coupling with all other exciton states (also referred to as Förster coupling) is possible. However, the diagonal coupling is already rather small (10–100 μeV in GaAs QWs), and a significant effect requires exciton states within this energy range which are spatially closer than the light wavelength λ. Coming back to the diagonal coupling, we note that the linearly polarized eigenstates can have widely distributed orientations due to the partly random nature of the disorder potential. Consequently, the optical excitation will in general excite a superposition of both eigenstates. The time evolution of the eigenstate doublets then induces a polarization dynamics involving also the polarization orthogonal to the excitation polarization. Since this is a coherent process, the coherence degree of the emission cross-polarized and co-polarized to the excitation is comparable, and the spin relaxation in the ensemble does not imply a loss of coherence. A model considering localized exciton states in an anisotropically Gaussian-correlated potential landscape gave a consistent description of the experimental observations for a 4 nm GaAs quantum well [18–20]. An anisotropy along the $[1\bar{1}0]$ direction with correlation lengths of 28 nm along and 17 nm perpendicular were deduced, leading to an average doublet splitting of 29 μeV .

5.3 Spatial Correlation and Level-Repulsion

As compared to a spatially uncorrelated level statistics assumed in the simple oscillator model, the exciton as quantum-mechanical particle of finite mass in the disorder potential shows a spatial correlation of its states by two effects: (i) level-repulsion, i.e. a reduction in the probability of finding states close both in energy and real space due to their orthogonality [21]. (ii) a spatial correlation of the disorder potential due to the properties of the growth process [22]. Effect (i) leads to an initial reduction of the RRS intensity of some 20%, that recovers on a time-scale that is longer than the initial rise time of the RRS, and is given by the inverse width of the level repulsion energy [22, 23]. Effect (ii) results in an initially increased RRS intensity due to super-radiance (i.e. in-phase emission) of excitons spatially separated by less than the disorder correlation length, which decays to the long-time value with a strongly damped oscillation. The relative importance of the two effects is determined by the ratio $\sigma_\mathrm{v}/E_\mathrm{c}$, where σ_v is the variance of the disorder potential, and E_c is the exciton COM kinetic energy $\hbar^2/(2M\xi^2)$ corresponding to the potential correlation length ξ. For $\sigma_\mathrm{v}/E_\mathrm{c} < 1$, the quantum mechanical level-repulsion dominates the level statistics, while the spatial correlation dominates for $\sigma_\mathrm{v}/E_\mathrm{c} \gg 1$.

We have investigated the influence of the exciton level statistics on the RRS dynamics of GaAs SQWs [24]. The described interplay of level-repulsion and spatial correlation in the calculated RRS dynamics is shown in Fig. 7a for different values of $\sigma_\mathrm{v}/E_\mathrm{c}$ on a dimensionless time-scale σt, where σ is corresponding to the width of the RRS spectrum. Experimentally measured RRS dynamics, corrected for the microscopic dephasing rate γ, are given in b) and c). For the non-growth-interrupted 15 nm GaAs SQW, the dynamics is comparable to a simulation with a small disorder correlation length $\sigma_\mathrm{v}/E_\mathrm{c} = 0.66$. This value is consistent with an estimation using the structural parameters of the samples: ξ is given by the exciton Bohr radius $a_\mathrm{B} \approx 10\,\mathrm{nm}$ and the exciton mass is $M = 0.25\, m_\mathrm{e}$. With the value of $\sigma = 0.11\,\mathrm{meV}$ measured in the RRS spectrum, we find $\sigma/E_\mathrm{c} = 0.08$, corresponding to $\sigma_\mathrm{v}/E_\mathrm{c} = 0.36$, in good agreement with the value range describing the measured dynamics.

For a growth-interrupted 12 nm GaAs SQW, the RRS spectrum is influenced by the islands of long correlation length on the growth-interrupted GaAs surface, resulting in a monolayer splitting. The RRS dynamics is compared in Fig. 7c with the calculated dynamics using a monolayer disorder model [24]. Agreement of both RRS dynamics and spectra is reached using a short range disorder σ_sr=0.18 meV, a monolayer energy of σ_ml=0.6 meV and an island size of ξ_ml=56 nm. The value of σ_ml is in agreement with the observed monolayer splitting energy in the RRS spectrum, and the monolayer island size ξ_ml is also comparable with the analysis of emission spectra of a series of growth-interrupted samples [25].

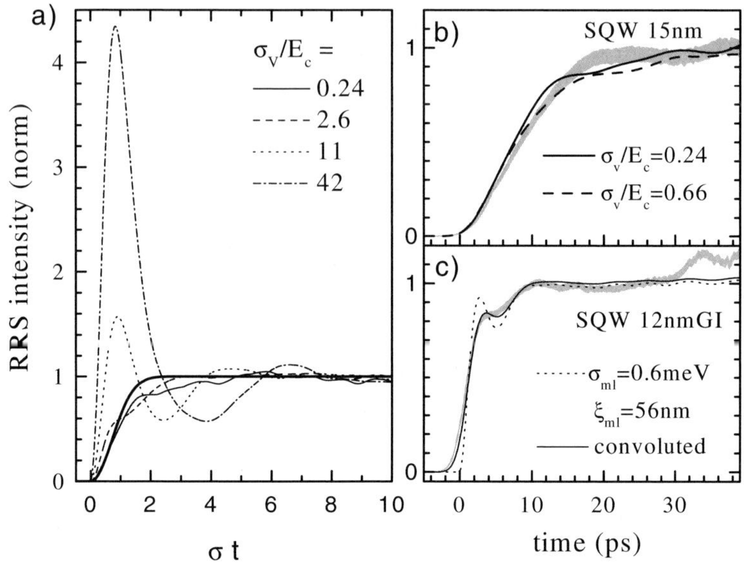

Fig. 7. (a) Calculated RRS intensity versus reduced time σt after excitation, for different values of $\sigma_{\mathrm{v}}/E_{\mathrm{c}}$ as indicated, and $\gamma = 0$. The RRS intensity is normalized at long times. The thick solid curve is the uncorrelated RRS dynamics of the simple oscillator model. (b) Comparison of the measured RRS dynamics of a non-growth-interrupted 15 nm SQW (grey shaded region) with the ones calculated for $\sigma_{\mathrm{v}}/E_{\mathrm{c}} = 0.24, 0.66$ convoluted with the experimental time resolution. (c) Comparison of the measured RRS dynamics of a 12 nm growth-interrupted SQW with the one calculated for a monolayer disorder with parameters as given (dotted line), and convoluted with the experimental time resolution (solid line). From Ref. [24].

5.4 Effects of Long-Range Correlations: Sloped Speckles

Until now we assumed that the exciton potential has only a disordered component with a correlation length short with respect to the excited area of the sample. Long-range variations of the average exciton energy $\hbar\omega_{\mathrm{X}}(\boldsymbol{R})$, that within the excited spot can be described by their gradient and eventually by their curvature were considered in Ref. [26]. Such potentials have been deliberately created by application of spatially inhomogeneous mechanical stress. The resulting energy shift was described by $k \cdot p$ theory with input from linear continuum elasticity calculations. An example for a 15 nm GaAs SQW is given in Fig. 8a. In the presence of an energy gradient, a moving speckle pattern of the resonantly excited Rayleigh scattering in time-momentum space is observed (Fig. 8b), which implies also a time-dependent reflection direction.

The temporal and directional speckles contain thus information on the local energy gradients in the sample. For a quantitative analysis we consider the speckle correlation

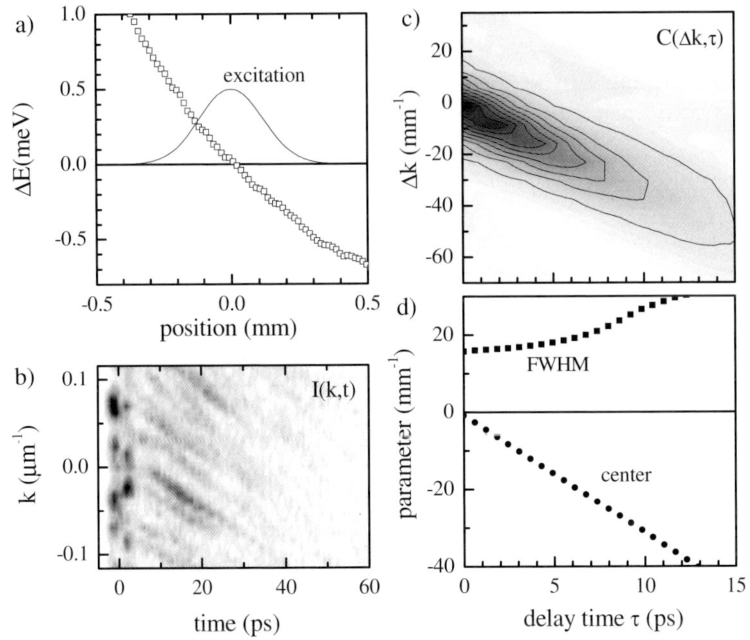

Fig. 8. (a) Measured spatially resolved average exciton transition energy of a strained 15 nm GaAs SQW, relative to 1.5376 eV. The intensity profile of the excitation is also shown. (b) Time- and directionally resolved SE after pulsed excitation. (c) Experimental intensity correlation $C(\Delta \boldsymbol{k}, \tau)$ on a linear grey scale. (d) Delay-time dependent width and center position of $C(\Delta \boldsymbol{k}, \tau)$. Taken from Ref. [26].

$$C(\Delta \boldsymbol{k}, \tau) = \frac{\langle I(\boldsymbol{k}, t) I(\boldsymbol{k} + \Delta \boldsymbol{k}, t + \tau) \rangle_{\boldsymbol{k},t}}{\langle I(\boldsymbol{k}, t) \rangle_{\boldsymbol{k},t} \langle I(\boldsymbol{k} + \Delta \boldsymbol{k}, t + \tau) \rangle_{\boldsymbol{k},t}} - 1 \,. \tag{36}$$

A moving speckle pattern translates into a tilted correlation with a maximum moving along $\Delta \boldsymbol{k} = -\hbar \tau \nabla_{\boldsymbol{R}} \omega_{\mathrm{X}}(\boldsymbol{R})$, as can be seen in Fig. 8c, and is analyzed in Fig. 8d. In this way it is possible to determine local transition energy gradients by speckle analysis. Additionally, the time-dependence of the width of the correlation determines the local curvature of the transition energy [26]. The model of localized excitons in a center-of-mass potential discussed in Sect. 3 was extended to systematic spatial variations of the potential, giving an accurate modelling of the observed speckle correlations.

5.5 Polaritonic Effects in Multiple Quantum Wells

In multiple quantum wells (MQWs), the radiative coupling of the exciton states is of increased importance. In a structure with N wells, the coupling leads to the formation of N exciton-polariton modes, which are extended over all quantum wells, and have different radiative renormalizations and radiative lifetimes [27, 28]. The sum of the renormalization is conserved in the coupling, which for $\boldsymbol{k}$ within the radiative cone is renormalization zero and radiative decay rate $N\Gamma_0$, with the radiative polarization decay rate Γ_0 of a single well [29], In the specific case of a double quantum well [28, 30], at $\lambda/2$ distance both modes have no renormalization, and only one mode has a radiative decay rate $2\Gamma_0$. At $\lambda/4$ distance instead, both modes have equal radiative damping, and opposite renormalization $\pm\Gamma_0$, as was also experimentally observed [31]. Stacking many quantum wells with $\lambda/2$ distance, a Bragg structure is created, in which all radiative renormalizations are zero, and only one (super-radiant) mode has the radiative decay rate $N\Gamma_0$, as was shown theoretically and experimentally [28, 32–35].

The RRS dynamics from MQWs is consequently more complex. In case of a weak disorder that results in a polariton scattering in $\boldsymbol{k}$ space, but not in a significant change of the exciton energies compared to the radiative renormalizations, the RRS dynamics is dominated by the dynamics of the polariton modes. Their polarization is decaying according to their radiative rates, showing an additional time-dependent interference due to their different radiative renormalizations. This limit was approached in experiments on 100 period MQW Bragg structures [36], or 10 period GaAs MQWs of 16 nm width [12, 37]. With increasing disorder, only the polariton modes with radiative renormalizations larger than the disorder-induced exciton linewidth retain their character [38], and are dominating the initial dynamics of the RRS [37]. The observed fast initial dynamics from MQWs in previous works [39–44] is most likely due to such polaritonic effects. We will show some results on MQWs in Sect. 7.

6 Spectrally Resolved Speckle Analysis

In spectrally resolved speckle analysis, the SE intensity is resolved in direction and in frequency, and the spectral coherence degree $c(\omega)$ and the total spectral emission intensity $I(\boldsymbol{k},\omega)$ are determined. Spectrally resolved measurements can be experimentally easier than time-resolved measurements, depending on the respective characteristic rates γ_r, γ_p and σ. In time-resolution, it is relatively easy to obtain a large time range, giving the lower limit of the detectable rates, while it can be complicated to have a high time resolution, giving the upper limit for the rates. In spectral resolution, instead, it is relatively easy to obtain a large spectral range, giving the upper limit of the detectable rates, while it can be complicated to have a high spectral resolution, giving the lower limit for the the rates. Apart from the experimental differences, the results of spectral and time-resolved speckle analysis are complementary. Additionally to the spectral coherence degree, the spectral speckle correlation

$$C(\omega,\Delta\omega) \equiv \frac{\langle I(\omega+\Delta\omega/2)I(\omega-\Delta\omega/2)\rangle_{\boldsymbol{k}}}{\langle I(\omega+\Delta\omega/2)\rangle_{\boldsymbol{k}}\langle I(\omega-\Delta\omega/2)\rangle_{\boldsymbol{k}}} \approx \frac{1}{(2\gamma)^2+\Delta\omega^2} \tag{37}$$

gives a direct access to the polarization decay rate γ (homogeneous linewidth) of the resonances if the inhomogeneous distribution is much broader than γ. As an example, experimental results on a 5 nm GaAs SQW are given in Fig. 9. The speckle formation in direction-frequency space is clearly visible in (a). The spectrum shows the spectrally narrow fluctuations, which determine the width of the spectral speckle correlation shown in (b). The resulting spectrally resolved emission intensity, its coherence degree, and the homogeneous linewidth are given in (c). The increasing coherence degree with increasing photon energy is due to the low lattice temperature ($kT \approx 100\,\mu\text{eV}$), at which the coherence loss is dominated by energy relaxation due to phonon-emission, piling up the incoherent density at lower exciton energies. The increase of the homogeneous linewidth with energy is due to the increase of the final state density in the phonon emission process. More details can be found in [10, 13, 45].

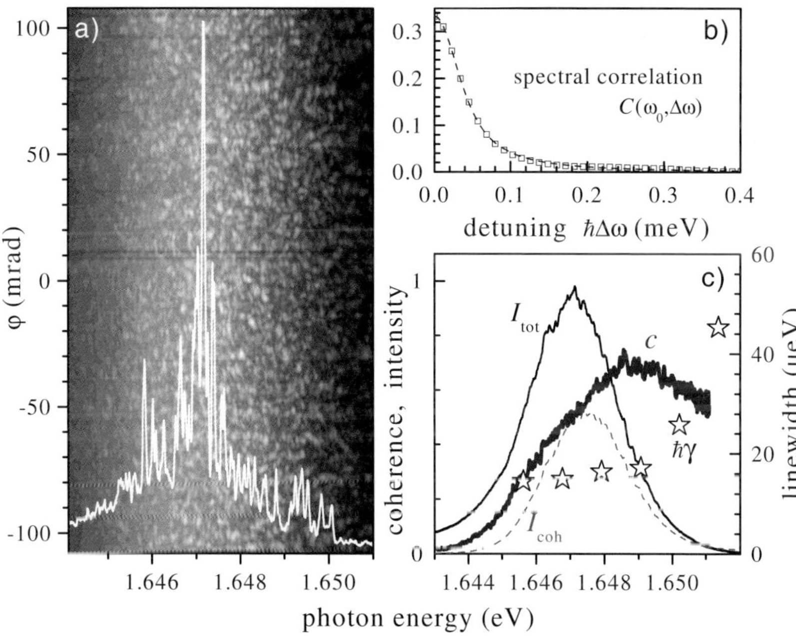

Fig. 9. (a) Directionally and spectrally resolved emission intensity (2 decades logarithmic grey scale) of a 5 nm GaAs SQW at 1.5 K. The white line is a spectrum at a fixed emission direction. (b) Measured spectral speckle correlation $C(\omega_0, \Delta\omega)$ at $\hbar\omega_0 = 1.6476\,\text{eV}$. The dashed line is a Lorentzian fit convoluted with the system response. (c) Spectral dependence of the SE intensity (I_{tot}), its coherence c (thick line), the deduced RRS part (I_{coh}), and the homogeneous line width $\hbar\gamma(\omega)$ (stars). From Ref. [45].

7 Spectral Interferometry

In spectral interferometry (SI), the coherent field amplitude $E(\boldsymbol{k},\omega) \propto P(\boldsymbol{k},\omega)$ is retrieved in amplitude and phase from the spectral interference between the secondary emission and a reference field using pulsed excitation [46]. The exciting pulse is split into one part exciting the sample, and a second part of field $E_0(\omega)$ and intensity $I_0(\omega)$ that is recombined with the secondary emission by a beam-splitter. The spectrally resolved detected intensity $I_{\mathrm{det}}(\mathbf{k},\omega)$ is given by

$$\begin{aligned} I_{\mathrm{det}}(\boldsymbol{k},\omega) &= |E_0(\omega)+E(\boldsymbol{k},\omega)|^2 + I_{\mathrm{inc}}(\boldsymbol{k},\omega) \\ &= I_0(\omega)+I(\boldsymbol{k},\omega)+\Re(2E_0^*(\omega)E(\boldsymbol{k},\omega))\,. \end{aligned} \tag{38}$$

The values of $I_0(\omega)$ and $I(\boldsymbol{k},\omega)$ are determined by blocking the secondary emission and the reference, respectively, so that one can deduce the interference term $S(\boldsymbol{k},\omega) = \Re(2E_0^*(\omega)E(\boldsymbol{k},\omega))$. The inverse Fourier transform of $S(\boldsymbol{k},\omega)$ is the sum of two convolution integrals:

$$F^{-1}(S(\boldsymbol{k},\omega)) = \int \mathrm{d}t' E_0(-t')E(\boldsymbol{k},t-t') + \int \mathrm{d}t' E_0(t')E(\boldsymbol{k},t'-t)\,. \tag{39}$$

The excitation field $E_0(t)$ is chosen to lead the SE field $E(\boldsymbol{k},t)$ in time with no temporal overlap, which one can achieve for pulsed excitation by an appropriate delay. In this case the second convolution integral is zero for $t>0$, and one finds

$$F(\Theta(t)F^{-1}(S(\boldsymbol{k},\omega))) = E_0^*(\omega)E(\boldsymbol{k},\omega)\,. \tag{40}$$

Knowing the excitation field $E_0(\omega)$, which is strong and can be characterized by non-linear methods like frequency resolved optical gating, one can determine the complex SE field $E(\boldsymbol{k},\omega)$, and its time-resolved counterpart $E(\boldsymbol{k},t) = F^{-1}(E(\boldsymbol{k},\omega))$. The coherent emission intensity is thus determined both in frequency and in time domain. Additionally the spectrally resolved incoherent emission intensity can be deduced by $I_{\mathrm{inc}}(\boldsymbol{k},\omega) = I(\boldsymbol{k},\omega) - |E(\boldsymbol{k},\omega)|^2$, while the time-resolved incoherent intensity is not accessible.

Care has to be taken in the comparison of the secondary emission intensity $I(\boldsymbol{k},\omega)$ with the retrieved electric field $E(\boldsymbol{k},\omega)$ in the determination of $I_{\mathrm{inc}}(\boldsymbol{k},\omega)$ when effects of speckle averaging are present. In $I(\boldsymbol{k},\omega)$ the speckle averaging leads to a reduction of the speckle contrast, while the speckle-averaged intensity is unchanged. The same speckle averaging leads to an averaging of the fields in the interference term $S(\boldsymbol{k},\omega)$, which results in a reduction of the field strength and thus of the measured coherent intensity $|E(\boldsymbol{k},\omega)|^2$. The measured coherent speckle intensity distribution, however, is still made from a superposition of fully coherent random fields, so that it is exponentially distributed. Deducing the spectral coherence degree from the spectral interferometry is thus problematic.

Spectral interferometry requires principally the same detection as the spectral speckle analysis. The difference is the interference with an optical reference pulse, which has to be phase stable relative to the emission of the sample. The phase stability can be reached either passively or actively. Using passive phase stability the relative path lengths of signal and reference light must not fluctuate by more than $\approx \lambda/10$ during the time of the measurement. Such stability requires not only a stable mechanical position of the optical elements, including the sample, but also a stability of the gas temperature along the light path. A simple way of reaching such stability is to use the passive phase stability present when the specular reflection of the sample is used as reference pulse [47]. An aperture in the far-field (i.e. in $\boldsymbol{k}$ space) transmits either only the specular direction, only secondary directions, or both. The transmitted light is then imaged to the near field, at which the light intensity is detected. In this way, reference and signal are passing along a similar path through the same optical elements, and a sufficient passive phase stability can be achieved. Additionally, a perfect spatial overlap between reference and signal is present. In the data analysis one has to take into account that the specular reflection, besides the non-resonant surface reflection, also contains the resonant specular response of the sample.

Previous reports of spectral interferometry of RRS were taken on MQWs on single speckles [39] or on averages over some ten speckles [40, 41]. Results with a large speckle average and time-range as well as results SQWs were not reported. The RRS intensities of SQWs and MQWs of various thicknesses measured in this work by SI averaged over about 1000 speckles are shown in Fig. 10. The excitation angle was 9 degrees to the sample normal, and the RRS was collected in a solid angle range of 9 degrees diameter close to the sample normal. The high resolution of the employed spectrometer allowed for a large SI time window of 180 ps.

The difference in the initial dynamics between SQW and MQWs is obvious. When comparing with the dynamics of an uncorrelated level statistics (dotted lines), the SQW shows an initial RRS suppression due to the effect of level repulsion [22, 24], while the MQWs show an initial RRS enhancement, with pronounced oscillations for MQWs of wide well width, i.e. small disorder. These results clarify the polaritonic origin of the fast initial RRS dynamics from MQWs [39–41, 48], which was never observed in SQWs. Additionally, one can observe a reduced decay rate γ in the MQW as compared to the SQW of similar inhomogeneous broadening [24]. Phonon-scattering only contributes about 8 ns^{-1} to the decay [24], so that the reduction should be attributed to the formation of the polaritonic modes, in which the decay rates of the individual QWs are redistributed, but the total rate is conserved. The super-radiant modes with large rates decay fast, and dominate the initial RRS. Later, the sub-radiant modes dominate, which have rates smaller than the ones of SQWs.

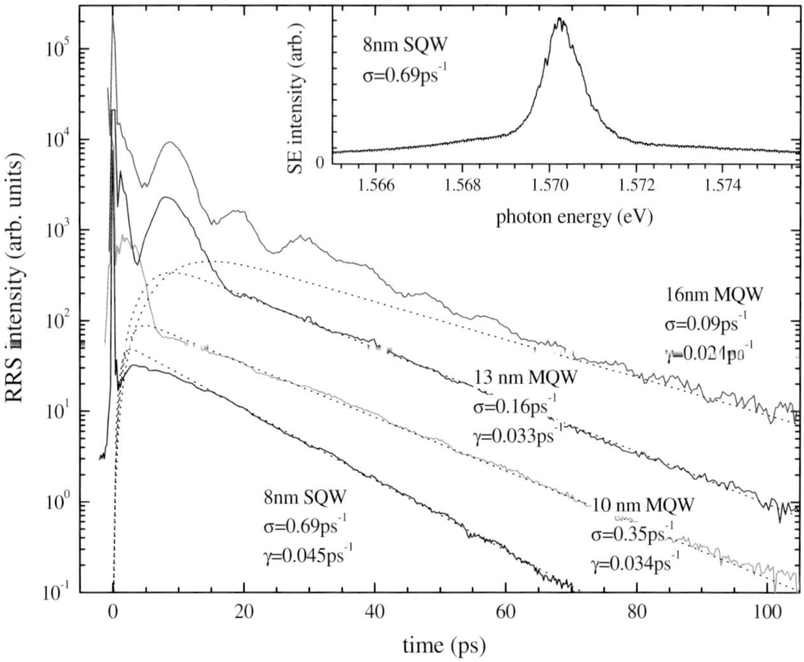

Fig. 10. Time-resolved, speckle-averaged RRS intensity dynamics measured by spectral interferometry of different MQWs of 10 periods and an 8 nm SQW at T=5 K (solid lines). The dotted lines are theoretical curves for uncorrelated level statistics for a polarization decay γ fit to the observed decay. The inset shows the ensemble-averaged SE intensity spectrum of the 8 nm SQW, from which the inhomogeneous broadening σ of the level distribution was deduced.

The knowledge of the scattered field in phase and amplitude allows us to access the time-dynamics of the emission spectrum, which can be analyzed using a speckle-averaged Wigner–Ville transform. This is a Fourier transform of the speckle-averaged field auto-correlation centered at the time t:

$$I_{\mathrm{wv}}(t,\omega) = \int \mathrm{d}\tau\, \mathrm{e}^{-\mathrm{i}\omega\tau} \left\langle E\left(\boldsymbol{k}, t+\frac{\tau}{2}\right) E^*\left(\boldsymbol{k}, t-\frac{\tau}{2}\right)\right\rangle_{\boldsymbol{k}} \tag{41}$$

It has the property that its time-integration yields the coherent emission spectrum $I_{\mathrm{coh}}(\omega) = \int \mathrm{d}t I_{\mathrm{wv}}(t,\omega)$, and its spectral integration yields the time-resolved coherent emission $I_{\mathrm{coh}}(t) = \int \mathrm{d}\omega I_{\mathrm{wv}}(t,\omega)$. A temporal windowing, which is usually introduced into the Wigner–Ville distribution in order to define the temporal and thus also spectral resolution, and was applied for the analysis of a single speckle [36] is not used here. Instead, an effective time windowing is created by the speckle-averaging, which sets the time-window to the temporal correlation of the field over the speckle. In this way, the time-resolution is time-dependent, adapting itself to the dynamic spectral content

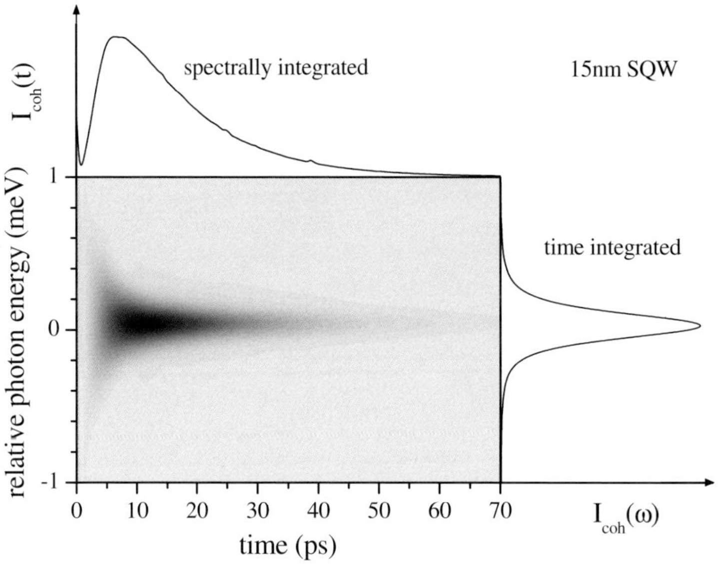

Fig. 11. Speckle-averaged Wigner–Ville distribution $I_{\mathrm{WV}}(t, \hbar\omega)$ of the coherent secondary emission from a 15 nm SQW on a linear grey scale from -0.1 (white) to 1 (black). About 1000 speckle were used.

of the scattered field. An example is given in Fig. 11, where the experimental result for the 15 nm SQW is shown. The obvious initial reduction of the spectral width with increasing time is due to the changing time-window, which stabilizes to the spectral distribution of the contributing exciton resonances after the RRS intensity is fully developed ($\sigma t > 1$). No additional spectral dynamics is present, showing that the emission can be described by an ensemble of uncoupled localized excitons with spectrally rather uniform polarization decay times. This result is typical for SQWs. Only for large inhomogeneous broadenings (> 1 meV) the gradient of phonon-scattering rates within the linewidth [45] is relevant, which introduces a temporal red-shift of the emission observed for a 8 nm SQW (not shown).

The situation is different for multiple quantum wells, in which the polaritonic coupling between the excitons of different wells leads to the formation of new polaritonic modes. Experimental data for a 16 nm MQW of ten periods and a barrier width of 10 nm are given in Fig. 12. A pronounced spectral dynamics due to several polaritonic modes of different energies and lifetimes is visible. The inhomogeneous broadening determines the time-scale on which the few super-radiant polaritonic modes are dominating the emission [37]. Due to the small inhomogeneous broadening of 0.11 meV FWHM, three polaritonic modes are visible at longer times, with a interference of 10 ps period between the modes corresponding to their splitting of 0.4 meV. The two modes of lower energy are nearly degenerate, but show a slightly different

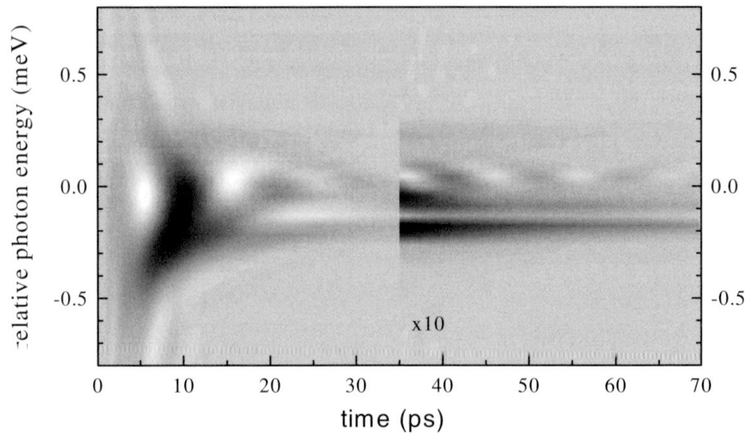

Fig. 12. $I_{\mathrm{wv}}(t, \hbar\omega)$ from a 16 nm GaAs/10 nm $Al_{0.3}Ga_{0.7}As$ 10-period MQW on a linear grey scale from -0.2 (white) to 1 (black). The data for $t > 35$ ps is scaled by a factor of 10. About 1000 speckle were used.

decay time, so that their destructive interference is changing with time. A microscopic theoretical description of the polariton formation in presence of disorder is up to now not available. Present models are treating the light-coupling fully, but are not taking into account the disorder in a microscopic way [37, 49] or making simplifications in the numerical treatment like an azimuthal average [36, 50]. In the investigated MQW the wells have a distance of about $\lambda/10$, for which, without disorder, the polariton modes with the large radiative coupling are on the low energy side of the resonance [35]. This is also found in the measured data. Both the initial dynamics and the spectrum of the RRS thus show a strong influence of the radiative coupling in MQWs on the RRS.

8 Conclusion and Outlook

The novel technique of speckle analysis of resonant secondary emission can be used as a standard tool for coherent solid state spectroscopy. The different flavors of the technique, time-resolved, spectrally resolved, or spectral interferometry, give access to complementary information on the excited optical resonance. The dynamics of the coherent and incoherent emission is measured, determining in the simplest case coherence and density decay times. Furthermore, speckle cross-correlation of the emission for varied external parameters gives a tool that measures small energy splittings of the individual scatterers in a dominantly inhomogeneously broadened ensemble, with a resolution limited only by the microscopic linewidths. An example is the long-range exchange splitting of localized excitons [18]. Such studies can be expanded to the influence of external electric, magnetic or strain fields. With the continuous improvement of the available multichannel detection systems and data processors, the main drawback of speckle analysis, that is the requirement of a large number of independent measurements, is of decreasing importance. Speckle analysis can thus become a widely used tool for coherent spectroscopy of inhomogeneously broadened ensembles, just as four-wave mixing or spectral hole burning.

References

1. W. Demtröder: *Laser Spectroscopy* (Spinger, Berlin 1998)
2. J. Shah: *Ultrafast Spectroscopy of Semiconductors and Semiconductor Nanostructures* (Springer, Berlin 1996)
3. E. Runge, R. Zimmermann: Phys. Rev. B **61**, 4786 (2000)
4. R. Zimmermann, E. Runge, V. Savona: Theory of Resonant Secondary Emission: Rayleigh Scattering versus Luminescence. In: *Quantum Coherence, Correlation and Decoherence in Semiconductor Nanostructures*, ed by T. Takagahara (Elsevier Science, USA 2003) p 89
5. H. Castella, J.W. Wilkins: Phys. Rev. B **58**, 16186 (1998)
6. R. Zimmermann, E. Runge: Optical Lineshape and Radiative Lifetime of Excitons in Quantum Structures with Interface Roughness. In *Proc. 22nd Int. Conf. on the Physics of Semiconductors*, ed by D. Lookwood (World Scientific, Singapore 1995) p 1424
7. R. Zimmermann, E. Runge, F. Große: Optical Spectra of Quantum Structures: Influence of Interface and Alloy Disorder. In: *Proc. 23rd Int. Conf. on the Physics of Semiconductors* ed by M. Scheffler, R. Zimmermann (World Scientific, Singapore 1996), vol 3, p 1935
8. R. Zimmermann, F. Große, E. Runge: Pure & Appl. Chem. **69**, 1179 (1997)
9. E. Runge, R. Zimmermann: Optical Properties of Localized Excitons in Nanostructures: Theoretical Aspects. In: Advances in Solid-State Physics, vol 38, ed by B. Kramer (Vieweg, Braunschweig 1999) p 251
10. W. Langbein: Speckle-Analysis of Resonant Light Emission from Solids. Habilitationsschrift, Universität Dortmund, Dortmund (2002)
11. W. Langbein, J.M. Hvam, R. Zimmermann: Phys. Rev. Lett. **82**, 1040 (1999)
12. W. Langbein, J.M. Hvam: phys. stat. sol. (a) **178**, 13 (2000)
13. W. Langbein: phys. stat. sol. (b) **234**, 84 (2002)
14. W. Langbein, J.M. Hvam, R. Zimmermann: Time-Resolved Speckle Analysis: Probing the Coherence of Excitonic Secondary Emission. In: *Proc. 24th Int. Conf. on the Physics of Semiconductors* ed by D. Gershoni (World Scientific, Singapore 1999)
15. W. Langbein, J.M. Hvam: Excitonic Coherence in Semiconductor Nanostructures Measured by Speckle Analysis. In: *Advances in Solid State Physics*, vol 39, ed by B. Kramer (Vieweg, Braunschweig 1999) pp 463–472
16. W. Langbein, K. Leosson, J.R. Jensen, J.M. Hvam, R. Zimmermann: Phys. Rev. B **61**, R10555 (2000)
17. D. Gammon, E.S. Snow, B.V. Shanabrook, D.S. Katzer, D. Park: Phys. Rev. Lett. **76**, 3005 (1996)

18. W. Langbein, R. Zimmermann, E. Runge, J.M. Hvam: phys. stat. sol (b) **221**, 349 (2000)
19. R. Zimmermann, W. Langbein, E. Runge, J.M. Hvam: Physica E **10**, 40 (2001)
20. R. Zimmermann, W. Langbein, E. Runge, J. Hvam: Speckle Correlation Spectroscopy on Localized Spin-Split Exciton States in Quantum Wells. In: *Proc. 25th Int. Conf. on the Physics of Semiconductors*, ed by N. Miura (Springer, Berlin 2001)
21. E. Runge, R. Zimmermann: Annalen der Physik **7**, 417 (1998)
22. V. Savona, R. Zimmermann: Phys. Rev. B **60**, 4928 (1999)
23. V. Savona, E. Runge, R. Zimmermann, S. Haacke, B. Deveaud: phys. stat. sol. (b) **221**, 365 (2000)
24. G. Kocherscheidt, W. Langbein, U. Woggon, V. Savona, R. Zimmermann, D. Reuter, A.D. Wieck: Phys. Rev. B **68**, 085207 (2002)
25. K. Leosson, J.R. Jensen, W. Langbein, J.M. Hvam: Phys. Rev. B **61**, 10322 (2000)
26. G. Mannarini, F. Grosse, R. Zimmermann, S. Kaßbohm, W. Langbein: Phys. Rev. B **69**, 15th February (2004)
27. D.S. Citrin: Phys. Rev. B **51**, 14361 (1995)
28. T. Stroucken, A. Knorr, P. Thomas, S.W. Koch: Phys. Rev. B **53**, 2026 (1996)
29. L.C. Andreani, F. Bassani: Phys. Rev. B **41**, 7536 (1990)
30. D.S. Citrin: Phys. Rev. B **49**, 1943 (1994)
31. W. Langbein, C. Mann, U. Woggon, M. Klude, D. Hommel: phys. stat. sol. (a) **190**, 861 (2002)
32. M. Hübner, J. Kuhl, T. Stroucken, A. Knorr, S.W. Koch, R. Hey, K. Ploog: Phys. Rev. Lett. **76**, 4199 (1996)
33. G.R. Hayes, J.L. Staehli, U. Oesterle, B. Deveaud, R.T. Phillips, C. Ciuti: Phys. Rev. Lett. **83**, 2837 (1999)
34. M. Hübner, J.P. Prineas, C. Ell, P. Brick, E.S. Lee, G. Khitrova, H.M. Gibbs, S.W. Koch: Phys. Rev. Lett. **83**, 2841 (1999)
35. J.P. Prineas, C. Ell, E.S. Lee, G. Khitrova, H.M. Gibbs, S.W. Koch: Phys. Rev. B **61**, 13863 (2000)
36. J.P. Prineas, J. Shah, B. Grote, C. Ell, G. Khitrova, H.M. Gibbs, S.W. Koch: Phys. Rev. Lett. **85**, 3041 (2000)
37. G. Malpuech, A. Kavokin, W. Langbein, J.M. Hvam: Phys. Rev. Lett. **85**, 650 (2000)
38. A.V. Kavokin, J. Baumberg: Phys. Rev. B **57**, R12697 (1998)
39. D. Birkedal, J. Shah: Phys. Rev. Lett. **81**, 2372 (1998)
40. D. Birkedal, J. Shah, A. Shchegrov, L. Pfeiffer: phys. stat. sol. (a) **178**, 5 (2000)
41. G.R. Hayes, B. Deveaud, V. Savona, S. Haacke: Phys. Rev. B **62**, 6952 (2000)
42. S. Haacke, S. Schaer, B. Deveaud, V. Savona: Phys. Rev. B **61**, 5109 (2000)
43. M. Wörner, J. Shah: Phys. Rev. Lett. **81**, 4208 (1998)
44. S. Haacke, R. Taylor, R. Zimmermann, I. Bar-Joseph, B. Deveaud: Phys. Rev. Lett. **78**, 2228 (1997)
45. G. Kocherscheidt, W. Langbein, G. Mannarini, R. Zimmermann: Phys. Rev. B **66**, 161314(R) (2002)
46. L. Lepetit, G. Chériaux, M. Joffre: J. Opt. Soc. Am. B **12**, 2467 (1995)
47. G. Kocherscheidt, W. Langbein, V. Savona: phys. stat. sol. (b) **238**, 486 (2003)
48. V. Savona, S. Haacke, B. Deveaud: Phys. Rev. Lett. **84**, 183 (2000)
49. A. Kavokin, G. Malpuech, W. Langbein: Solid State Commun. **120**, 259 (2001)
50. B. Grote, C. Ell, S.W. Koch, H.M. Gibbs, G. Khitrova, J.P. Prineas, J. Shah: Phys. Rev. B **64**, 045330 (2001)

Donor-Related Exciton Luminescence in Wide-Bandgap Semiconductors: Diamond, Zinc Oxide, and Gallium Nitride

Rolf Sauer and Klaus Thonke

1 Introduction

The article addresses the topical issue of donors and their luminescence characterization in wide-gap semiconductors. In epitaxial diamond layers grown by chemical vapor deposition on synthetic diamond substrates there is now – after many years of intense research into this issue – firm evidence for phosphorus being a substitutional donor. Conclusive data include Hall measurements, photocurrent and infrared absorption spectroscopies, donor-acceptor pair (DAP) luminescence, and – very recently – bound exciton luminescence showing all optical signatures expected for such excitonic transitions. We will particularly discuss these novel phosphorus-related exciton luminescence spectra. ZnO and GaN are naturally n-type. High-quality samples show in the near-bandgap region multiple, narrow bound exciton recombination transitions due to a variety of different yet not fully identified donors. These spectra are replicated at lower energies, with the excitation of the donors in the luminescence final states, often called bound exciton two-electron satellites. The correlation of these two types of spectra for individual donors is made manifest by studying different samples and by Zeeman spectra in magnetic field as high as 28 Tesla. Once such correlations have been established for individual donors one can use the information contained in the donor final state excitations to determine their ionization energies alone from the photoluminescence spectra.

2 Diamond

2.1 The Donor Problem

Standard elemental impurities in natural diamond are B and N. They give rise to the usual classification of natural diamonds into four principal classes (which is also adopted for high pressure-high temperature (HPHT) synthetic diamonds): Type Ia with a high concentration of aggregated N, type Ib with a much lower concentration of single substitutional N atoms, type IIa which is relatively free from nitrogen (exhibiting no infrared absorption due to N), and type IIb which is semiconducting due to the incorporation of B acceptors [1]. Isolated N atoms generate deep donor states having an ionization energy of $E_\mathrm{d} = 1.7\,\mathrm{eV}$ [2] with little technical applicability. Optically, isolated N atoms are inactive with respect to excitonic luminescence. Instead, N atoms form a variety of complexes with defects, as e.g. vacancies, emitting vibronic point-defect spectra at photon energies extending over a broad range from below 1 eV up to the bandgap of diamond, $E_\mathrm{g} = 5.49\,\mathrm{eV}$ [1, 3]. Nitrogen A aggregates form a very deep donor level at $\approx 4\,\mathrm{eV}$ below the conduction band [4]. Boron acts as a substitutional acceptor with an ionization energy $E_\mathrm{a} = 0.37\,\mathrm{eV}$ [5, 6]. In infrared excitation and photocurrent spectroscopic measurements it exhibits many dipole allowed transitions from the ground state to electronic excited states. A classification of these states in terms of the effective-mass theory turns out to be not straight-forward [1, 7], and this is at least partially due to the lack of unambiguous and sufficiently exact values of the valence band parameters A, B, and C [8]. Neutral B acceptors bind free excitons forming an (A^0,X) complex. Upon radiative recombination, it emits a luminescence spectrum as expected from a 'shallow' Coulomb impurity exhibiting no-phonon (NP) and $\boldsymbol{k}$-vector-conserving TA-, TO-, LO-, and ($\mathrm{TO}+\mathrm{O}^\Gamma$)-phonon replicas. The spectrum has been studied in detail in cathodoluminescence (CL) as early as in the 1960s [9]. (The acceptor binding the exciton and being responsible for the p-type electrical conductivity of type IIb diamonds was at that time identified with Al). Recent advanced studies of the boron-bound exciton ($\mathrm{BE_B}$) spectrum refer to isotope shifts in ^{12}C and ^{13}C isotopically pure diamond crystals [10–12] and to ample fine structure observed in high-resolution measurements due to a fourfold splitting of the luminescent (A^0,X) exciton state and a doublet splitting of the A^0 acceptor ground state [11–13]. In diamond layers grown by microwave-assisted chemical vapor deposition (CVD) the effective B concentration has been varied from $\approx 10^{16}\,\mathrm{cm}^{-3}$ up to $10^{20} - 10^{21}\,\mathrm{cm}^{-3}$ and a concomitant reduction of the ionization energy from 0.37 eV to zero has been observed [14, 15]. This has made possible the fabrication of a number of electronic devices such as field effect transistors using high, nearly δ-doped CVD diamond layers that provide almost temperature-independent hole concentrations [16].

In contrast to the acceptor case there is no natural donor in diamond with ionization energy E_d comparable in size with the value of B, $E_\mathrm{a} = 0.37\,\mathrm{eV}$. Such a 'shallow' donor would be highly desirable for electronic devices working with pn-junctions. Intense efforts over the past years have been directed towards solving the problem of identifying atomic species which exhibit practicable donor properties. Theoretically, a number of donors have been suggested: Li and Na should be interstitial donors with $E_\mathrm{d} = 100$ and 300 meV, respectively, and P should be a substitutional donor with E_d between 200 and 1090 meV [17–19]. Other workers predict P to be a donor with $E_\mathrm{d} = 400\,\mathrm{meV}$ [20]. However, it is expected that Li is a fast diffusor and that in all cases the solid solubilities are very low. Hence it seems unlikely that these elements can be incorporated into the diamond lattice in practicable amounts by diffusion or during the growth. Theoretical works motivated by recent experimental studies on S-doped CVD diamond predict a wide range of ionization energies of a S donor from 150 meV to 1.2 eV, either for neutral or ionized S atoms. Experimental data are contradictory and do not yield unambiguous results. The present activities on the search for donors are summarized in a recent review [21].

Experimental attempts to n-dope diamond by implantation followed by annealing were not just successful. Since diamond is not the stable phase of carbon, this technique is basically problematic. Implantation introduces damage, and the resulting defects may cause electrical conductivity persisting even at high temperatures if the damage is incompletely annealed out. This conductivity can be due to nearest-neighbor or variable-range hopping between the defect states to be distinguished from the conduction band conductivity of carriers which have been exponentially activated from the donor states. Such behavior has e.g. been observed in Li-implanted natural type IIa diamonds [22] and in synthetic HPHT diamonds [23] after annealing. Thermal activation energies for the n-type conductivities from 80 to 270 meV were found and could be firmly ascribed to the presence of Li but the temperature dependencies were typical for a variable-range hopping mechanism. There are many other experimental investigations on the above mentioned donor candidates in the literature; credit is given to them in a number of comprehensive papers, as e.g. the survey on ion-implanted diamond [24] with a recent update on selected issues in ion implantation [25] by Prins, the more general articles on prospective n-type impurities and methods of diamond doping by Popovici and Prelas [26], and on the search for donors in diamond by Kalish [21].

2.2 Phosphorus Donors in CVD Diamond

Attempts to n-dope diamond with P by a technique other than ion implantation [27, 28] were concerned with the addition of P compounds to the gas mixture in the growth of diamond layers either by hot-filament CVD or by microwave-assisted CVD, and with the use of P as a catalyst in the direct phase transformation from graphite to diamond in an HPHT process. In the latter case, the individual diamond grains of cubic appearance that were obtained were too small (sizes of ~ 1 mm) to perform conductivity measurements [29]. In the growth of CVD diamond layers on Si substrates with PH_3 addition by the hot-filament method, preference was given to morphological questions, growth rates, crystal quality, and related issues but the layers were not electrically characterized [30]. In alternative work on hot-filament diamond CVD on Si substrates with the addition of P_2O_5 n-type conductivity was achieved with electron mobilities up to 50 $cm^2/V\,s$ and concentrations in the $10^{15} cm^{-3}$ range at room temperature; however thermal carrier activation energies were only in the 100 meV range, and the authors could not exclude

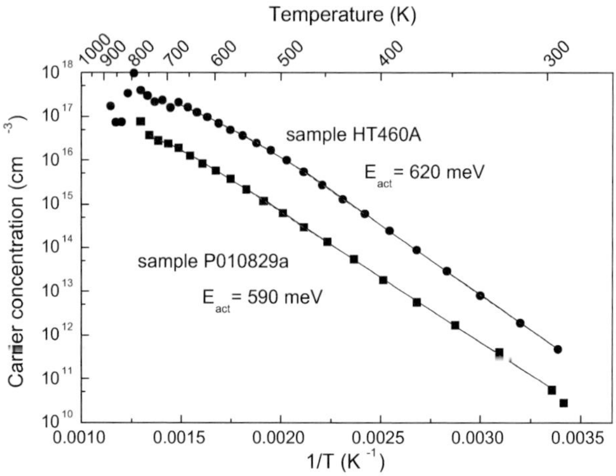

Fig. 1. Temperature dependence of electron concentrations from Hall measurements of two P-doped CVD diamond layers with doping levels of $N_d \approx 4\times10^{17}\,\mathrm{cm}^{-3}$ (sample P010829a) and $N_d \approx 1 \times 10^{19}\,\mathrm{cm}^{-3}$ (sample HT 460A). (From [48].)

the possibility that defects and the presence of oxygen played a major role in their electrical measurements [31].

A breakthrough in n-type doping was finally made by growing diamond layers homoepitaxially on commercially available HPHT synthetic type Ib diamond substrates by microwave-assisted CVD with the addition of PH_3 to the H_2/CH_4 gas mixture [32]. Initially, the electron concentrations were thermally activated with an energy of around 0.34–0.5 eV in a limited temperature range $T \geq 500\,\mathrm{K}$ with Hall mobilities of 20–25 $\mathrm{cm}^2/\mathrm{V\,s}$ at 450–500 K [32, 33]. Also, Si was incorporated into the deposited films, and at room temperature hopping conduction appeared to dominate the electrical measurements. These drawbacks were thought to be mainly due to the use of a quartz-tube reactor in the CVD plasma system. Consequently, in succeeding work, the quartz plasma cavity was replaced by a stainless steel chamber; the amount of CH_4 in the H_2 carrier gas stream – recognized to have to be small for good n-type layers – was reduced to 0.15%, and other parameters were also optimized [34]. With these improvements, state-of-the-art P-doped CVD layers exhibit reproducibly electron concentrations which are exponentially activated in a large temperature range with $E_{\mathrm{act}} \approx 600\,\mathrm{meV}$ and have room temperature mobilities up to 240 $\mathrm{cm}^2/\mathrm{V\,s}$. Donor concentrations from below 10^{17} to above $10^{19}\,\mathrm{cm}^{-3}$ have been obtained. Figure 1 shows an example of such electrical data. A remaining problem is that the PH_3 concentrations, with respect to the H_2 carrier gas, are in the ppb range so that residual gases, such as water vapor or oxygen, can significantly influence the surface chemical reactions during the film growth and lead to badly reproducible values of the doping concentrations. Optical data to be discussed below also show there is a partial compensation of the P donors by acceptors identified with B atoms.

2.3 Photocurrent and Infrared Absorption Spectroscopy

The availability of good P-doped, n-type CVD diamond layers motivated optical studies to investigate the electronic structure of the P donor. Important contributions came from two methods, photocurrent spectroscopy and infrared absorption. Photocurrent (and related optical cross section) measurements on a variety of P-doped CVD diamonds identified two levels at 0.56 eV and 0.81 eV optical excitation energy (Fig. 2) [35, 36]. While the latter level remained unexplained and appeared unrelated with P, the first level scaled in signal amplitude with the P doping concentration, and its energy was remarkably close to the thermal activation energy E_{act} of the carrier concentration. Infrared absorption ("excitation spectroscopy") revealed consistently up to three lines at 523 meV, 562 meV, and 583 meV which were identified with transitions from the P donor ground state of $1s(A_1)$ symmetry to excited $2p_{\pm}$, $2p_0$, and $3p_{\pm}$ states, respectively (Fig. 3) [37, 38].

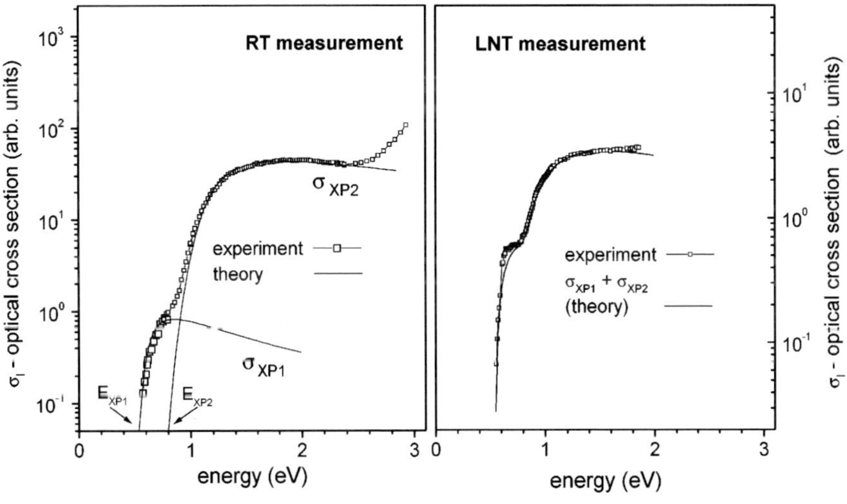

Fig. 2. Optical photoionization cross section σ vs photon energy for a P-doped epitaxial diamond layer. Full lines are fits using standard theory. The activation energy of the dark conductivity was found to be $E_a = (0.55 - 0.60)$ eV (after [35]).

It is expected that the effective-mass theory (EMT) models the binding energies of the p states rather accurately. Indeed, it was possible to match the experimental structure of the excited states with EMT very satisfactorily when the transverse and longitudinal electron masses, m_t and m_l, were slightly modified as compared to literature data obtained in quite different ways (Fig. 4). Addition of the observed spectral transition energies to the theoretical binding energies delivers the energy position of the $1s(\mathrm{A}_1)$ donor ground state which itself is extremely badly described by the theory due to large central cell corrections. The donor ionization energy E_d obtained in this way was initially (600 ± 20) meV [37]. The most recent value is $E_\mathrm{d} = 604$ meV with an apparently much lower error not explicitly specified [38].

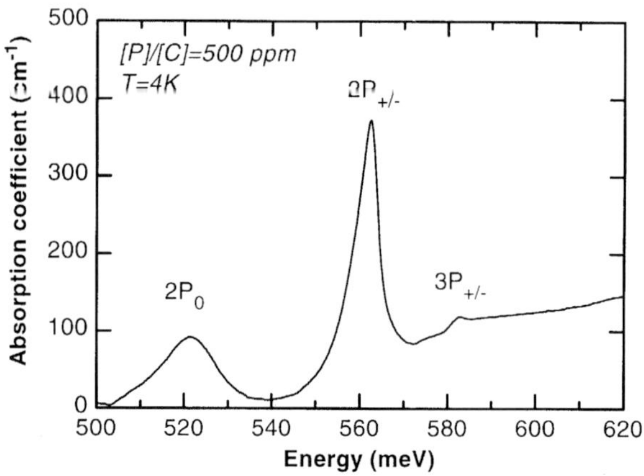

Fig. 3. Infrared absorption of a P-doped CVD diamond layer grown with a [P]/[C] ratio of 500 ppm in the vapor phase exhibiting dipole transitions from the donor P ground state $1s(\mathrm{A}_1)$ to excited p-like states. (From [38].)

Comparing this value of the ionization energy with the thermal activation energy of the electron concentration from Hall measurements, $E_\mathrm{act} \approx 600$ meV, it is clear that there must be a partial compensation of the donors according to standard carrier statistics to yield $E_\mathrm{act} = E_\mathrm{d}$; if no compensation were assumed the electrical measurements would yield $E_\mathrm{act} = E_\mathrm{d}/2$.

2.4 Cathodoluminescence Measurements

Cathodoluminescence measurements on diamonds doped with P were performed since about 1995. The expectation was that a donor P would give rise to a characteristic bound exciton ($\mathrm{BE_P}$) emission spectrum similar in appearance to that emitted by acceptor B-doped, p-type conducting diamonds [9] or to the photoluminescence spectra related with the 'shallow' acceptors B, Al, Ga, In and the 'shallow' donors P, As, Sb, Bi in Si [39]. B-doped natural and HPHT diamond crystals exhibit phonon-free and $\boldsymbol{k}$-vector-conserving phonon-assisted exciton transitions. The NP transition is a close doublet at 5.368 eV and 5.356 eV, and the phonon replicas are shifted by 141 meV (TO-phonon), 163 meV (LO-phonon) and 308 meV (combined TO and O^Γ-phonons). More complex fine structure has been recognized in recent

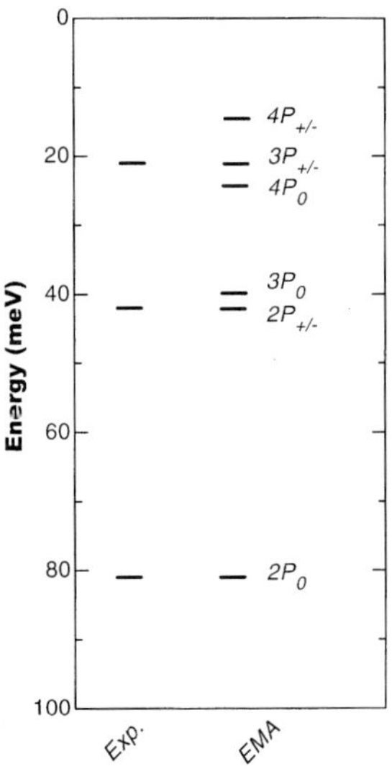

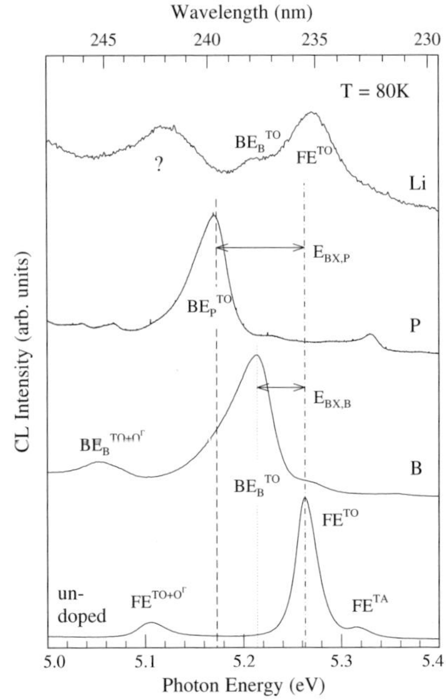

Fig. 4. (Left figure.) Binding energies for donor electrons calculated by the effective-mass theory (EMA) and experimental values. The zero of energy corresponds to the conduction band edge. (From [38].)

Fig. 5. (Right figure.) Cathodoluminescence spectra at $T = 80\,\mathrm{K}$ of polycrystalline diamond films doped with B, P, and Li. Bottom spectrum: Undoped film showing free exciton emission with the participation of $\boldsymbol{k}$-vector-conserving TA-, TO-, and (TO+O^{Γ})-phonons. B doping: Spectrum as known from the literature. Li doping: Association of line at 5.12 eV with Li unclear. P doping: Identification of lines at 5.33 eV and 5.175 eV with P-bound excitons $\mathrm{BE}_{\mathrm{P}}^{\mathrm{NP}}$ and $\mathrm{BE}_{\mathrm{P}}^{\mathrm{TO}}$, respectively. Hot-filament CVD layer grown by S. Bohr and R. Haubner, Technical University of Vienna. (From [41].)

high-resolution CL measurements consisting of eight NP transitions [11, 13] but will not be discussed here. The bound exciton spectrum is shifted from the low-energy threshold of the free exciton radiation ($\mathrm{FE}^{\mathrm{TO}}$ at 5.266 eV) by 51 meV, which is the localization energy E_{loc} of the exciton at the acceptor.

First indications of a donor-exciton spectrum were obtained 1996. The samples were polycrystalline CVD diamond layers grown by the hot filament technique on Si: B substrates doped with P by the addition of suitable compounds or grown on Si: P substrates. P incorporation was demonstrated by SIMS measurements. Observed were new peaks in the CL spectra close to 5.175 eV and – very weakly – 5.32 eV ascribed tentatively to bound exciton

TO-phonon and NP-transitions, respectively, related with the P doping [40]. Since at that time P was predicted theoretically to be a donor with ionization energy from $\sim$ 200 meV up to more than 1090 meV it was not evident whether these CL lines could be associated with isolated P atoms or with a P complex incorporating also other impurity species or structural defects. The luminescence data could be substantiated in subsequent investigations: Nearly 20 P-doped samples showed the lines as in Fig. 5 with intensities related to their P doping concentrations [41].

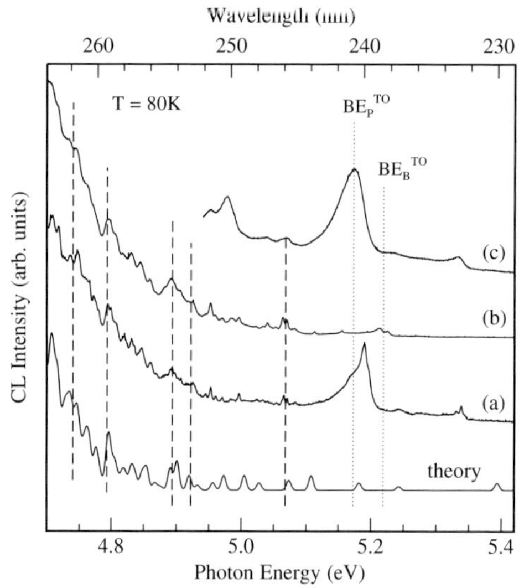

Fig. 6. Donor-acceptor pair (DAP) spectra ($h\nu \leq 5$ eV) due to P donors and B acceptors, and bound exciton spectra ($h\nu > 5$ eV) in three P-doped diamond samples. Sample (a) grown epitaxially by S. Koizumi, NIMS (Tsukuba), using microwave CVD; sample (b) grown by HPHT using P as a catalyst by M. Akaishi and D. Michau, NIMS (Tsukuba), and sample (c) grown on a Si substrate using the hot-filament technique by S. Bohr and R. Haubner, Technical University of Vienna. Bottom spectrum: Simulation following standard DAP theory, see text. (From [42].)

Unambiguous optical evidence for P as a donor was contributed by line-resolved donor-acceptor pair (DAP) spectra [42]. A number of samples showed rich line spectra in the region from $\sim$ 5 eV down to $\sim$ 4.6 eV, and in all samples these spectra were essentially identical (Fig. 6). The spectra were observed from CVD layers grown homoepitaxially on type Ib synthetic diamond substrates which exhibited the P-related exciton lines close to 5.175 eV and 5.32 eV; they were also observed from HPHT crystals where this near-band-edge exciton radiation was absent. Crystal growers have learnt that B is often unintentionally introduced into diamond layers even when B doping was not performed in previous growth experiments or when B could not originate from B-doped Si substrates or from plasma etching of the silica walls of the CVD reactor. Hence it was natural to assume that the DAP line spectra were due to recombination transitions between electrons at P donors and holes at B acceptors [42]. Following the procedure of evaluating DAP spectra in GaP, Si and in other traditional cases, theoretical model spectra were generated assuming for sufficiently distant pairs only Coulomb pair interactions to yield photon energies

$$h\nu = E_{\mathrm{g}} - (E_{\mathrm{d}} + E_{\mathrm{a}}) + \frac{e^2}{4\pi\varepsilon\varepsilon_0 R_i}. \quad (1)$$

R_i is the (discrete) donor-acceptor distance which in an fcc elemental crystal can take the values $R_i(m) = a_0\sqrt{(m-b)/2}$ with $b = 0$ for DA pairs on the same sublattice and $b = 5/8$ for DA pairs on different sublattices, and m is the "shell number" [43]. The lattice constant of diamond is $a_0 = 3.568$ Å, the dielectric constant $\varepsilon = 5.7$, E_{g} at low temperatures equals 5.49 eV, and $E_{\mathrm{a}} = 370$ meV was assumed for the anticipated acceptor B. The spectral line intensities were assumed to reflect in a certain range of R_i values the statistical degeneracy factors of equivalent pairs occurring at R_i. A model spectrum is shown in Fig. 6 simulating quite satisfactorily a number of characteristic groups of lines for photon energies below 5.0 eV. A second model spectrum – though considered less probable in describing the experimental spectra – was also taken into account finally resulting in a value for the donor ionization energy of $E_{\mathrm{d}} = (630 \pm 50)$ meV [42]. If the theoretical spectrum of Fig. 6 is taken and only the line groups close to 4.8 eV ($m = 26$) or 4.9 eV ($m = 14$) are considered, the resulting ionization energy is $E_{\mathrm{d}} = (600 \pm 3)$ meV in very good agreement with the value deduced from the infrared absorption measurements [44].

As mentioned earlier, an important improvement in the deposition of P-doped CVD diamond layers on type Ib synthetic diamond substrates was made by the replacement of the quartz-tube reactor by a stainless steel metal chamber. This had also bearing on the quality of the near-gap CL spectra. In such layers, bound exciton spectra were observed [45] confirming on one hand the previous data [40, 41] but revealing also new transitions. The spectrum consists of peaks at 5.32 eV, 5.18 eV, and 5.02 eV identified with P-induced $\mathrm{BE}^{\mathrm{NP}}$, $\mathrm{BE}^{\mathrm{TO}}$, $\mathrm{BE}^{\mathrm{TO+O^{\Gamma}}}$ bound exciton transitions, respectively. The $\mathrm{BE}^{\mathrm{TO}}$ peak has as a linewidth of ≈ 44 meV comparable to the value of ≈ 38 meV in [41]. In addition, free exciton transitions [9] at 5.27 eV ($\mathrm{FE}^{\mathrm{TO}}$) and 5.11 eV ($\mathrm{FE}^{\mathrm{TO+O^{\Gamma}}}$) were resolved whose appearance documents the good quality of these layers. The $\mathrm{BE}^{\mathrm{NP}}$ and $\mathrm{BE}^{\mathrm{TO}}$ peaks have slightly increased energy positions compared to the initial data [40, 41]. This is possibly a consequence of the lower doping level achieved in these more recent growth runs. Low-energy shifts of bound exciton transitions for increasing doping levels due to exciton-exciton exchange interactions are well known for many semiconductors and have particularly for B-doped CVD diamond been documented [46].

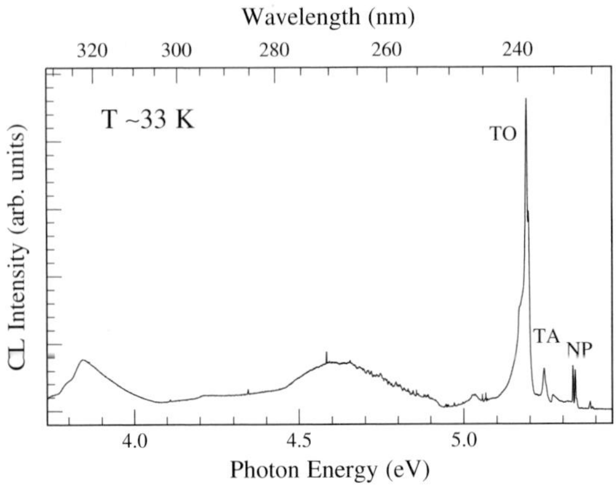

Fig. 7. Survey CL spectrum of a P-doped epitaxial diamond layer grown by microwave-assisted CVD on a synthetic type Ib diamond (Sample P 010829a from Fig. 1). Bands below 5.0 eV are due to unintentional B doping, lines above 5.0 eV are P-bound exciton transitions. (From [48].)

The new P-bound exciton spectra were studied as a function of temperature and nicely showed the interplay between free and bound excitons by exciton trapping at and releasing from the binding donors [45].

Very recently, again a big step forward has been made [47, 48]. CVD samples doped from 10^{17} to some $10^{18}\,\mathrm{cm}^{-3}$ with P donors revealed exorbitantly narrow bound exciton transitions and, at the same time, well resolved DAP spectra. Fig. 7 is a survey CL spectrum exhibiting at $h\nu > 5.0$ eV the P-bound exciton ($\mathrm{BE_P}$) spectrum. The spiky part of the spectrum between 4.5 eV and 5.0 eV is no noise but the DAP spectrum. When one zooms into this part it is exactly identical to the spectra shown in Fig. 6. Another sample shows the $\mathrm{BE_P}$ spectrum much more weakly but has high intensities in the DAP spectrum which is again absolutely identical to that depicted in Fig. 6 except for better resolution due to narrower linewidths. In Fig. 7 there are two broad bands centered at ≈ 3.8 eV and 4.6 eV with a weak maximum at ≈ 4.2 eV in between. A further broad band at 2.3 eV is not shown in the figure. These bands have been shown to be characteristic of B-doped HPHT and CVD diamond [49]. The 2.3 eV band is part of a very broad emission covering the green (2.3 eV) and blue (2.9 eV) spectral ranges. It was traditionally labelled 'band A' recombination until it was shown by a number of works that the green portion of the band correlates with B incorporation and the blue portion is due to radiative recombination at dislocations. From these broad bands it is obvious that the CVD layers contain unintentionally introduced B acceptors. Hence it is more than natural to conclude that the thermal activation energy E_{act} of the electron concentration in these n-conducting samples

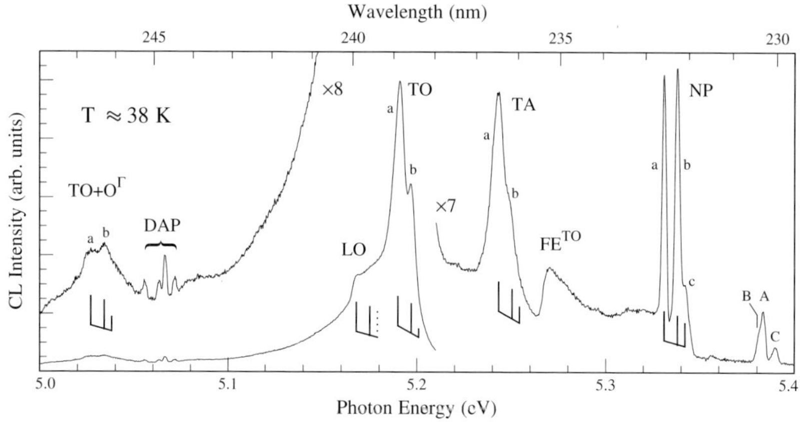

Fig. 8. High-resolution P-bound exciton (BE_P) spectrum showing narrow split NP lines (a, b, c) and $\boldsymbol{k}$-vector-conserving phonon replicas. The fine structure is at least partially recognized in all phonon satellites (see combs). B, A, C denotes an independent CL spectrum appearing very closely below the excitonic gap energy of $E_{gx} = 5.4060\,\mathrm{eV}$. (From [48].)

is identical to the donor ionization energy E_d due to the partial compensation of the P donors by B acceptors.

The high-energy portion of the spectrum in Fig. 7 is displayed on an expanded scale in Fig. 8. It exhibits a very sharp NP transition followed by $\boldsymbol{k}$-vector-conserving TA-, TO-, LO-, and ($TO+O^{\Gamma}$)-phonon replicas which are all unambiguously identified following previous investigations of free and bound exciton spectra in natural and HPHT diamonds [9]. In particular, the phonon energies are 88 meV (TA), 140 meV (TO), 162.4 meV (LO), and 304.2 meV ($TO+O^{\Gamma}$) so that a value of 164.2 meV results for the O^{Γ} (Raman) phonon. The FWHM linewidth in the NP transition is only 2 meV, and this exorbitantly narrow width allows to resolve a triplet line structure a, b, c. This fine structure is recognizable in the various phonon replicas although washed out due to their larger widths estimated to be $\leq 9\,\mathrm{meV}$. The whole BE spectrum is as nearly perfect in appearance as the BE spectra obtained from donor- or acceptor-doped Si at impurity concentrations comparable to the present CVD diamond values. The nature of the fine structure remains presently unexplained. The splitting is in the luminescence initial, excitonic state as the line components b, c are thermally activated compared to component a at increasing temperature (Fig. 9). Also, additional line components (a', b') come up yielding in total a complex split bound exciton state. Potential reasons for the splitting could be the coupling of the angular momenta of the two electrons and the hole in the (D^0,X) complex to form different values of the total angular momentum, or valley-orbit splitting effects of the bound electrons. It has also to be taken into account that in diamond the spin-orbit splitting of holes is small, amounting to 6 meV (experimentally) [50] and 13 meV (theoretically) [51]. Bound holes often exhibit a reduction of the values for free holes so that part of the observed fine structure could come from the spin-orbit effect. The multiplicity of the (D^0,X) spectrum is reminiscent of the eightfold fine structure observed in the (A^0,X) spectrum where it also could only partially be explained [11].

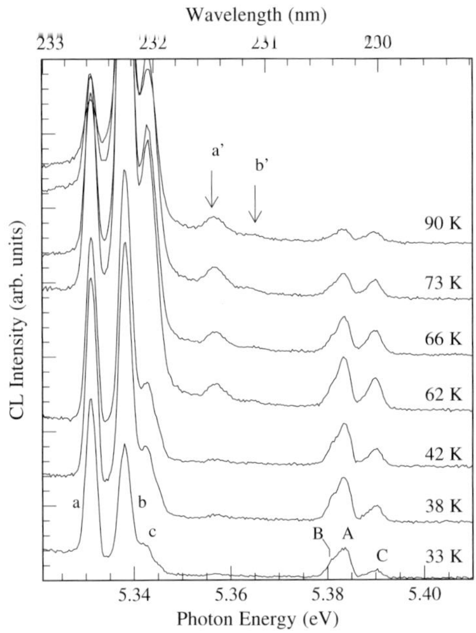

Fig. 9. Lines a, b, c from excited exciton states at increasing temperatures and appearance of additional lines a', b'. Temperature dependence of the B, A, C triplet. (From [48].)

Two features need to be discussed which are characteristic of 'shallow' bound excitons and support the assignment of the present new spectrum to isolated substitutional P donors as the binding centers. The exciton localization energy E_{loc} at the impurity is given by the spectral shift of the bound exciton transition from the low-energy threshold of the free exciton. The new spectrum related to the P donors with $E_{\mathrm{d}}(\mathrm{P}) = 604\,\mathrm{meV}$ yields $E_{\mathrm{loc}}(\mathrm{BE_P}) = 75\,\mathrm{meV}$ from Fig. 7 whereas for the well-known boron-bound exciton with $E_{\mathrm{a}}(\mathrm{B}) = 370\,\mathrm{meV}$ a value of $E_{\mathrm{loc}}(\mathrm{BE_B}) = 51\,\mathrm{meV}$ is obtained. Hence, $E_{\mathrm{loc}}(\mathrm{BE_P})/E_{\mathrm{d}}(\mathrm{P}) = 0.124$ and $E_{\mathrm{loc}}(\mathrm{BE_B})/E_{\mathrm{a}}(\mathrm{B}) = 0.138$. On average, one may take $E_{\mathrm{loc}} \approx 0.13E_i$ (with $E_i = E_{\mathrm{d}}$ or E_{a}) which constitutes an analogue of Haynes' rule for donors and acceptors in silicon, $E_{\mathrm{loc}} \approx 0.1E_i$

[52]. The second characteristic feature refers to the relative strength of the NP transition in the BE spectrum to the dominant TO-phonon replica. As excitons are more strongly localized at P donors than at B acceptors their wavefunctions are more widely spread in $\boldsymbol{k}$-space in the first case yielding higher oscillator strength in the quasi-direct NP transition. This expectation is experimentally fully satisfied: The relative NP intensities, $I(\mathrm{NP})/I(\mathrm{TO})$, are ≈ 0.03 for P-bound excitons and ≈ 0.002 for B-bound excitons. Summing up our discussion of the new spectrum in Fig. 7 it is evident from this optical signature that P behaves as a typical 'shallow' donor in diamond and that many features are very similar to the classical cases of shallow donors and acceptors in silicon.

A novel puzzling luminescence feature arises in all P-doped samples at high photon energies very close to the excitonic band edge. This is the triplet of lines labelled B, A, C in Figs. 8 and 9. The B line appears as a shoulder on the A line and is positioned at 5.38 eV, only $E_{\mathrm{loc}} = 26$ meV below the exciton threshold E_{gx} at 5.406 eV. The lines are rather narrow (in terms of CVD diamond) at FWHM of 3–4 meV. The spectrum has not been observed previously and has to be distinguished from a number of luminescence lines which in different publications were reported to emerge at high photon energies up to 5.361 eV. Interesting in the present context is a line pair C' (5.374 eV) and D' (5.361 eV) in type IIb diamond clearly related with the boron doping of such diamonds [53]. In our case, Fig. 9 shows that line C is thermally activated compared to line A with an energy $E_{\mathrm{act}} \approx 6$ meV which is close to the spectral spacing of the two lines of 7 meV. The whole triplet is quenched at increasing temperature with $\approx 20 - 25$ meV consistent with $E_{\mathrm{loc}} = 26$ meV, the shift of the spectrum from E_{gx}. Taken together these properties prove the triplet to originate in the radiative recombination of an exciton bound very loosely at a center X of yet unknown nature.

It can be argued that X is not a simple Coulombic impurity since the acceptor B ($E_{\mathrm{a}} = 370$ meV) is the shallowest known Coulomb impurity in diamond, and no other elemental species is expected to have lower ionization energy. The triplet shows no $\boldsymbol{k}$-vector-conserving phonon replicas (the very weak peak at ≈ 5.295 eV in Fig. 8 may be an indication of a faint replica), again excluding a very shallow Coulomb binding center. The small exciton localization energy of only 26 meV appears to discard also a model for X in terms of a structural center for which one expects a much stronger exciton localization. An interesting suggestion has been made in terms of an isoelectronic center [47, 48]. The optical signature exhibited by the triplet is characteristic of excitons at axial centers of isovalent nature. Among many examples documented in the literature for a variety of semiconductors there is one in Si representative in its optical properties for this kind of centers. This is the A, B, C line triplet at 1.223 eV whose luminescence was studied intensely (preferentially in Al-doped Si) proving that a $\langle 111 \rangle$-oriented isoelectric complex is the binding center [54] before succeeding work brought evidence for the center to be formed by next-nearest Al-N pairs [55]. The present triplet was named after this Si example. The isoelectronic complex as a neutral center binds an exciton by core potential differences to the lattice atoms or size-induced local strain, and the total angular momenta made up by $j_e = 1/2$ (electron) and $j_h = 3/2$ (hole) are J (exciton) $= 2$ and 1 split into states $|J, M_J\rangle = |2, \pm 2\rangle$, $|2, \pm 1\rangle$, $|2, 0\rangle$, $|1, \pm 1\rangle$, and $|1, 0\rangle$ by the strain.

The $|2, \pm 2\rangle$ exciton state lies lowest in energy and dipole transitions are forbidden to the "vacuum" ground state $J = 0$ unless there is a perturbation relaxing the selection rule. A weakly allowed transition would yield the B line. Line A is due to the dipole-allowed transition from the $|2, \pm 1\rangle$ state, and the excited state C line can be related with one of the remaining states which are possibly overlapping. While this model of an isoelectronic center is fully consistent with the observed spectral features, potential atomic species making up the pair suggest themselves: The CVD diamond layers incorporate P as donors and – well documented as discussed earlier in this article – B as unintentionally introduced compensating acceptors; these two can form close isoelectronic P-B pairs. Also consistent with this model is the very small exciton binding energy of 26 meV when one remembers the situation found for isoelectronic centers in GaP. Here, the binding energy is small for centers whose atomic species fit well into the host lattice, as for N_P ($E_{loc} = 11.5$ meV) but can became very large when there is a significant lattice mismatch, as for Bi_P ($E_{loc} = 98.5$ meV); $Li_{Ga} - Li_{Ga} - O_P$ ($E_{loc} = 238.1$ meV); $Cd_{Ga} - O_P$ ($E_{loc} = 420.2$ meV) [56]. Given B and P atoms in diamond, being either isochoric with C or close to C in the periodic table, respectively, one would expect a very small excitonic localization energy for P-B pairs as observed. Hence, all experimental data are consistent with the model of an isoelectronic B-P pair. It is an interesting question whether the model can be further substantiated by future experimental data.

Returning to the donor problem it has now firmly been established by the CVD growth of diamond epitaxial layers and various measuring techniques that P is a donor in diamond, and the present article has illustrated the optical signatures of excitonic radiative recombination related with it.

3 Donors in ZnO and GaN: Determination of Ionization Energies from Photoluminescence (PL) Spectra

In modern optoelectronics, the direct band structure, high-band-gap material gallium nitride (GaN) plays a significant role for light emitters in the blue to ultraviolet spectral region; similarly, it is vitally important in electronics, and first high power/high-frequency devices are available [57]. Although now being widely used, there is still an ongoing dispute on the nature and the exact binding energies of shallow donors which are either intentionally added or omnipresent as background doping species [58, 59].

Recently, there is renewed interest in the possible competitor material zinc oxide (ZnO) which has basic properties very similar to GaN: The direct bandgap energy is just slightly smaller (ZnO: $\approx 3.3\,\mathrm{eV}$, GaN: $\approx 3.4\,\mathrm{eV}$ at room temperature), and the crystal structure is also hexagonal with similar lattice constants. Since in earlier efforts some 30 years ago, the problem of doping ZnO p-type was never solved [60] (again very similar to the GaN case), the interest in this material faded away for a long time. With state-of-the-art crystal growth methods like MOCVD and MBE, new promising attempts in obtaining p-type conductivity – though partially still low and related with high compensation degrees – were reported [60–63].

In low-temperature photoluminescence spectra of nominally undoped (effectively low n-type) GaN and ZnO the near-bandgap spectral emission is dominated by the decay of bound excitons, i.e. the radiative recombination of an electron-hole pair bound by exchange forces to a neutral donor or acceptor. Due to differences of the core potentials, the total energy of the bound electron-hole pair varies slightly for different donors or acceptors. For both materials, these bound exciton lines are very sharp for samples with sufficiently good material quality but fall into a very narrow energy interval making their identification difficult. Whereas it has mostly been clarified for GaN which lines are donor-related and which lines have to be assigned to acceptors, there are contradictory reports for ZnO: Older work, based on only partially resolved magneto-PL spectra, claims the majority of the lines to be acceptor-related [64], however, more recent publications agree that the commonly observed lines in state-of-the-art bulk ZnO crystals are donor-related [65, 66].

3.1 Two-Electron Satellites

The assignment of a bound-exciton (BE) line to being donor- or acceptor-related can be most firmly based on magneto-PL data. In a favorable case, such data can deliver a complete set of allowed and weakly forbidden Zeeman components together with their thermalization behavior which serves to distinguish splittings occurring in the transition initial and final states. Only when comparison to model calculations for the splittings is satisfactory and the effective g-values derived have sizes consistent with those known for free particles one can consider a possible assignment reliable.

Another reliable indicator for bound exciton emission to be donor-related exists in the so-called "two electron satellites" (TES). The TES lines occur when the electron of the neutral donor D^0 is excited to $2s$, $2p$, etc. final states during the bound exciton recombination process whereas it is left in its $1s$ ground state for the principal BE transition. Hence, the TES lines appear as low-energy sidebands of the principal BE transitions with spacings specific for individual donors. Since the partial energy transfer from the exciton to the final-state bound electron is less probable than the direct exciton recombination to the donor ground state, the TES satellites are much weaker than the main (D^0,X) transition. The spacing of a TES line from its parent main (D^0,X) transition delivers the energy difference between the donor ground state and the corresponding excited state. Thus, TES satellites allow a rather direct determination of donor single particle binding energies with a precision comparable to that of donor infrared absorption or photocurrent measurements.

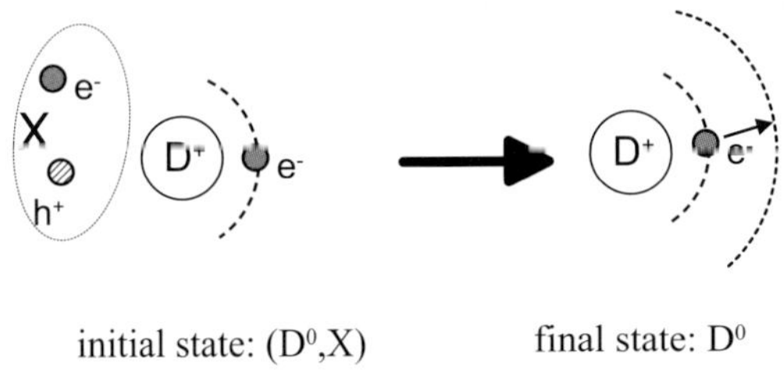

Fig. 10. Recombination scheme for a donor-bound exciton. Left: Initial state incorporating an electron-hole pair (exciton "X") bound to the neutral donor (positive core D^+ with electron e^-). Right: Final state incorporating core D^+ and electron e^-. In the normal (D^0,X) recombination, the donor electron remains in the ground state 1*s*. In the TES transition, it is left in an excited state 2*s*, 2*p*, 3*s* etc.

Analogous considerations hold for acceptor-bound excitons for which two-hole transitions can occur. Since for acceptors in GaN and ZnO the hole binding energies are significantly larger than the donor electron binding energies, two-hole satellites are much more widely spaced from their parent bound-exciton transition than in the case of donors .

The first TES transitions were identified experimentally by Dean et al. for GaP [67], where a whole excitation series of the bound electron from $2p_0$ to $5p_\pm$ was resolved. Similar spectra were found in donor-doped silicon [68] which allowed a reliable comparison to experimental energy values from other measurements and to theoretical predictions. Also for a variety of other materials such as CdS [69, 70], CdSe [71], InP [72], GaAs [73–75], and CdTe [76] TES lines were studied in much detail. For GaN, TES lines were first reported by Skromme et al. [80].

3.2 Two-Electron Satellites for ZnO

For ZnO, TES transitions of donor-bound excitons were reported as early as 1969 by Reynolds et al. [77] and later by Klingshirn et al. [78] and Reynolds et al. [79]. In state-of-the-art ZnO bulk crystals grown by the chemical vapor transport (CVT) technique or by hydrothermal processes, the PL spectra show at low sample temperatures extremely sharp bound exciton lines with FWHM linewidths as low as $\approx 110\,\mu eV$ [66, 81], accompanied by several TES lines. An example is shown in Figs. 11 and 12 for as-grown CVT material.

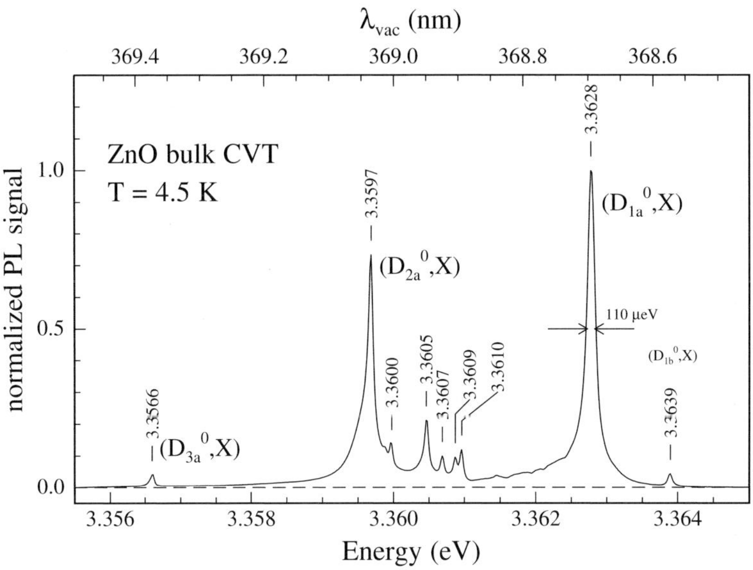

Fig. 11. High-resolution PL spectrum of a typical CVT ZnO substrate as received (Eagle Picher). The spectral resolution was ≈ 30 μeV.

The dominant features are the donor-bound exciton lines labelled (D^0_{1a},X) at 3.3628 eV and (D^0_{2a},X) at 3.3597 eV. The first line is most probably identical to the line labelled "I4" in earlier work [64, 82] and was related to a very shallow donor which incorporates hydrogen [83]. It disappears after annealing above 600°C. The line labelled (D^0_{1b},X) at 3.3639 eV, i.e. 1.1 meV above the (D^0_{1a},X) line, is a transition from an excited state of the donor-bound exciton as proved by temperature-dependent PL measurements [81] and Zeeman data. Several bound excitons typically observed by excitation spectroscopy in older ZnO crystals reveal luminescence from excited states ≈ 5 meV above the BE lines which was related to holes from the next higher valence band [84]. Groups of lines from excited states at 1.5 meV, 5 meV, and 12 meV excess energy were found for the dominant BE transitions by Gutowski et al. [85], who interpreted the majority of them as acceptor-related.

The (D^0_{2a},X) line remains stable after annealing and was recently assigned to Ga contaminants [86]. The intermediate sharp lines are, according to our Zeeman data, either excited state transitions or related to further, independent donors. The line (D^0_{3a},X) at 3.3566 eV (presumably the same line as labelled "I9" earlier [64, 82]) was reported to increase after annealing [65] and might be related to indium contaminations [86].

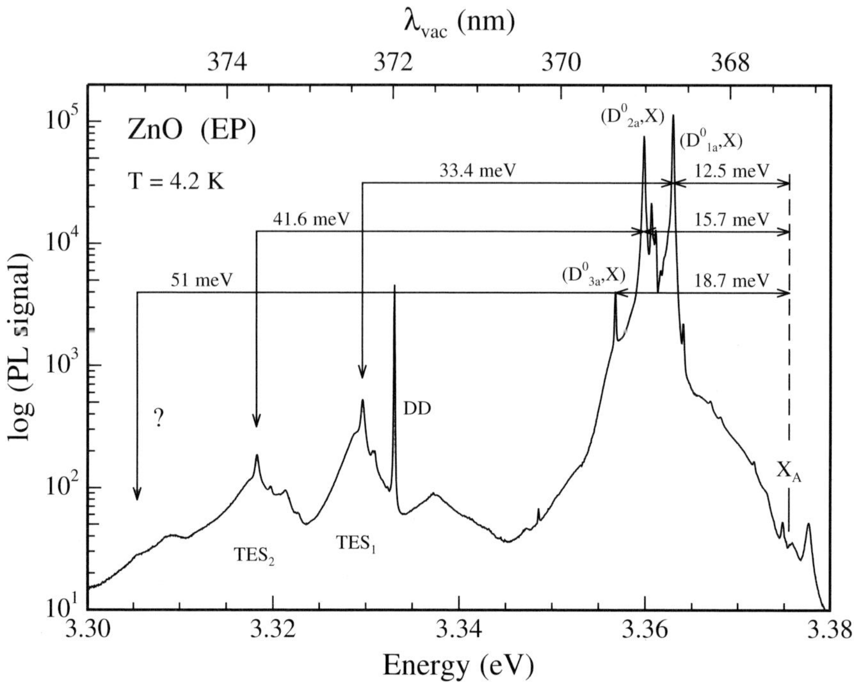

Fig. 12. Overview PL spectra in the range of bound exciton lines and their related two-electron satellites (logarithmic intensity scale). "X_A" marks the energy position of the A valence band free exciton as determined from reflectance measurements on the same sample, and "DD" is the emission from an exciton bound to a deep donor.

For each of the strong and sharp (D^0,X) transitions we find at 30–50 meV lower energy a group of two-electron satellite lines labelled "TES_1", "TES_2", ... (Fig. 12). They scale in intensity with their parent (D^0,X) line when different samples are investigated. Also Zeeman spectra to be discussed below directly prove this relationship.

In a series of Zeeman experiments, the splitting of the (D^0,X) lines in magnetic fields up to 28 T was analyzed [81]. Details of line splittings for (D^0_{1a},X) and (D^0_{1b},X) are shown in Fig. 13. The other strong lines in the range from 3.356 eV to 3.362 eV including the "DD" line reveal essentially the same splitting patterns, and also the splitting rates are very similar. We discuss the Zeeman spectra in some detail for (D^0_{1a},X) and (D^0_{1b},X). The main transition (D^0_{1a},X) splits into two strong inner Zeeman components and two much weaker, quasi-forbidden outer components. The excited transition (D^0_{1b},X) splits twofold. The fact that both transitions, (D^0_{1a},X) and (D^0_{1b},X), are due to one donor manifests itself in strong nonlinear splitting rates of some components when approaching at increasing field strength. Specifically, we

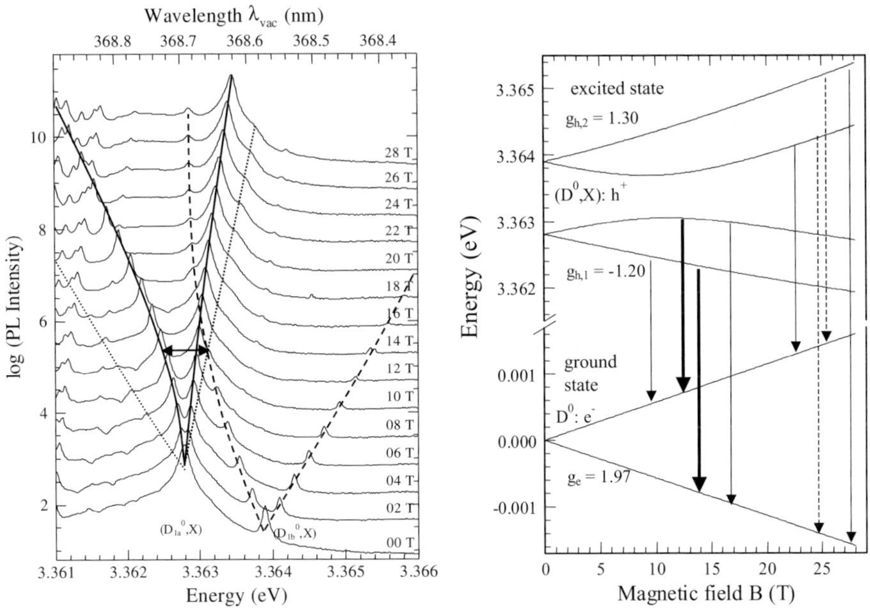

Fig. 13. Left: Details of the Zeeman splitting of the line (D^0_{1a},X) and its related transition from an excited initial state (D^0_{1b},X) for fields from 0 T to 28 T. The solid line marks the splitting of the strong inner components of (D^0_{1a},X), the dotted line that of the much weaker, quasi-forbidden outer components. For (D^0_{1b},X) we see a twofold splitting (dashed lines). Around $B = 10\,\mathrm{T}$, a clear anti-crossing behavior of the low-energy components of both lines is observed as marked by the double arrow "↔". Right: Schematics of the simple model which fits the observed line splittings.

find a pronounced anti-crossing behavior between those components marked by a double-arrow in Fig. 13.

The data can be simulated by a simple model developed by Hopfield and Thomas [87, 88] and Wheeler and Dimmock [89] for excitons in an anisotropic crystal. One assumes that the two electrons in the (D^0,X) complex couple to a $S = 0$ singlet state with dipole-allowed transitions of the bound exciton to the D^0 final state, and to a $S = 1$ triplet state with forbidden dipole transitions. Thus, referring to luminescence, in the exciton complex only the hole splits in a magnetic field, and in the final state only the remaining donor electron splits. In our case, this model has to be extended by the inclusion of an interaction between the holes in the (D^0_{1a},X) and (D^0_{1b},X) complexes. Group-theoretical analysis shows, that the interaction observed for $\boldsymbol{B}\|\boldsymbol{c}$ (i.e. Faraday configuration, with $\boldsymbol{c}$ $\|$ hexagonal axis) can only take place between states of character Γ_7. The Hamiltonian describing this interaction reads as

$$\left(\begin{array}{c|cc} & \Psi_1 & \Psi_2 \\ \hline \Psi_1 & \pm g_{\mathrm{h1}}\mu_{\mathrm{B}}B/2 + qB^2 + E_{1,0} & c\mu_{\mathrm{B}}B \\ \Psi_2 & c\mu_{\mathrm{B}}B & \pm g_{\mathrm{h2}}\mu_{\mathrm{B}}B/2 + qB^2 + E_{2,0} \end{array}\right) \quad (2)$$

where $E_{1,0}$ and $E_{2,0}$ are the zero-field energies of the lower- and higher-energy hole states, respectively, $g_{\mathrm{h1,2}}$ are the effective hole g-factors, c is the factor weighting the field-induced linear interaction between the hole states, and q is a term describing the quadratic diamagnetic shift. The inner, allowed components of a $\Gamma_7 \rightarrow \Gamma_7$-transition are expected to split in the linear range with a rate $\pm\frac{1}{2}(g_{\mathrm{e}} - g_{\mathrm{h}})\mu_{\mathrm{B}}B$, whereas the outer (forbidden) components split with a rate $\pm\frac{1}{2}(g_{\mathrm{e}} + g_{\mathrm{h}})\mu_{\mathrm{B}}B$. We find for the electron a g-factor of 1.97 ± 0.02 in close agreement with the value 1.957 found in EPR experiments [83]. To obtain the correct interaction, one of the g-factors for the two hole states has to be set negative. Based on the selection rules, we derive from our best fit $g_{\mathrm{h},1} = -1.20 \pm 0.05$ and $g_{\mathrm{h},2} = +1.30 \pm 0.05$. The absolute value of the hole g-factors is close to the value found for free excitons [90]. For the excited hole state, with a transition strength lower by a factor of ≈ 5, the symmetry of the sub-states is inverted, since now the outer components are visible and the inner ones are forbidden. The interaction term is obtained as $c = 0.45$, and the diamagnetic term as $q \approx 0.3\,\mu\mathrm{eV/T}^2$ close to the theoretically expected value $q = 0.44\,\mu\mathrm{eV/T}^2$ [89]. The negative sign of the g-factor of the lower hole state together with its Γ_7 character is a strong indication that indeed for ZnO the ordering of the valence bands is inverted relative to most other hexagonal semiconductors. This was suggested by Lambrecht et al. [91] to occur due to hybridization of the p-orbitals with the d-orbitals of Zn.

After the precise fit of the (D^0,X) Zeeman splitting patterns, the TES satellites can be analyzed. There, the full patterns of (D^0_{1a},X) and (D^0_{1b},X) are replicated but superimposed on the splittings and shifts (relative to the 1s donor ground state) of the $(2p, 2s)$ single electron D^0 states. Since for the excited donor electron states the magnetic field-induced splittings become similar in size to their spacings, mixing effects of orbitals have to be taken into account. This problem was treated earlier for the analogous problem of hydrogen atoms in very high magnetic fields by Roesner et al. [92] and can be scaled to the present case. Another complication which has to be accounted for is the anisotropy of both the electron mass and the dielectric constant for hexagonal semiconductors like ZnO and GaN. This leads to a splitting of the $n = 2$ electron state into three substates $2s$, $2p_0$, and $2p_\pm$, and furthermore, to a shift of the $1s$ ground state. Depending on the sign of the mass anisotropy, the ordering of the states can be different. In semiconductors with partial ionic binding the electrons cause a lattice distortion. The coupling of the electrons to phonons has to be correctly described in terms of "polarons" with a higher effective mass than calculated for a rigid lattice from the $E(k)$ dispersion relation [93]. The energies of donor-bound polarons in an anisotropic crystal were calculated by perturbation theory by Sak [94]

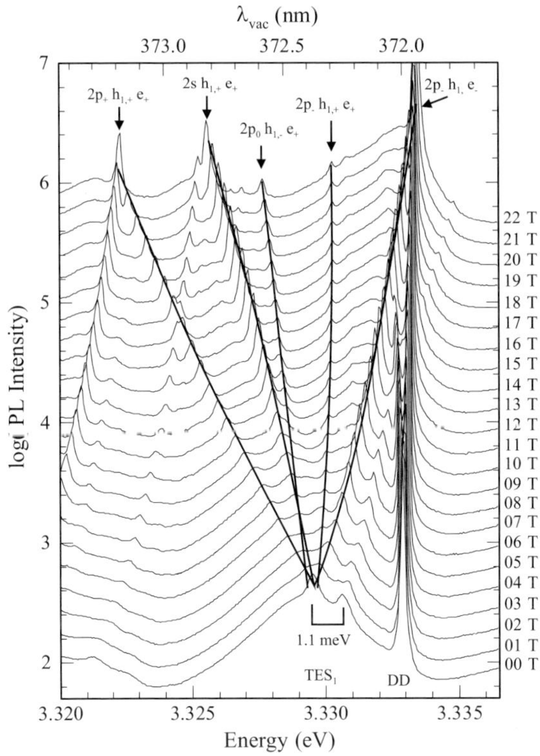

Fig. 14. Details of the Zeeman splitting of the TES_1 lines related to (D^0_{1a},X) and (D^0_{1b},X) for fields from 0 T to 22 T. The spacing of 1.1 meV between these lines is reproduced in the TES for $B = 0$. For fields $B > 0$, the splitting of the two components is correctly described by the combined splittings of the initial (D^0,X) and the final D^0 states. The solid lines mark components identified by this simple model. Labels indicate the final electron states and the initial hole states. A very similar splitting pattern is found for TES_2 related to (D^0_{2a},X) and (D^0_{2b},X).

and Larsen [95], and numerically by Engineer and Tzoar [96]. A numerical estimate of the $n = 2$ splitting for the effective-mass-like donor in ZnO orders the $2p_0$ state at the lowest energy followed by $2s$ and $2p_\pm$ with sequential equal spacings of ≈ 0.4 meV. The $2s$ state energy is subject to a further "chemical shift" or central-cell correction specific for the real donor under consideration. Typically, the $2s$ state chemical shift is 1/8 of the chemical shift of the $1s$ donor ground state, and thus expected to be less than 1 meV for the present case.

Once the Zeeman components of the TES lines have been identified, the effective polaron electron mass can be directly calculated from the energetic spacing of the $2p_+$ and $2p_-$ components, which are split apart by twice the cyclotron frequency

$$\Delta E = 2\omega_c = \frac{2eB_z}{m_{\text{eff}}} \tag{3}$$

In the ZnO case, the Zeeman splitting pattern in Faraday configuration, i.e. with $\boldsymbol{B}\|\boldsymbol{c}$, yields a value for $m_{\text{eff},\perp}$ of 0.23 [97]. Using the values for the dielectric constant from Ref. [98], $\epsilon_{\|} = 8.49$ and $\epsilon_{\perp} = 7.4$, the effective electron polaron Rydberg Ry^* for the donor-bound electron can be calculated with the formulae given by Sak [94] and Larsen [95]. This results in $Ry^* = 52\,\text{meV}$ for the ideal, effective-mass-like donor without central-cell correction [97]. The $2p_\pm$ state is then 12.9 meV below the ionization limit. Since the wavefunction of this state has only a very small overlap with the donor core potential, it is expected to be unaltered for real donors with core potentials deviating from the simple screened Coulomb potential assumed in the effective mass theory.

With this knowledge, one can proceed in calculating the single particle ionization energies for the donors responsible for the various (D^0,X) emission lines. From the spectrum in Fig. 12 the final state excitations $\Delta E_{1s\leftrightarrow 2p\pm} = E_{2p_\pm} - E_{1s}$ can be read to be $\approx 33.4\,\text{meV}$. Adding to this value the binding energy of the $2p_\pm$ state calculated for the effective-mass-like donor yields an ionization energy $E_{\text{D1}} = (33.4 + 12.9)\,\text{meV} = 46.3\,\text{meV}$ for the shallowest donor. This is less than predicted by the EMT and thus requires a negative chemical shift of the $1s$ ground state of $\approx (52 - 46)\,\text{meV} = 6\,\text{meV}$. It is believed that this particular donor is introduced by interstitial hydrogen [83, 99], and a repulsive donor core potential is consistent with interstitial donors in other semiconductors, as e.g., lithium in silicon [100].

In a similar way, for the second prominent donor responsible for the (D_2^0,X) emission lines, a value of $E_{\text{D2}} = (41.6 + 12.9)\,\text{meV} = 54.5\,\text{meV}$ results. Taking account of the uncertainties in the values for the dielectric constant and the effective masses, E_{D2} should be correct within $\pm 1\,\text{meV}$. For the third deeper donor involved in the (D_3^0,X) transitions, the related TES $p_\pm$ peak is more difficult to identify. Most likely, a value for the $1s \leftrightarrow 2p$ energy difference of 51 meV applies resulting in a total binding energy of $E_{\text{D3}} = (51 \pm 2 + 12.9)\,\text{meV} \approx (64 \pm 2)\,\text{meV}$. These values are in the range of the (less precise) energies determined for similar samples by Hall effect measurements [101].

3.3 Haynes' Rule and the Two-Electron Satellites for ZnO

Haynes [52] first described a correlation between the donor (or acceptor) ionization energies E_{D} (or E_{A}) and the exciton localization energies E_{loc}. In a generalized form, this linear dependence reads as

$$E_{\text{loc}} = aE_i + b \tag{4}$$

with $E_i = E_\mathrm{D}$ or E_A. From the discussion above it follows for donors

$$E_\mathrm{D} = E_{1s \leftrightarrow 2p\pm} + E_{2p_\pm} . \quad (5)$$

Thus, (4) can be rewritten as

$$E_\mathrm{loc} = a(E_{1s \leftrightarrow 2p\pm} + E_{2p_\pm}) + b = aE_{1s \leftrightarrow 2p\pm} + b' \quad (6)$$

with $b' = b + aE_{2p_\pm}$.

For different donors in a sample, (6) relates the variation in their $1s \leftrightarrow 2p_\pm$ excitation energies to the corresponding variation of their exciton localization energies

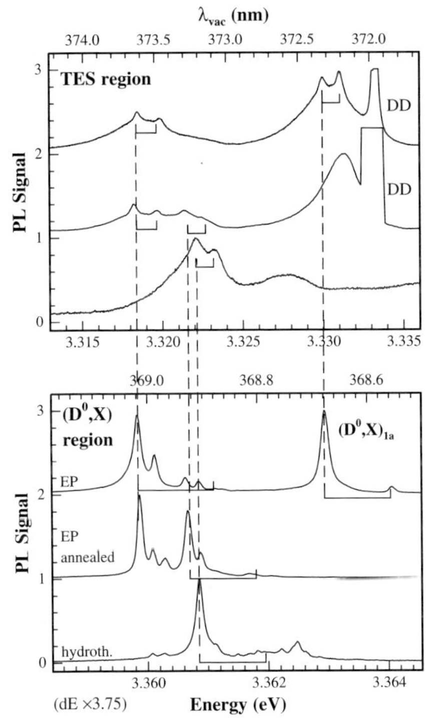

Fig. 15. Upper panel: PL spectra in the TES region for three different ZnO substrate samples: CVT grown as-received (labelled "EP" after the supplier Eagle Picher, upper trace), CVT annealed (middle trace), and hydrothermally grown ZnO (lower trace). Lower panel: The (D^0,X) region of the same samples, shown on an energy scale stretched by a factor of $s = 3.75$ relative to the upper panel. This value yields optimum coincidence of the spectra in the (D^0,X) and the TES regions as marked by the dashed lines and the square brackets. (From [81].)

$$\Delta E_{\mathrm{loc}} = a \Delta E_{1s \leftrightarrow 2p\pm} \tag{7}$$

The absolute position of a $1s \rightarrow 2p_{\pm}$ TES luminescence PL line is displaced from the free exciton energy E_{X} by an energy

$$E_{\mathrm{TES}} = E_{\mathrm{X}} - E_{\mathrm{loc}} - E_{1s \leftrightarrow 2p\pm} . \tag{8}$$

The variation of this energy for different donors is then

$$| \Delta E_{\mathrm{TES}} | = \Delta E_{\mathrm{loc}} + \Delta E_{1s \leftrightarrow 2p\pm} = \Delta E_{1s \leftrightarrow 2p\pm} \times (a + 1) \tag{9}$$

and hence, using (7),

$$\frac{| \Delta E_{\mathrm{TES}} |}{\Delta E_{\mathrm{loc}}} = \frac{a + 1}{a} \equiv s . \tag{10}$$

This equation implies that the energy spacings found in the TES range for different donors should be the same as those observed in the (D^0,X) range, but expanded by a "stretching factor" s which is directly related to the factor a in the generalized Haynes' rule (4).

The best coincidence between the dominant (D^0,X) lines and their related $2p_{\pm}$ TES satellites is found for a stretching factor $s = (3.7 \pm 0.1)$ (Fig. 15). The corresponding factor in (4) is $a = (0.37 \pm 0.01)$, much larger than the commonly quoted value of ≈ 0.2.

In Fig. 16 the interrelation between the exciton localization energies E_{loc} and the $E_{1s \leftrightarrow 2p\pm}$ energy differences is depicted for the major (D^0,X) lines. As expected from (6), the data points can be fitted by a straight line with slope $a = 0.37$ confirming the validity of Haynes' rule. The offset $b' =$

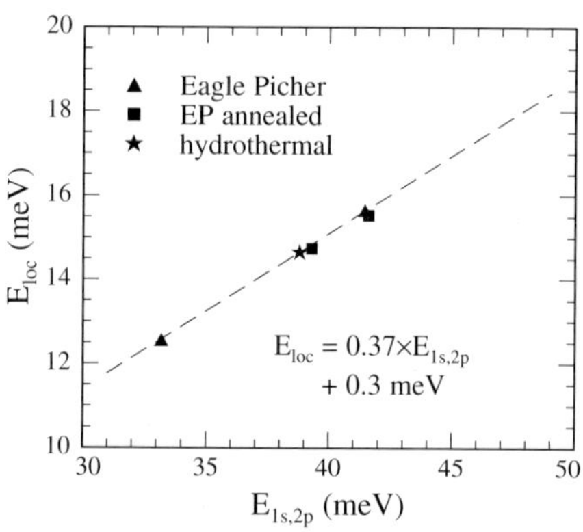

Fig. 16. Interrelation between the localization energies E_{loc} of the donor-bound excitons and the $E_{1s \leftrightarrow 2p\pm}$ energy as determined from the spacing between the (D^0,X) and TES lines in the PL spectra. A linear fit to the data using $a = 0.37$ as determined from the "stretching factor" $s = 3.75$ yields an abscissa of (0.3 ± 0.4) meV. (From [81].)

(0.3 ± 0.4) meV corresponds to a constant shift $b = b' - a \times E_{1s \leftrightarrow 2p\pm} = (0.3 - 4.8 \pm 0.5)$ meV $= (-4.5 \pm 0.5)$ meV according to (6). Thus, we find a final correlation

$$E_{\mathrm{loc}} = (0.37 \pm 0.01) E_{\mathrm{D}} - (4.5 \pm 0.2)\,\mathrm{meV} \tag{11}$$

or vice versa

$$E_{\mathrm{D}} = (2.7 \pm 0.1) E_{\mathrm{loc}} + (12.7 \pm 0.6)\,\mathrm{meV}. \tag{12}$$

Haynes' rule in the form of (12) allows to determine the ionization energy for further unknown donors by just measuring the localization energy of the related (D^0,X) emission line.

3.4 Haynes' Rule and the Two-Electron Satellites for GaN

The low-temperature PL spectrum of high-quality GaN layers shows a close analogy to that of ZnO bulk crystals as presented above. High-quality samples are, e.g., MOVPE-grown homoepitaxial GaN layers on (lower quality) GaN bulk substrate crystals [102], free-standing HVPE-grown GaN layers detached from their original sapphire substrate [58, 59], or so-called ELOG structures where a sapphire substrate pre-structured by SiO_2 stripes is laterally overgrown by GaN.

Since nominally undoped GaN is usually n-type ($n \approx 10^{16} - 10^{17}\,\mathrm{cm}^{-3}$), the spectra are dominated by donor-bound exciton (D^0,X) lines. They are located at ≈ 3.47 eV (Fig. 17). For samples with good crystal quality, several close (D^0,X) recombination lines are observed. They are introduced by different donors, and again part of them originate from excited exciton states (Fig. 18). The FWHM width of the individual lines is in excellent samples below 100 μeV and governed by lifetime broadening. Some 5 meV lower in energy acceptor-bound exciton (A^0,X) transitions are observed. Towards higher energy, the broader emission from free excitons (more precisely: "exciton-polaritons", [93]) is observed. They incorporate either holes from the "A" valence band (X_A) or from the "B" valence band (X_B) plus their excited states. The free excitons can also be identified in reflectance measurements, whereas bound excitons do not contribute to such spectra due to their comparably low density of states.

Since the (D^0,X) lines from different donors are energetically very close and strain conditions always vary slightly from sample to sample, a straightforward comparison and identification of donors present in different samples is very difficult. Assistance in the identification of (D^0,X) lines can again come from the analysis of the related TES lines. For the donor-bound excitons in GaN, also two-electron satellite lines show up with similar relative intensities as in ZnO.

A PL spectrum for free-standing GaN is shown in Fig. 19. The authors observe numerous lines which partially thermalize and reveal intensities which are sample dependent and correlate with specific (D^0,X) lines. Based on these

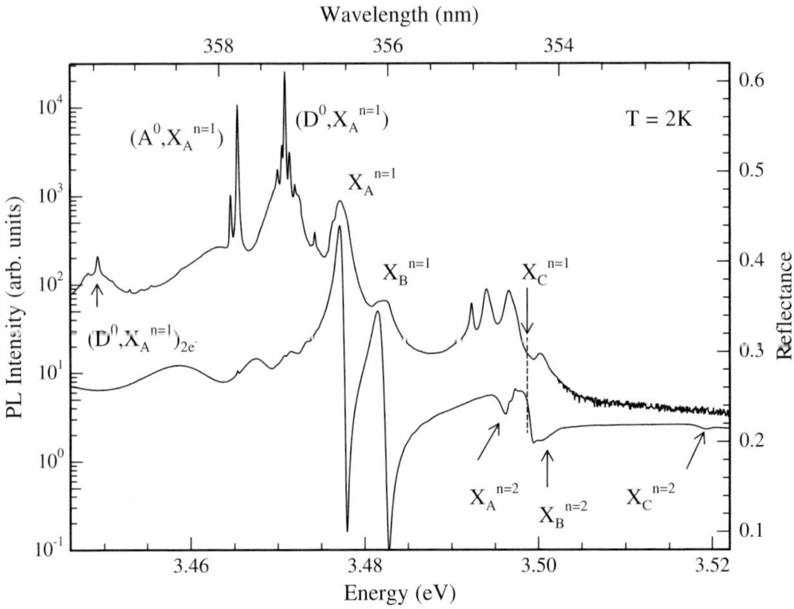

Fig. 17. Low-temperature PL spectrum (upper trace) and reflectance spectrum (lower trace) of a homoepitaxial MOVPE-grown GaN layer in the bandgap region. The resolution is 0.12 meV. (From [102].)

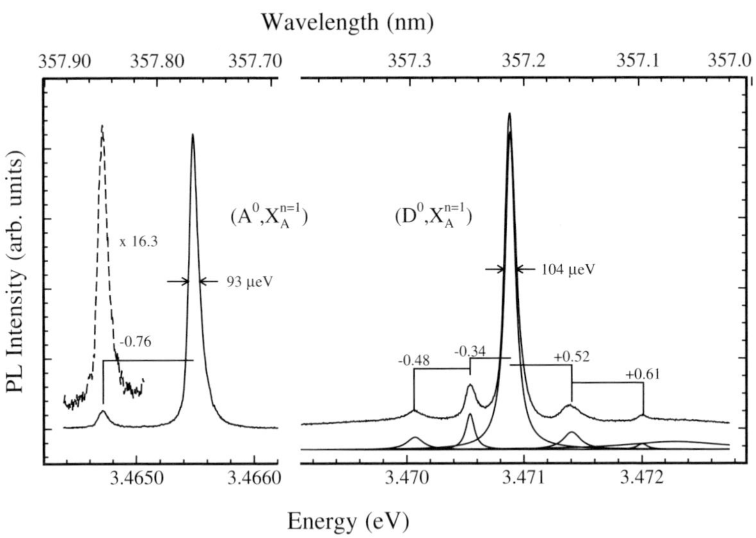

Fig. 18. Highly resolved low-temperature PL spectrum of a homoepitaxial MOVPE-grown GaN layer showing details of the (D^0,X) and (A^0,X) lines. (From [103].)

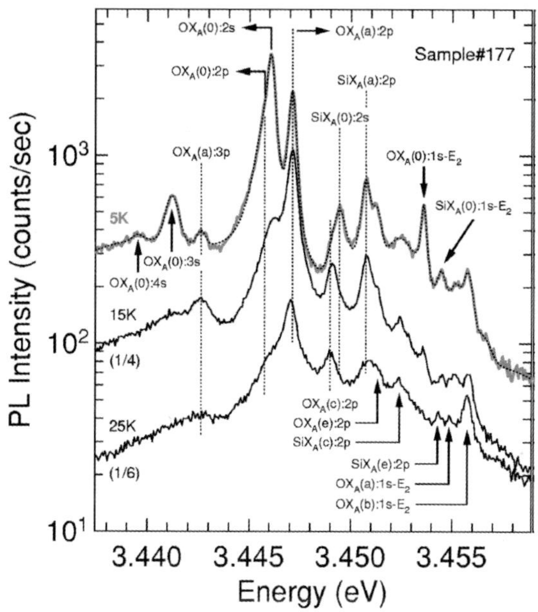

Fig. 19. PL spectra of the TES region of free-standing GaN recorded at different sample temperatures. Contributions from various excited states of the excitons bound to either oxygen or silicon donors and different final states of the bound excitons were identified. (From [58].)

ratios, the TES lines can be tentatively linked to their parent (D^0,X) transitions.

Magneto-PL measurements are also again a valuable tool to assign observed TES lines to specific final state excitations of the donors. As an example, Fig. 20 shows the result of such measurements. The authors succeeded in resolving not only the sub-components of the $n = 2$ electron state, but also observed higher final state excitations up to $5g$. They could fit the data rather well with a model based on calculations of a hydrogen atom in high magnetic fields. For the highest fields, an interaction of the donor-bound electron with optical phonons has to be taken into account for a precise description [59, 74, 76]. Again as in the ZnO (D^0,X) spectra, the GaN (D^0,X) recombination includes transitions from an excited initial state $\approx 1.4\,\mathrm{meV}$ higher in energy. However, this transition is only observed in the TES satellites but not in the principal parent (D^0,X) recombination [58].

After correlating TES lines to their parent (D^0,X) transitions and identifying in the TES lines the p-like states, one can directly compare the (D^0,X)-to-TES energy separations to those $1s \leftrightarrow (2p, 3p, ...)$ transition energies which for known donors have been observed in infrared absorption or in photoconductivity experiments. This way, Freitas et al. [58] assigned the (D^0,X) peak

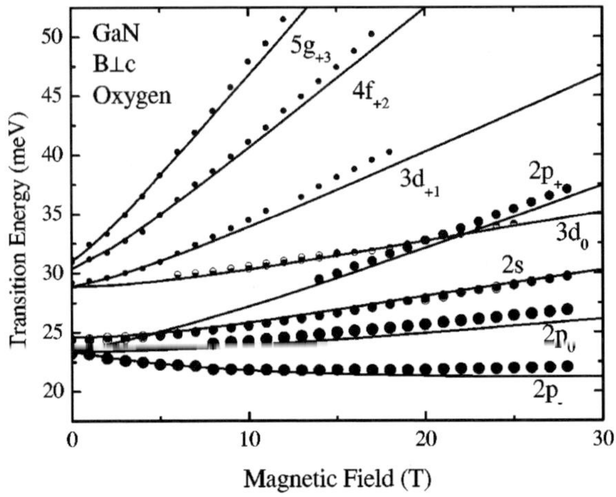

Fig. 20. Splitting of the TES lines for donors in GaN in magnetic fields $\boldsymbol{B} \perp \boldsymbol{c}$. Energies are given as differences to the $1s$ ground state. Excitations up to the $5g$ state are detected. The solid lines mark theoretical fits. (From [59].)

at 3.4714 eV in their free-standing GaN samples to the oxygen donor (O_N), and the peak at 3.4722 eV to the silicon (Si_{Ga}) donor [105]. To determine the ionization energy of the donors involved, on must know again the ideal effective-mass energy of the $2p_\pm$ states. This in turn requires knowledge of the anisotropic effective mass and dielectric constant. The polaron masses can be directly evaluated from the splitting of the $2p_\pm$ states in the Zeeman-PL measurements, yielding $m_\| = m_\perp = 0.222\, m_0$ [59]. The dielectric constants for GaN are $\epsilon_\| = 10.4$ and $\epsilon_\perp = 9.5$ [106]. With the expressions given in [94, 95] one finds a binding energy of the $2p_\pm$ states of 7.8 meV. Thus, the total donor ionization energies can be calculated to be $(25.7 + 7.8)\,\text{meV} = 33.5\,\text{meV}$ for the oxygen donor O_N, and $(22.8 + 7.8)\,\text{meV} = 30.6\,\text{meV}$ for the silicon donor Si_{Ga}, where the values for the $1s \leftrightarrow 2p_\pm$ energy spacings were taken from [58].

For the case of GaN, it is also possible to determine the prefactor a for the generalized Haynes' rule (4) from the stretching factor s (10). For this sake it is necessary to correlate the (D^0,X) peaks with their associated TES satellites. Fig. 21 shows the situation for a homoepitaxial sample [107]. For different excitation spots on the sample, different (and partially unknown) donors D_1 ... D_3 dominate the spectrum in the (D^0,X) spectral range. Their relative intensities are reflected in specific TES groups so establishing a correlation between them. The best coincidence of energies is obtained, when the (D^0,X) range is spread in energy by a factor of $s = (4.13 \pm 0.1)$, yielding $a = (0.32 \pm .01)$. To obtain the offset b in the generalized Haynes' rule we have to represent (D^0,X) localization energies E_{loc} versus $E_{1s} \leftrightarrow E_{2p\pm}$ spacings, similar to Fig. 16. These data are shown in Fig. 22.

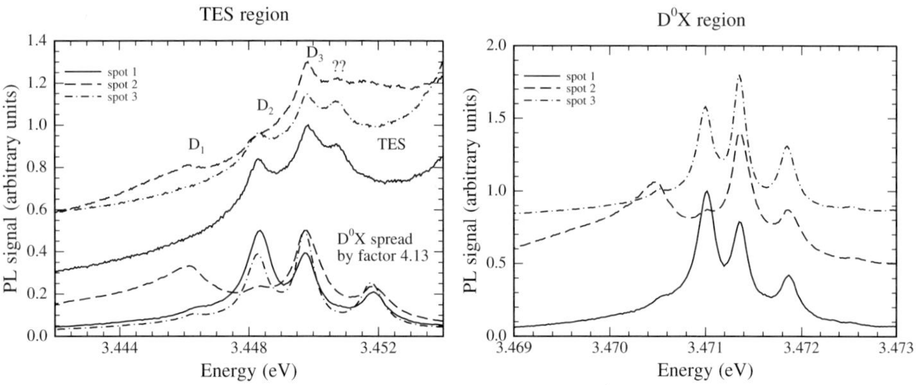

Fig. 21. Right: PL spectra of homoepitaxial GaN recorded from different sample spots at 5 K. Depending on the excitation spot, the unintentionally incorporated donors $D_1 - D_3$ have different (D^0,X) signals. '?' marks a line presumably due to an excited state transition of D_3. Left: The corresponding spectra in the TES range (upper traces). Optimum coincidence of the spectral energy structure in the TES range with that in the (D^0,X) range is obtained for a spreading factor $s = 4.13$ (lower trace).

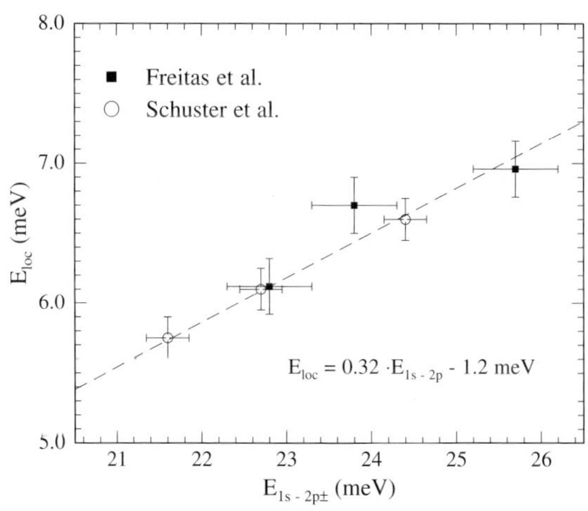

Fig. 22. Exciton localization energies versus energy spacings of the TES lines from their parent (D^0,X) lines. Square symbols: Data from [58]. Open circles: Data from [107]. The dashed line is a fit to the latter data points keeping the slope fixed at 0.32 (see text).

Assuming the validity of (7), a linear fit with a slope kept fixed at 0.32 as determined via the stretching factor s yields an offset parameter $b' = -1.2\,\text{meV}$. Using (6), we finally obtain an expression relating exciton localization energies to the donor ionization energies

$$E_{\text{loc}} = (0.32 \pm 0.01)E_{\text{D}} - (3.7 \pm 0.2)\,\text{meV} \tag{13}$$

or vice versa

$$E_{\text{D}} = (3.13 \pm 0.1)E_{\text{loc}} + (11.6 \pm 0.6)\,\text{meV}. \tag{14}$$

Within the range of donors which show up experimentally, Haynes' rule appears to be valid. The prefactor a turns out to be much larger than generally assumed (e.g. $a = 0.214$ in [58] or $a = 0.20 \pm 0.01$ in [108]).

In summary, our discussion has shown how the parameters in the generalized Haynes' rule for ZnO and GaN can be determined. Once these parameters are known they can serve to determine rather precisely the ionization energies of further donors solely from their bound exciton PL spectra.

References

1. J. Walker: Rep. Prog. Phys. **42**, 1605 (1979)
2. R.G. Farrer: Solid State Commun. **7**, 685 (1969)
3. G. Davies, N.B. Manson. In: *Properties and Growth of Diamond*, EMIS Datareview Series No. 9, ed by G. Davies (INSPEC, IEE, London 1994) p 173
4. A.T. Collins, E.C. Lightowlers. In: *The Properties of Diamond*, ed by J.E. Field (Academic Press, New York 1979) p 79
5. R.M. Chrenko: Phys. Rev. B **7**, 4560 (1973)
6. E.C. Lightowlers, A.T. Collins: J. Phys. D: Appl. Phys. **9**, 951 (1976)
7. A.T. Collins, E.C. Lightowlers, Phys. Rev. **171**, 843 (1968)
8. See the listings of A, B, and C values in: L. Reggiani, D. Waechter, S. Zykotynski: Phys. Rev. B **28**, 3550 (1983) and M. Willatzen, M. Cardona, N.E. Christensen: Phys. Rev. B **50**, 18054 (1994)
9. P.J. Dean, E.C. Lightowlers, D.R. Wight: Phys. Rev. **140**, A352 (1965)
10. A.T. Collins, S.C. Lawson, G. Davies, H. Kanda: Phys. Rev. Lett. **65**, 891 (1990)
11. H. Sternschulte, S. Wahl, K. Thonke, R. Sauer, T. Ruf, M. Cardona, T.R. Anthony: Mater. Sci. Forum **258-263**, 757 (1997)
12. T. Ruf, M. Cardona, H. Sternschulte, S. Wahl, K. Thonke, R. Sauer, P. Pavone, T.R. Anthony: Solid State Commun. **105**, 311 (1998)
13. S.J. Sharp, A.T. Collins, G. Davies, G.S. Joyce: J. Phys: Condens. Matter **9**, L451 (1997)
14. J.C. Bourgoin, J. Krynicki, B. Blanchard: phys. stat. solidi (a) **52**, 293 (1979)
15. J.-P. Lagrange, A. Deneuville, E. Gheeraert: Diam. Rel. Mater. **7**, 1390 (1998) and references therein

16. A. Aleksov, M. Kubovic, N. Kaeb, U. Spitzberg, A. Bergmaier, G. Dollinger, Th. Bauer, M. Schreck, B. Stritzker, E. Kohn: Diam. Rel. Mater. **12**, 391 (2003) and further references therein
17. K. Jackson, J.R. Pederson, J.G. Harrison: Phys. Rev. B **41**, 12641 (1990)
18. S.A. Kajihara, A. Antonelli, J. Bernholc: Mater. Res. Soc. Symp. Proc. **162**, 315 (1990)
19. S.A. Kajihara, A. Antonelli, J. Bernholc, R. Car: Phys. Rev. Lett. **66**, 2010 (1991)
20. D. Saada, J. Adler, R. Kalish: Appl. Phys. Lett. **77**, 878 (2000)
21. R. Kalish: Diam. Rel. Mater. **10**, 1749 (2001)
22. S. Prawer, C. Uzan-Saguy, G. Braunstein, R. Kalish: Appl. Phys. Lett. **63**, 2502 (1993)
23. R. Job, M. Werner, A. Denisenko, A. Zaitsev, W.R. Fahrner: Diam. Rel. Mater. **5**, 757 (1996)
24. J.F. Prins. In: *The Properties of Natural and Synthetic Diamond*, ed by J.E. Field (Adacemic Press, London 1992) p 301
25. J.F. Prins: Diam. Rel. Mater. **10**, 1756 (2001)
26. G. Popovici, M.A. Prelas: Diam. Rel. Mater. **4**, 1305 (1995)
27. G.S. Sandhu, C.T. Kao, M.L. Swanson, W.K. Chu: Mater. Res. Soc. Symp. Proc. **162**, 321 (1990)
28. R. Kalish, A. Reznik, C. Uzan-Saguy, S. Koizumi, M. Kamo. In: *Advanced Materials '98. The 5th NIRIM Internat. Symp. Advanced Materials (ISAM '98)*, ed by M. Kamo et al. (National Institute for Research in Inorganic Materials 1998) p 105
29. M. Akaishi, H. Kanda, S. Yamaoka: Science **259**, 1592 (1993)
30. S. Bohr, R. Haubner, B. Lux: Diam. Rel. Mater. **4**, 133 (1995)
31. K. Okano, H. Kiyota, T.Kurosu, M. Iida: Diam. Rel. Mater. **3**, 35 (1993)
32. S. Koizumi, M. Kamo, Y. Sato, H. Ozaki, T. Inuzuka: Appl. Phys. Lett. **71**, 1065 (1997)
33. S. Koizumi: phys. stat. solidi (a) **172**, 71 (1999)
34. S. Koizumi, T. Teraji, H. Kanda: Diam. Rel. Mater. **9**, 935 (2000)
35. M. Nesladek, K. Meykens, K. Haenen, L.M. Stals, T. Teraji, S. Koizumi: Phys. Rev. B **59**, 14852 (1999)
36. K. Haenen, K. Meykens, M. Nesladek, G. Knuyt, C. Quaeyhaegens, L. Stals, S. Koizumi, E. Gheeraert: phys. stat. solidi (a) **174**, 53 (1999)
37. E. Gheeraert, S. Koizumi, T. Teraji, H. Kanda, M. Nesladek: phys. stat. solidi (a) **174**, 39 (1999)
38. E. Gheeraert, N. Casanova, S. Koizumi, T. Teraji, H. Kanda: Diam. Rel. Mater. **10**, 444 (2001)
39. P.J. Dean, J.R. Haynes, W.F. Flood: Phys. Rev. **161**, 711 (1967)
40. H. Sternschulte, T. Albrecht, K. Thonke, R. Sauer, M. Griesser, H.M. Grasserbauer: Mat. Res. Soc. Symp. Proc. **423**, 693 (1996)
41. H. Sternschulte, K. Thonke, R. Sauer: phys. stat. solidi (a) **172**, 37 (1999)
42. H. Sternschulte, K. Thonke, R. Sauer, S. Koizumi: Phys. Rev. B **59**, 12924 (1999)
43. J.J. Hopfield, D.G. Thomas, M. Gershenzon: Phys. Rev. Lett. **10**, 162 (1963); D.G. Thomas, M. Gershenzon, F.A. Trumbore: Phys. Rev. Lett. **133**, A269 (1964)
44. R. Sauer. In: *Thin Film Diamond, Series 'Semiconductors and Semimetals'*, ed by C.E. Nebel, J. Ristein (Academic Press 2004), Chapter 8, p 379

45. K. Nakazawa, K. Tanabe, M. Tachiki, H. Kawarada, S. Koizumi: Phys. Rev. B **64**, 235203 (2001)
46. H. Sternschulte, T. Albrecht, K. Thonke, R. Sauer. In: *The Physics of Semiconductors*, ed by M. Scheffler, R. Zimmermann (World Scientific, Singapore 1996) p 169
47. N. Teofilov, R. Sauer, K. Thonke, S. Koizumi: Physica B **340-342**, 99 (2003)
48. R. Sauer, N. Teofilov, K. Thonke, S. Koizumi: Diam. Rel. Mater. (in press)
49. S.C. Lawson, H. Kanda, H. Kiyota, T. Tsutsumi, H. Kawarada: J. Appl. Phys. **77**, 1729 (1995)
50. C.J. Rauch: *Proc. Int. Conf. Phys. Semicond., Exeter, 1962*, (Institute of Physics and the Physical Society, London, 1963) p 276
51. M. Willatzen, M. Cardona, N.E. Christensen: Phys. Rev. B **50**, 18054 (1994)
52. J.R. Haynes: Phys. Rev. Lett. **4**, 361 (1960)
53. P.J. Dean, I.H. Jones: Phys. Rev. **133**, A1698 (1964)
54. J. Weber, W. Schmid, R. Sauer: Phys. Rev. B **21**, 2491 (1980)
55. H.Ch. Alt, L. Tapfer: J. Electron. Mater. **14a**, 833 (1985)
56. P.J. Dean: J. Lumin. **7**, 51 (1973) and further references therein
57. J.I. Pankove, T.D. Moustakas: "GaN I" and "GaN II". In: *Semiconductors and Semimetals*, vol **50** and **57**, ed by R.K. Willardson, E.R. Weber, (Academic Press, San Diego 1998 and 1999)
58. J.A. Freitas, W.J. Moore, B.V. Shanabrook, G.C.B. Braga, S.K. Lee, S.S. Park, J.Y. Han: Phys. Rev. B **66**, 233311 (2002)
59. A. Wysmolek, K.P. Korona, R. Stepniewski, J.M. Baranowski, J. Bloniarz, M. Potemski, R.J. Jones, D.C. Look, J. Kuhl, S.S. Park, S.K. Lee: Phys. Rev. B **66**, 24317 (2002)
60. D.C. Look, B. Claflin: phys. stat. solidi (b) **241**, 624 (2004)
61. D.C. Look, D.C. Reynolds, C.W. Litton, R.L. Jones, D.B. Eason, G. Cantwell: Appl. Phys. Lett. **81**, 1830 (2002)
62. K.K. Kim, H.-S. Kim, D.-K. Hwang, J.-H. Lim, S.-J. Park: Appl. Phys. Lett. **83**, 63 (2003)
63. J.F. Rommeluere, L. Svob, F. Jomard, J. Mimila-Arroyo, A. Lusson, V. Sallet, Y. Marfaing: Appl. Phys. Lett. **83**, 287 (2003)
64. For a collection of data, see Landolt–Börnstein: *Numerical Data and Functional Relationships in Science and Technology – New Series III*, vol 41B, ed by U. Rössler (Springer, Berlin 1999) p 85 and references therein
65. D.C. Reynolds, D.C. Look, B. Jogai, C.W. Litton, T.C. Collins, W. Harsch, G. Cantwell: Phys. Rev. B **57** 12151 (1998)
66. K. Thonke, N. Kerwien, A. Wysmolek, M. Potemski, A. Waag, R. Sauer. In: *Proceedings of the 26th ICPS, Edinburgh (2002)*, ed by A.R. Long and J.H. Davies (IOP Bristol 2003) p 22
67. P.J. Dean, J.D. Cuthbert, D.G. Thomas, R.T. Lynch: Phys. Rev. Lett. **18**, 122 (1967)
68. R. Sauer: J. Lum. **12/13**, 495 (1976)
69. D.C. Reynolds, C.W. Litton, T.C. Collins: Phys. Rev. **174**, 845 (1968)
70. C.H. Henry, K. Nassau: Phys. Rev. B **2**, 997 (1970)
71. D.C. Reynolds, C.W. Litton, T.C. Collins: Phys. Rev. **177**, 1161 (1969)
72. D.C. Reynolds, C.W. Litton, R.J. Almassy, S.B. Nam, P.J. Dean, R.C. Clarke: Phys. Rev. B **13**, 2507 (1976)
73. D.C. Reynolds, C.W. Litton, T.C. Collins, S.B. Nam: Phys. Rev. **12**, 5723 (1975)

74. J.P. Cheng, B.D. McCombe, J.M. Shi, F.M. Peeters, J.T. Devreese: Phys. Rev. B **48**, 7910 (1993)
75. V.A. Karasyuk, D.G.S. Becket, M.K. Nissen, A. Villemaire, T.W. Steiner, M.L.W. Thewalt: Phys. Rev. B **49**, 16381 (1994)
76. M. Grynberg, S. Huant, G. Martinez, J. Kossut, T. Wojtowicz, G. Karczewski, J. M. Shi, F. M. Peeters, J. T. Devreese: Phys. Rev. B **54**, 1467 (1996)
77. D.C. Reynolds, T.C. Collins: Phys. Rev. **185**, 1099 (1969)
78. C. Klingshirn: phys. stat. solidi (b) **71**, 547 (1976)
79. D.C. Reynolds, D.C. Look, B. Jogai, C.W. Litton, T.C. Collins, W. Harsch, G. Cantwell: Phys. Rev. B **57**, 19 (1998)
80. B.J. Skromme, H. Zao, B. Goldenberg, H.S. Kong, M.T. Leonard, G.E. Bulman, C.R. Abernathy, S.J. Pearton: Mat. Res. Soc. Symp. Proc. **449**, 713 (1997)
81. A. Schildknecht, R. Sauer, K. Thonke: Physica B **340**, 205 (2003)
82. D.C. Reynolds, C.W. Litton, T.C. Collins: Phys. Rev. **140**, A1726 (1965)
83. D.M. Hofmann, A. Hofstetter, F. Leitner, H. Zhou, F. Henecker, B.K. Meyer, S. Orlinski, J. Schmidt, P.G. Baranov: Phys. Rev. Lett. **88**, 45504 (2002)
84. H. Schrey, C. Klingshirn: Solid State Commun. **33**, 485 (1980)
85. J. Gutowski, N. Presser, I. Broser: Phys. Rev. B **38**, 9746 (1988)
86. M. Strassburg, A. Rodina, M. Dworzak, U. Haboeck, I.L. Krestnikov, A. Hoffmann, O. Gelhausen, M.R. Phillips, H.R. Alves, A. Zeuner, D.M. Hofmann, B.K. Meyer: phys. stat. solidi (b), 607 (2004)
87. J.J. Hopfield, D.G. Thomas: Phys. Rev. **122**, 2135 (1961)
88. D.G. Thomas, J.J. Hopfield: Phys. Rev. **128**, 2135 (1962)
89. R.G. Wheeler, J.O. Dimmock: Phys. Rev. **125**, 1805 (1962)
90. G. Blattner, G. Kurtze, G. Schmieder, C. Klingshirn: Phys. Rev. **25**, 7413 (1982)
91. W.R.L. Lambrecht, A.V. Rodina, S. Limpijumnong, B. Segall, B.K. Meyer: Phys. Rev. B **65**, 75207 (2002)
92. W. Roesner, G. Wunner, H. Herold, H. Ruder: J. Phys. B **17**, 29 (1984)
93. See, e.g., C.F. Klingshirn: *Semiconductor Optics*, 2nd corr. printing (Springer, Berlin Heidelberg New York 1997) p 133
94. J. Sak: Phys. Rev. B **3**, 3356 (1971)
95. D.M. Larsen: Phys. Rev. B **9**, 823 (1974)
96. M. Engineer, N. Tzoar: Phys. Rev. B **8**, 702 (1973)
97. K. Thonke et al.: To be published
98. H. Yoshikawa, S. Adachi: Jpn. J. Appl. Phys., Part 1 **36**, 6237 (1997)
99. C.G. Van de Walle: Phys. Rev. Lett. **85**, 1012 (2000)
100. B. Szablak, M. Altarelli: Solid State Commun. **37**, 341 (1981)
101. D.C. Look, D.C. Reynolds, J.R. Sizelove, R.L. Jones, C.W. Litton, G. Cantwell, W.C. Harsch: Solid State Commun. **105**, 399 (1998)
102. K. Kornitzer, T. Ebner, K. Thonke, R. Sauer, C. Kirchner, V. Schwegler, M. Kamp, M. Leszczynski, I. Grzegory, S. Porowski: Phys. Rev. B **60**, 1471 (1999)
103. K. Kornitzer, M. Grehl, K. Thonke, R. Sauer, C. Kirchner, V. Schwegler, M. Kamp, M. Leszczynski, I. Grzegory, S. Porowski: Physica B **273-274**, 66 (1999)
104. A. Wysmolek: Reported on the MRS Fall Meeting, Boston (2002)

105. Currently a discrepancy exists in the literature between the assignment of TES lines to specific donors as advanced by Freitas et al. [58] which is based on the relative line intensities, and that of Wysmolek et al. [59] which is based on a comparison of PL lifetimes.
106. A.S. Barker Jr., M. Ilegems: Phys. Rev. B **7**, 743 (1973)
107. T. Schuster, K. Thonke et al.: To be published
108. B.K. Meyer: Mater. Res. Soc. Symp. Proc. **449**, 497 (1997)

Spectroscopy of Biexcitons and Trions in II–VI Quantum Dots

Ulrike Woggon

An overview will be given of similarities and differences in the optical properties of two kinds of quantum-confined II-VI semiconductors, nanocrystals and epitaxially grown islands. In both systems, Coulomb correlation energies have been observed which are up to ten times larger than those in III-V semiconductor materials. This peculiarity makes II-VI semiconductor quantum dots interesting objects for studying few-particle states. In this review we focus on binding energies and recombination dynamics of the elementary excitations exciton, biexciton and trion in three-dimensionally confined CdSe and CdS.

1 Introduction

1.1 Biexcitons in Bulk Semiconductors

In semiconductors, excitons and biexcitons are electronic states created as a result of Coulomb interaction between electrons and holes. The trion, another elementary excitation, is a positively or negatively charged exciton. When more than one electron-hole pair inside a semiconductor is excited, we necessarily have to deal with electron-electron, hole-hole and electron-hole interactions which may result in new binding states at energies below the exciton energy within the band-gap of the bulk semiconductor. For example, a biexciton is created in case of non-vanishing biexciton binding energy $\Delta E_{\mathrm{XX}}^{\mathrm{B}}$ which is defined by the energy difference

$$\Delta E_{\mathrm{XX}}^{\mathrm{B}} = 2E_{\mathrm{X}} - E_{\mathrm{XX}} \tag{1}$$

with E_{X} the exciton energy and E_{XX} the biexciton energy. The dispersion relation of the biexciton is

$$E_{\mathrm{XX}}(K) = 2E_{\mathrm{X}} - \Delta E_{\mathrm{XX}}^{\mathrm{B}} + E_{\mathrm{exch}} + \frac{\hbar^2 K_{\mathrm{XX}}^2}{4M_{\mathrm{X}}} \tag{2}$$

with K_{XX}^2 the wave vector of the center of mass motion of the biexciton, M_{X} the exciton mass and E_{exch} the electron-hole exchange interaction consisting of the short- and long-range contribution

$$E_{\text{exch}} = C\left(\Delta_{\text{ST}} + \frac{1}{3}\Delta_{\text{LT}}\right) . \tag{3}$$

Δ_{ST} is the singlet-triplet splitting, Δ_{LT} the longitudinal-transverse splitting energy and the constant C is determined by the symmetry of the valence band.

Biexcitons can be formed, for example, after interband excitation and energy relaxation of electrons and holes, in pump-probe experiments after absorption of one pump- and one probe-photon, or directly by two-photon absorption (TPA). The biexciton ground state is a state with total angular momentum $J = 0$, i.e. in TPA it can only be populated by electrons and holes with antiparallel spin. The coupled exciton and biexciton transitions can be represented by the so-called diamond scheme which illustrates the selection rules in the presence of circularly polarized light (see Fig. 1). A photon with σ^+ polarization creates a spin-up electron and spin-down hole and vice versa in case of σ^--photons. The two-photon absorption process with two photons having both the same σ^+ (σ^-) polarization is optically forbidden.

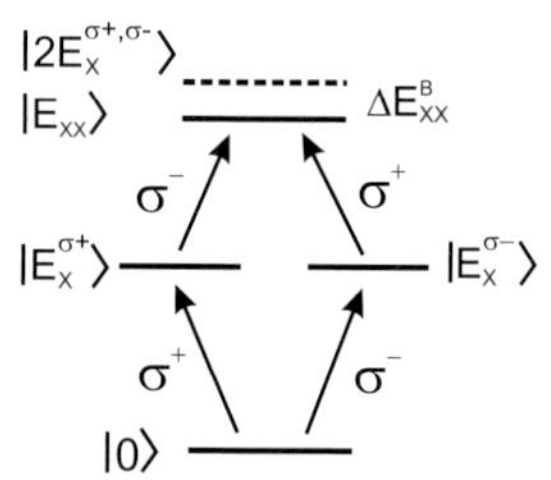

Fig. 1. Scheme of the exciton and the biexciton states and the allowed optical transitions in two-photon absorption using circularly polarized light. The horizontal lines denote the crystal ground state $|0\rangle$, the exciton states $|E_{\text{X}}^{\sigma+}\rangle$ and $|E_{\text{X}}^{\sigma-}\rangle$, and the biexciton state $|E_{\text{XX}}\rangle$. The diagonal arrows represent a transition caused by a σ^+ (σ^-) photon. The dashed line shows the energy of two uncorrelated excitons. $\Delta E_{\text{XX}}^{\text{B}}$ is the biexciton binding energy. The biexciton ground state can be populated only by electrons and holes with antiparallel spin.

Intensive experimental and theoretical studies revealed a strong dependence of the biexciton binding energy $\Delta E_{\text{XX}}^{\text{B}}$ on the electron and hole mass ratio $\sigma = m_{\text{e}}/m_{\text{h}}$ which strongly varies in the different semiconductor materials (for an overview about biexciton properties in bulk semiconductors see e.g. [1–7]). Observations in bulk semiconductors showed that $\Delta E_{\text{XX}}^{\text{B}}$ can hardly exceed the value $\Delta E_{\text{XX}}^{\text{B}} \approx 0.3\,E_{\text{Ryd}}$ where E_{Ryd} denotes the bulk exciton Rydberg energy. Examples for biexciton binding energies $\Delta E_{\text{XX}}^{\text{B}}$ in bulk wide- and narrow-gap semiconductors are, e.g., 28 meV for CuCl, 20.5 meV for CuBr, 19 meV for ZnO, 5.7 meV for CdS, 2 to 5 meV for CdSe, 2.2 meV for ZnSe, 1.5 meV for ZnTe, and 0.8 meV for GaAs. In general, bulk II-VI semiconductors have biexciton binding energies in the range between 2 and

5 meV [8, 9]. Quantum confinement of biexcitons in quantum wells, wires or quantum dots drastically modifies binding energies, oscillator strengths, and other optical properties. For a more general overview about biexcitons and charged excitons in II-VI semiconductor quantum wells we refer to the contribution of J. Gutowski et al. in this book.

1.2 Biexcitons in Quantum Dots

Compared with bulk semiconductors, theory predicted for biexciton states in three-dimensionally confined quantum structures (i) a confinement-induced enhancement of the biexciton binding energy, (ii) an observation of both ground and excited two-pair states in the optical spectra because of the absence of bulk-typical continuum states and (iii) transitions to excited two-pair states which were originally forbidden and occur now due to the change of selection rules caused by the Coulomb potential [10–17]. Therefore, the two-electron-hole pair or biexciton states are an important elementary excitation, especially in strongly confined quantum dots with dimensions below a_{B}, the bulk exciton Bohr radius. The biexciton ground state $|1s_{\mathrm{e}}1s_{\mathrm{e}}1S_{3/2}1S_{3/2}\rangle$ is a four-particle state with two quantum confined electrons ($1s_{\mathrm{e}}$) and holes ($1S_{3/2}$) being in the energetically lowest, discrete states given by the confinement potential. In quantum dots the Hamilton-operator to describe a four-particle system is

$$\mathbf{H} = \mathbf{H}_{\mathrm{e}} + \mathbf{H}_{\mathrm{h}} + \mathbf{V}_{\mathrm{ee}} + \mathbf{V}_{\mathrm{hh}} + \mathbf{V}_{\mathrm{eh}} + \Delta\mathbf{V}(\epsilon_1, \epsilon_2, \boldsymbol{r}_{\mathrm{e}}, \boldsymbol{r}_{\mathrm{h}}) + V_{\mathrm{e,h}}^{\mathrm{conf}} \tag{4}$$

where $\mathbf{H}_{\mathrm{e}}$ and $\mathbf{H}_{\mathrm{h}}$ are kinetic energies of electrons and holes, $\mathbf{V}_{\mathrm{ee}}$, $\mathbf{V}_{\mathrm{hh}}$ and $\mathbf{V}_{\mathrm{eh}}$ the Coulomb interaction terms describing the electron-electron, hole-hole and electron-hole interaction, $\Delta\mathbf{V}(\epsilon_1, \epsilon_2, \boldsymbol{r}_{\mathrm{e}}, \boldsymbol{r}_{\mathrm{h}})$ is the correction to the Coulomb potential resulting from the differences in the dielectric constants of the quantum dot and the barrier material, and $V_{\mathrm{e,h}}^{\mathrm{conf}}$ is the confining barrier potential. To solve the Schrödinger equation derived from (4), numerical matrix diagonalization methods have been applied in [10–12] and a confinement-induced enhancement of $\Delta E_{\mathrm{XX}}^{\mathrm{B}}$ is obtained. Beside the matrix diagonalization, also perturbation theory, variational calculations and Monte-Carlo techniques have been used to describe two-electron-hole pair states in quantum dots [13–17]. A treatment of biexciton binding energies for self-assembled quantum dots is presented, e.g., in [18]. Trion energies are calculated for spherical nanocrystals in [16] and the change in oscillator strength of the optical transitions as function of nanocrystal radius is discussed for different spin-orbit splitting energies. All theoretical treatments showed a strong sensitivity of the numerical value of the biexciton binding energy regarding the differences in the dielectric constants ϵ_1 and ϵ_2 of the quantum dot and the barrier material. A different value of ϵ outside the semiconductor quantum dot influences the screening of the Coulomb interaction between electrons and holes and thus their binding energy.

Alternative to methods based on effective mass approximation (EMA), the problems of charged exciton energies, formation and symmetry of multi-excitons, and exciton dissociation are studied in [19–21] using pseudopotential theory. The influence of external charges on the radiative recombination rate of an electron-hole pair in a CdSe quantum dot is investigated in [22] via atomistic empirical pseudopotential calculations. When a negative external charge is near the surface of a CdSe quantum dot, its Coulomb potential could be strong enough to pull the hole away from the electron with the consequence of a reduction of the radiative recombination rate.

1.3 Excitons and Biexcitons in Quantum Dots with Anisotropic Confinement Potentials

When discussing biexciton energies in quantum-confined structures, we have to take into account the fine structure splitting of the exciton. This is of particular importance in single-dot spectroscopy where the inhomogeneous broadening due to a quantum dot ensemble has been removed. In nanostructures of reduced symmetry, the electron-hole exchange interaction splits the fourfold degenerate heavy hole exciton with the pair angular momentum components $M = J_z + s_z = \pm 1, \pm 2$ by the splitting energy Δ_0 into a radiative doublet and two close lying nonradiative singlets (here $s_z = \pm 1/2$, $J_z = \pm 3/2$ are the electron and the heavy-hole momentum, respectively. For the nomenclature of the splittings see [23] and Fig. 2). The nonradiative "dark" exciton state consists predominantly of hh-excitons with angular momentum component of $M = \pm 2$ which do not couple to the radiation field. An anisotropic confinement potential (symmetry $< \mathrm{D}_{2d}$) reduces the point group symmetry further and the radiative doublet having the angular momentum $M = \pm 1$ is split by the energy Δ_1 into two "bright" exciton states which have linearly polarized optical transitions to the ground state. In [23] the splittings have been classified in Δ_0 for the energy splitting between bright and dark states, Δ_1 for the splitting of the bright exciton and Δ_2 for the energy difference between the dark states (see Fig. 2).

As can be seen from Fig. 2, the fine structure of the biexciton emission reproduces exactly the fine structure of the intermediate exciton transition [23–25]. The peak positions of the linearly polarized excitons and biexciton transitions show a mirror-like arrangement in the spectra, i.e. the biexciton to exciton transition has the inverted fine structure of the exciton to ground-state transition. The trion does not exhibit an exchange splitting. The trion ground state is a spin-singlet state and the wave function overlap of the two electrons (or holes) in the trion gives zero local spin density of the two electrons (holes).

According to [23, 26] the biexciton binding energy can be defined with respect to the energetically lowest lying state which is the dark exciton. For that definition, the quantum dot biexciton binding energy derived from single-dot photoluminescence experiments has to be corrected by the fine structure

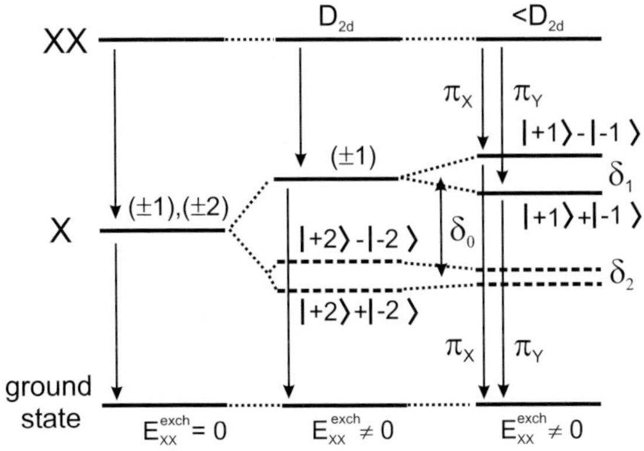

Fig. 2. Scheme of the exciton X and the biexciton XX states and the allowed optical transitions in quantum dots of different symmetry according to [23]. The energy splittings due to exchange interaction have been classified in Δ_0 for the energy splitting between bright and dark states, Δ_1 for the splitting of the bright exciton caused by anisotropies and Δ_2 for the energy separation between the two dark states. D_{2d} denotes the point group symmetry, ±1 and ±2 the pair angular momentum component M, and π_x and π_y are linearly polarized transitions.

splitting energies according to

$$\Delta E_{XX}^{B*} = \Delta E_{XX}^{B} - 2\Delta_0 \pm \Delta_1 - \Delta_2 \tag{5}$$

where $\pm\Delta_1$ stands for the two polarizations π_X and π_Y.

Since the prediction of the enhancement of the biexciton binding energy in quantum-confined semiconductors, a variety of experiments has been performed confirming that increase in ΔE_{XX}^{B} for localized, or quantum-confined, biexcitons [23, 27–36]. Experimental verification is obtained both for quantum dot ensembles and single quantum dots exploiting two-photon absorption, intensity-dependent photoluminescence, polarization-sensitive analysis of fine structure splittings, pump-and-probe and four-wave mixing experiments. Examples for measured biexciton binding energies in quantum-confined II-VI semiconductors are 14 meV for CdSSe nanocrystals ($R = 2.3$ nm) [30], 8 meV for CdS nanocrystals ($R = 7.5$ nm) [33], 18 to 22 meV for CdSe/ZnSe self-assembled islands [23, 34, 35], 38 meV for localized excitons in CdS/ZnS SQWs [37, 38]. The trion state is studied both experimentally and theoretically in two- and three-dimensionally confined semiconductor nanostructures, e.g. by magnetic field experiments and polarisation-sensitive micro-photoluminescence [25, 39–53]. Binding energies of a few meV are reported, in particular for epitaxially grown CdTe/ZnTe and CdSe/ZnSe islands.

In the following we present an analysis of exciton, trion, and biexciton states in both CdSe nanocrystals and self-organized CdSe/ZnSe and CdS/ZnS islands and compare and discuss biexciton and trion related data obtained in photoluminescence, differential absorption and four-wave mixing spectroscopy.

2 Biexcitons in Nanocrystals

2.1 Size-Dependence of the Biexciton Binding Energy

The theoretically predicted enhancement of biexciton binding energies has been first proven for II-VI nanocrystals embedded in a glass matrix. Widely used experimental techniques were pump-probe (or differential absorption) spectroscopy, both with nanosecond and femtosecond pulses, two-photon absorption spectroscopy and photoluminescence (PL) experiments under high optical excitations. First hints to the occurrence of excited two-pair states have been obtained from luminescence spectra of CdSSe quantum dots under high laser excitation [27]. Structures in the high energy wing of the PL spectrum were interpreted as transition arising from the decay of excited two-electron-hole pair states. By means of nonlinear absorption measurements (pump-probe spectroscopy) (see e.g. [12, 13, 28, 54–56] and Refs. therein), an induced absorption feature at energies above the exciton resonance was assigned to an absorption process involving one pump- and one probe photon resulting in the creation of excited two-pair states where one or both holes are not in their ground state. Accordingly, induced absorption at energies below the exciton was assigned to the formation of a biexciton in its ground state.

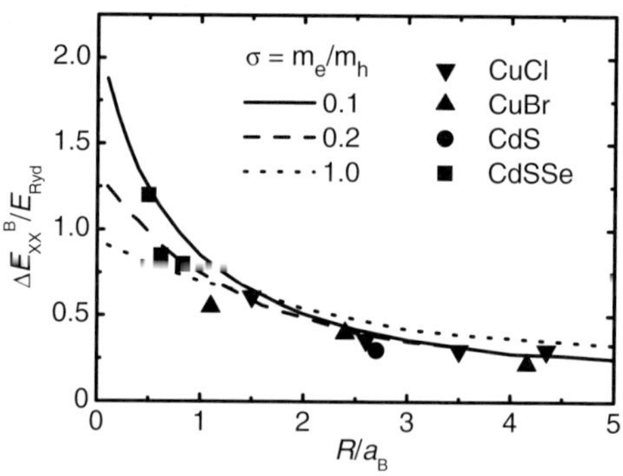

Fig. 3. Size dependence of the biexciton binding energy for spherical quantum dots with radius R. The curves are theoretical results obtained for different electron-hole mass ratios $\sigma = m_\mathrm{e}/m_\mathrm{h}$ and constant ratio $\epsilon_2/\epsilon_1 = 1$ for the dielectric constants of nanocrystal and host [10]. The experimental data are from [31] for CuCl, [32] for CuBr, [33] for CdS and [30] for CdSSe. $\Delta E^\mathrm{B}_\mathrm{XX}$ is the measured biexciton binding energy, E_Ryd is the bulk exciton Rydberg energy and a_B is the bulk exciton Bohr radius.

Figure 3 shows some of the presently known biexciton binding energies as a function of size measured for spherical quantum dots in glass matrices together with theoretical results from [10, 11]. Assuming similar confinement conditions (all nanocrystals have a transparent glass matrix with $E_{\mathrm{gap}} \approx 6\,\mathrm{eV}$ as confining barrier), the enhancement of biexciton binding energy depends predominantly on the quantum dot radius and the electron-hole mass ratio. The y-axis is given in units of the bulk excitonic Rydberg energy E_{Ryd} in order to discuss the enhancement compared with bulk semiconductors. This representation is widely used for discussing theoretical results and to relate them to experimental data.

2.2 Biexcitons in Optical Gain Processes

In II-VI nanostructures, the biexciton state gained its importance as an essential elementary excitation for modelling spectra and dynamics of optical gain. For strong pump power, optical gain develops as a result of stimulated decay of one- and two-pair electron-hole states and can occur in a broad spectral range including photon energies far below the absorption onset [57–60]. To model the gain mechanism and the femtosecond gain dynamics, a multi-exciton system is considered in [59–61]. The femtosecond dynamics of the optical gain is analyzed and a gain build-up time of 2 ps as well as a gain lifetime of 200 ps are found. The threshold pump power to achieve gain was $\sim 1\,\mathrm{mJ/cm^2}$ at a pump energy of 2.213 eV (560 nm), a pump pulse width of 80 fs and a temperature of $T = 10\,\mathrm{K}$ for a mean nanocrystal radius of $R = 2.5\,\mathrm{nm}$.

In Fig. 4 we illustrate typical absorption bleaching and gain spectra for CdSe nanocrystals embedded in glass and compare them with luminescence spectra measured at identical high excitation conditions. The average radius in the selected sample is about half the excitonic Bohr radius of $a_{\mathrm{B}} = 5.8\,\mathrm{nm}$ (bulk CdSe), i.e. a strong enhancement of the biexciton binding energy is expected. The differential absorption and emission spectra have been measured when pumping the samples above the absorption edge at 2.48 eV with nanosecond laser pulses. In the upper right part of Fig. 4, the level scheme of the confined exciton states in CdSe quantum dots according to experimental data published in [62] shows, to which states the pump energy is resonant: At 2.48 eV we excite $1P_{3/2}1p_{\mathrm{e}}$ exciton states with $\sim 0.5\,\mathrm{eV}$ excess energy with respect to the $1S_{3/2}1s_{\mathrm{e}}$ ground state as well as $2S_{3/2}1s_{\mathrm{e}}$ exciton states with $\sim 0.2\,\mathrm{eV}$ excess energy with respect to the ground state exciton. Assuming a complete relaxation of these excess energies within the size distribution, we thus excite with 2.48 eV surely quantum dot sizes having their ground state around 1.98 eV like in our example. The chosen sample consists of an ensemble of quantum dots with a size distribution centered around $R \sim 3.0\,\mathrm{nm}$, e.g. with our pump energy we can populate (within the experimental error) the majority of the quantum dots (the contribution from the small amount of

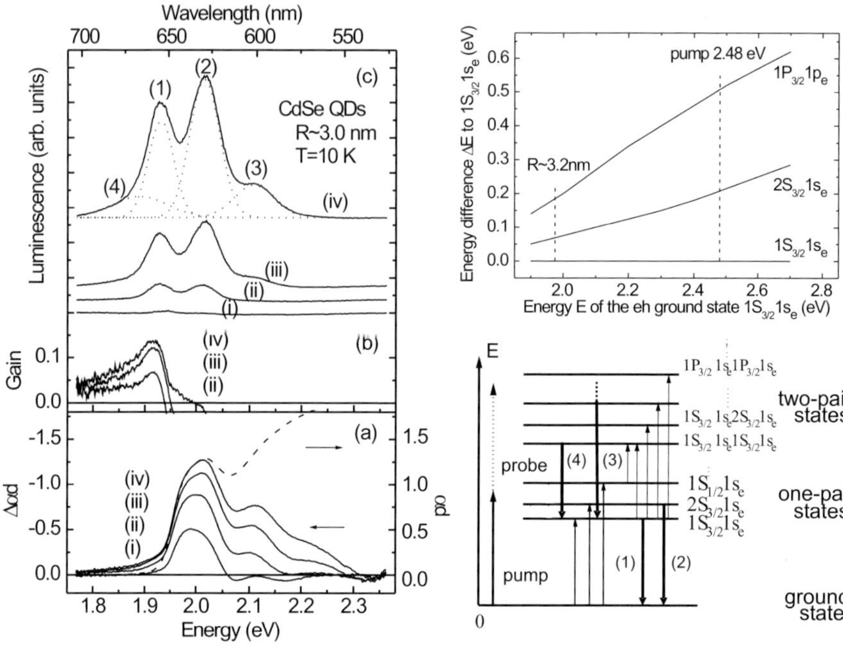

Fig. 4. Simultaneously measured photoluminescence and differential absorption (gain) spectra for CdSe nanocrystals embedded in a glass matrix ($T = 10\,\mathrm{K}$) [58]. (a) Differential absorption $\Delta\alpha d$ (solid lines) for pump intensities of (i) 10, (ii) 100, (iii) 400, and (iv) $1400\,\mathrm{kW/cm^2}$ (excitation energy 2.48 eV, pulse length 7 ns). For comparison the linear absorption spectrum αd is shown as a dashed line. (b) Gain spectra in units of $\Delta\alpha d$ calculated from the difference of the linear and differential absorption for pump intensities of (ii) 100, (iii) 400, and (iv) 1400 $\mathrm{kW/cm^2}$. (c) Evolution of the luminescence spectra (solid) for identical excitation intensities as in (a). The dotted lines show a deconvolution of curve (iv) into four luminescence peaks which are assigned to the exciton states $1S_{3/2}1s_\mathrm{e}$ (1), $2S_{3/2}1s_\mathrm{e}$ (2), and to the biexciton ground state $1S_{3/2}1S_{3/2}1s_\mathrm{e}1s_\mathrm{e}$ (4) (right lower scheme). For the energetically highest luminescence peak which appears at very high pump intensity, we suppose contributions from higher excited two-pair states (3) (biexcitons with holes populating excited states) and/or from an energetically higher exciton transition involving *P*-*p* type hole and electron states. Upper right part: Level scheme of the confined exciton states in CdSe quantum dots according to experimental data published in [62].

resonantly excited smaller quantum dots can be neglected because no emission from these sizes is detected). Consequently, the studied quantum dots should show a slightly Stokes-shifted emission from the ground state transition around 1.98 eV and, under high power and state filling conditions, emission from the excited exciton states $2S_{3/2}1s_\mathrm{e}$ and $1P_{3/2}1p_\mathrm{e}$ shifted by 0.07 eV and 0.185 eV to higher energies, respectively. Likewise, the absorption bleaching of the ground state should be centered at the energies of the $1S_{3/2}1s_e$ and $2S_{3/2}1s_e$ excitons around 1.98 eV and 2.05 eV, since these states are simultaneously bleached because of the common $1s_e$-electron.

The experimental data in Fig. 4 (a) show, besides the linear absorption spectrum αd, the development of the differential absorption $\Delta\alpha d$ with increasing pump intensity. For the lowest pump intensity of $10\,\mathrm{kW/cm^2}$ (i), the absorption is bleached around 2.0 eV indicating the fast relaxation of the excited electron-hole pairs to the lowest exciton states. The two hole levels belonging to the same electron level $1s_e$ are bleached simultaneously. For higher pump intensities (curves (ii)–(iv)), the bleaching signal is increased up to the same value as the maximum of the linear absorption which shows that indeed the whole size distribution has been excited. For the highest intensities the signal in differential absorption exceeds the linear absorption and yields the gain spectrum (Fig. 4 (b)). The gain is spectrally broad (nearly 200 meV) with a steeper decrease on the high energy side and a longer tail stretching to lower energies. With increasing intensity the gain increases and becomes spectrally broader. The maximum of the gain for the highest intensity amounts to about $6.5\,\mathrm{cm^{-1}}$.

The gain spectra are now compared with the luminescence taken under identical excitation conditions shown in Fig. 4 (c). For the lowest intensity (i) only one luminescence peak (1) near the absorption edge is detected at 1.93 eV. It can be assigned to the radiative recombination of electron-hole pairs after relaxation into the lowest quantum dot level $1S_{3/2}1s_e$ (the observed Stokes shift of 50 meV is in agreement with previous observations, e.g. in [62] and Refs. therein). At higher pump intensities (curves (ii)–(iv)), peak (1) grows and, at the same time, a second peak (2) at about 80 meV above peak (1) develops which finally even exceeds the first one originating from the decay of a one-pair state with the hole in the $2S_{3/2}$ level. Additionally, a third peak (3) about 180 meV above the first one emerges (clearly seen in curve (iv) of Fig. 4 (c)). Regarding this energetically highest luminescence peak (3) evolving at high pump intensity, contributions to this line can be expected from a P-p-type one-pair transition and/or from higher excited two-pair states (biexcitons with holes populating excited states, see scheme on the lower right part of Fig. 4). Furthermore, a shoulder (4) appears on the low-energy side of the first peak (1) in the spectral region where the gain is found. A careful deconvolution of the luminescence spectrum for the highest pump intensity shows that the spectrum can be excellently fitted with a sum over four Gaussians indicating that the low-energetic shoulder (4) arises from a peak centered about 40 meV below the first transition. This peak shows a much larger halfwidth (of ~100 meV) compared to peaks (1) and (2) (of ~50 and ~60 meV, respectively). From the spectral position of its appearance under high excitation, from the coincidence of its energetic position with the gain and from its large width we conclude that this new luminescence peak (4) arises from the decay of two-pair states (biexcitons) into a photon and an exciton. This hypothesis is supported by data from time-resolved luminescence of the four peaks: (1) and (2) exhibit a similar, almost mono-exponential decay of a few nanoseconds because of the coupled dynamics of two excitons sharing the same electron, peak (3) has a fast initial decay of a few hundreds of ps and decays much faster than peak (4) which shows a slight rise time followed by a fast component below 1 ns. A luminescence band centered around the gain maximum is likewise found in fs pump-probe experiments [61] (the spectral overlap of the broad emission bands prevents a more detailed analysis here, see single-dot experiments in Sect. 5).

A qualitative scheme of the transitions predicted theoretically for strongly confined CdSe quantum dots is shown in the lower right part of Fig. 4. Besides the transition between the lowest one-pair, or exciton, state and the quantum dot ground state, a whole ensemble of transitions from higher exciton states involving excited hole states become allowed due to the relaxation of the selection rules by e.g. the Coulomb interaction and valence band mixing effects. For the same reason, also biexciton states can very likely be formed and consequently transitions between these two-pair states, as well from their lowest level as from various excited levels, and the one-pair states will turn up. These transitions can therefore appear as well low- as high-energetically to the exciton-related peaks in the differential absorption spectrum. The biexcitonic origin of the optical gain in CdSe nanocrystals is supported by the good agreement between experiment and theory, both for modelling the gain spectra and their femtosecond dynamics and in explaining spectral hole burning data for the gain spectrum [59–61]. The experimental data from differential absorption measurements in combination with the analysis of the emission spectra indicate a degeneracy in the energies of excited one-pair states with p-type symmetry and excited biexciton states, i.e. four-particle states with particles in excited states. To clarify this we will discuss in the next section differential absorption experiments using circularly polarized light.

2.3 Biexciton Dynamics in Pump-Probe Experiments Using Circularly Polarized Light

As outlined in the previous Sect. 2.2, the pump-probe dynamics at energies above the ground state exciton and biexciton is more complex because of the spectral overlapping of excited one-pair (exciton) states and two-pair (biexciton) states having holes in excited states. At high pair densities, the created two-pair states mediate a new relaxation path due to their stimulated decay into one photon and one electron-hole pair with lower energy [56]. This superposition of two relaxation (or recombination) pathways made investigations of one-pair relaxation processes, such as the search for the phonon-bottleneck, difficult. Experimentally such processes are manifested by the ultrafast recovery of the bleaching band around the pump energy when exciting at energies above the exciton ground state. Within the excited biexciton picture a possible explanation of the missing absorption bleaching around the pump is the induced absorption of probe photons due to the creation of biexcitons around the pump wavelength that compensates for the bleaching induced by the hot carrier population. However, if a spin-selective experiment could be done, i.e. a pump-probe experiment with copolarized light, then the only state possible with σ^+ pump and probe light would be a biexciton with two parallel electron spins and two parallel hole spins [63]. This state is forbidden by Pauli's exclusion principle, hence the induced absorption of the probe photon should be suppressed and now a pronounced spectral hole due to one-pair population should appear when using exclusively co-circularly polarized light. For counterpolarized pump-probe experiments, however, the σ^+ pump photon and the σ^- probe photon would create an allowed biexciton with antiparallel electron and hole spins. Therefore, the induced absorption due to the absorbed probe photon would appear in the probe spectrum and counteract the bleaching. The concept of pump-probe experiments with circularly polarized light is illustrated in the diagram shown in Fig. 5. The left diagram shows the copolarized pump-probe case, the right the counterpolarized case [63].

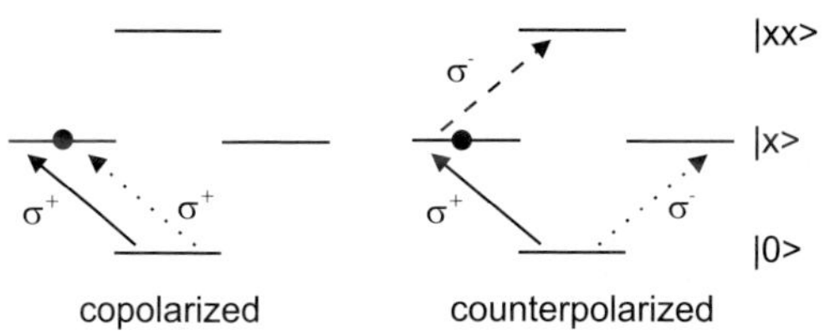

Fig. 5. Pump-probe experiment with circularly co- and counterpolarized beams [63]. Solid arrows show the allowed pump transitions. The dashed arrow shows the allowed probe transition under the creation of a biexciton in the counterpolarized case, giving an induced absorption signal. The dotted arrows show the forbidden probe transitions in the copolarized case, yielding a bleaching

To examine this experimental concept which would allow to distinguish between populations of excited exciton and biexciton states, a pump-probe experiment with circularly polarized laser beams has been performed in strongly confined CdSe quantum dots with radius of $0.5\,a_B$. Figure 6 shows the result of that pump-probe experiment: In Fig. 6 (a), showing the bleaching under copolarized conditions, the bleaching around the pump beam at 560 nm in the first 300 fs is clearly visible. Figure 6 (b) shows the same experiment with counterpolarized beams. No additional bleaching around the pump beam can be observed. Indeed, the biexciton formation by induced absorption of a probe photon results in a signal which compensates the absorption bleaching caused by the population of excited excitons.

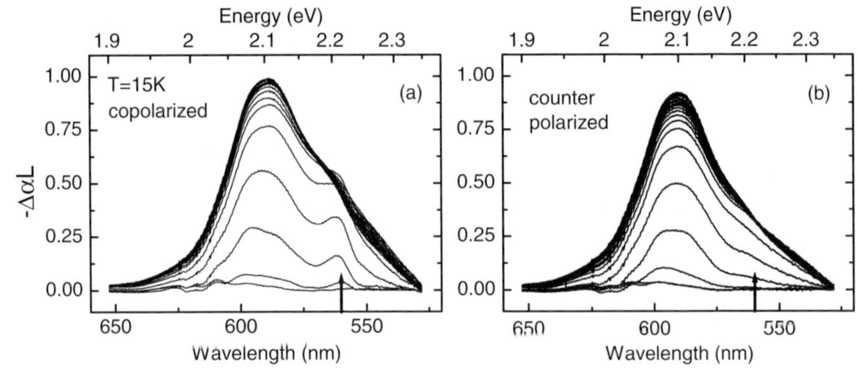

Fig. 6. Femtosecond pump-probe experiment using circularly polarized light in strongly confined CdSe nanocrystals ($R = 0.5\,a_{\rm B}$) [63]. Differential absorption (bleaching) spectrum for (a) circularly co-polarized and (b) counter-polarized pump and probe beams. The pump intensity is $7\,\mathrm{GW/cm^2}$, the pump wavelength is 560 nm (arrow), the pump pulse has a duration of 115 fs, the time delay increases from top to bottom in 80 fs steps, $T = 15\,\mathrm{K}$.

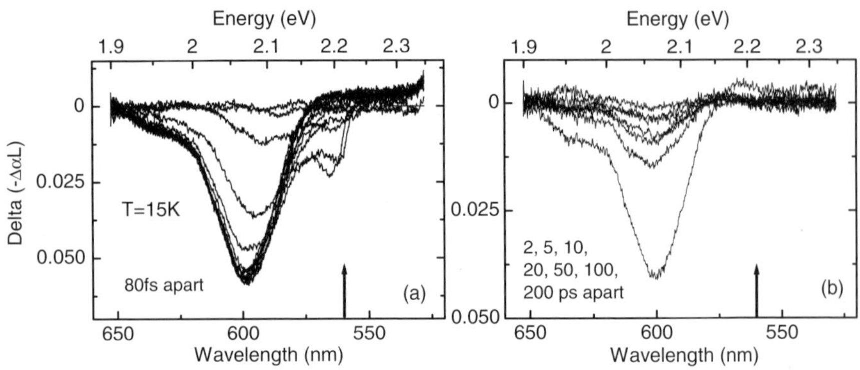

Fig. 7. Differential pump-probe experiment using co- and counter-circularly polarized light showing the dynamics of the biexciton population around the pump as an induced absorption (i.e. negative) signal in the difference spectrum between both polarizations [63]. The y-axis shows the difference of the bleaching for (counter minus co)-polarized beams. The pump intensity is $500\,\mathrm{kW/cm^2}$, all other parameters are identical to those in Fig. 6. (a) Signal dynamics in the early time range for pump delay times between -160 fs and 960 fs, (b) signal dynamics at later pump delay times between 2 ps and 200 ps.

According to the selection rules illustrated by the diamond level scheme in Figs. 1 and 5, the difference between the copolarized and counterpolarized differential absorption spectrum contains the direct information about the excited biexciton dynamics. Subtracting the copolarized from the counterpolarized diagram yields exactly the biexciton population and gives us access to its relaxation dynamics [63]. Fig. 7 shows the result of such a differential pump-probe experiment with two polarizations. Here the difference between the two differential absorption spectra measured for co- and counter-polarized pump and probe beams is plotted. A negative signal means net induced absorption and hence in our model biexciton population. At early times, the biexciton creation around the pump can be clearly seen. Simultaneously, at some lower energy there seems to be a further population of biexcitons being created. Subsequently, the biexcitons around the pump relax to lower energies, transform into one-pair states and after 500 fs there is only one big induced absorption peak visible. Its maximum at 598 nm is about 10 meV lower than the early bleaching peak. This is explained by the biexciton binding energy, which is on that order of magnitude. It is also interesting to note the pronounced shoulder around 640 nm which we assign to a population of biexcitons in their ground state. This ground state population reaches its maximum after about 2 ps and decays within 50 to 100 ps.

In Fig. 7 (b) the later dynamics in the picosecond range of the differential (counter-copolarized) bleaching spectra for times ranging from 2 to 200 ps is shown. The induced absorption decays to zero in about 100 to 200 ps. This recovery is strongly density dependent (not shown here). It takes places within 50 ps for intermediate pump intensity, and for high intensity within 20 ps [63]. Our spin-sensitive model explains this bleaching recovery by spin relaxation or spin flips, leading to the same spins after a certain spin flip time. The higher intensity seems to lead to faster spin flips, but also the influence of the surface or impurities in the dots could lead to spin flips. Additionally, the signal decays as a result of the decay of the excited biexciton population itself which we expect to be in the order of some picoseconds according to the build-up and decay time of the ground state biexciton population around 640 nm. A similar depolarization effect is observed recently in a study of biexciton and spin dephasing effects in CdSSe quantum dots by four-wave mixing and pump-and-probe spectroscopy [64]: The difference in the signals in the $\sigma^+\sigma^+$ and $\sigma^+\sigma^-$ geometries likewise decayed within about 5 ps.

The presented results from pump-probe experiments with circularly polarized light showed that excited exciton and biexciton populations can be distinguished and studied separately by exploiting spin-sensitive selection rules. The existence of two parallel relaxation channels of excitons and biexcitons which can be initiated at the same pump energy is proven and has to be taken into account in all studies of exciton population and polarization dynamics, which focus e.g. on such problems like phonon-bottleneck, gain mechanisms or Auger processes.

3 Biexcitons in Self-Organized Islands

3.1 Biexciton Binding Energies Determined by Four-Wave Mixing Experiments

As can be seen from Fig. 3, the enhancement of the biexciton binding energy $\Delta E_{\mathrm{XX}}^{\mathrm{B}}$ is experimentally proven in I-VII and II-VI nanocrystals in glass matrix, however, the experimentally found values for $\Delta E_{\mathrm{XX}}^{\mathrm{B}}$ typically do not exceed much the bulk exciton Rydberg energy E_{Ryd}. Since the biexciton binding energy crucially depends on the electron–hole mass ratio σ, on the ratio of dielectric constants and on the degree of quantum confinement, the interesting question arises about an upper limit for $\Delta E_{\mathrm{XX}}^{\mathrm{B}}$ and the possibility to obtain values considerably larger than the bulk exciton Rydberg energy E_{Ryd}. In this section we will show that for self-assembled quantum islands based on strained II-VI quantum structures, $\Delta E_{\mathrm{XX}}^{\mathrm{B}}$ may clearly reach and even exceed E_{Ryd} and binding energies similar to the thermal energy $k_{\mathrm{B}}T$ at room temperature can be observed.

We start with self-assembled CdSe/ZnSe-based nanostructures. The optical properties of excitons and biexcitons in CdSe quantum dots grown by migration-enhanced epitaxy (MEE) are studied by time-, energy-, and temperature-dependent photoluminescence, as well as by polarization dependent four-wave mixing (FWM) and two-photon absorption (TPA) experiments [34, 35]. The investigated nanostructures consist of a coherently strained $\mathrm{Zn_{1-x}Cd_xSe/ZnSe}$ quantum well with embedded islands of higher Cd content with sizes of a few nanometer due to strain-induced CdSe accumulation. The local increase in CdSe concentration results in a strong localization of the excitonic and biexcitonic wave functions.

FWM experiments have been performed to determine the biexciton binding energy from the coherent beating between the exciton and biexciton polarizations [34]. In Fig. 8 the spectrum of the time-integrated FWM signal is shown as a function of the delay time τ. The central frequency of the excitation is tuned in resonance with the localized excitons. The 80 fs pulses used ensure, by their large energy bandwidth, a simultaneous excitation of confined excitons and biexcitons. The signal reveals pronounced oscillations in the low energy tail arising from the beating between the exciton and biexciton polarizations which can be used to determine the biexciton binding energy. At a higher detection energy of 2.46 eV, where the signal is dominated by the excitonic response, the beat amplitude decreases. Experimental evidence of the biexcitonic origin of these beats is also obtained from an analysis of the phase shift of the beat signals in case of different polarizations of the incident laser beams. The biexcitonic beating starts with a minimum (maximum) at zero delay for co-linear (cross-linear) polarization [34, 65]. From the FWM signal a period between 200 and 260 fs is derived which corresponds to an energy splitting of $\sim$ 20 meV. Good agreement between the oscillation period and the spectral separation of the exciton and biexciton in FWM and

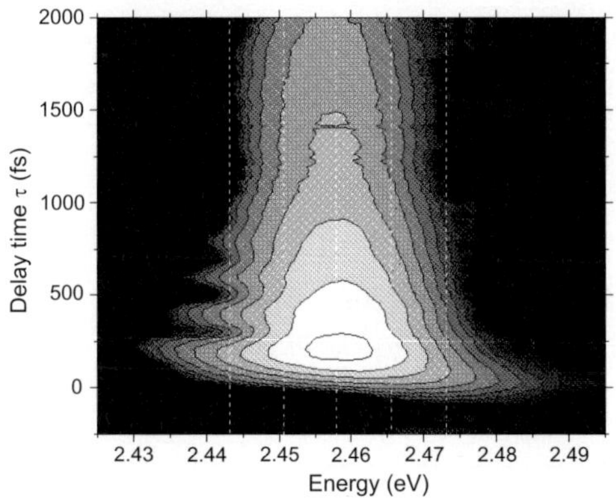

Fig. 8. Contour plot of the spectrally resolved, time-integrated Four-Wave Mixing (FWM) signal as a function of delay time τ measured for an ensemble of epitaxially grown CdSe/ZnSe quantum dots [34, 66]. The 80 fs-pulse excites resonantly both the exciton and biexciton transitions in a sample consisting of a 3.5 ML thick CdSe layer grown between 40 nm ZnSe substrate layer and a 20 nm ZnSe cap layer (amplified Ti:Sa laser combined with an optical parametric amplifier providing an excitation intensity of $100\,\mathrm{nJ/cm^2}$ per pulse, $T = 5\,\mathrm{K}$). From the signal oscillation on the low-energy side of the FWM spectrum the biexciton binding energy is derived (see Fig. 9).

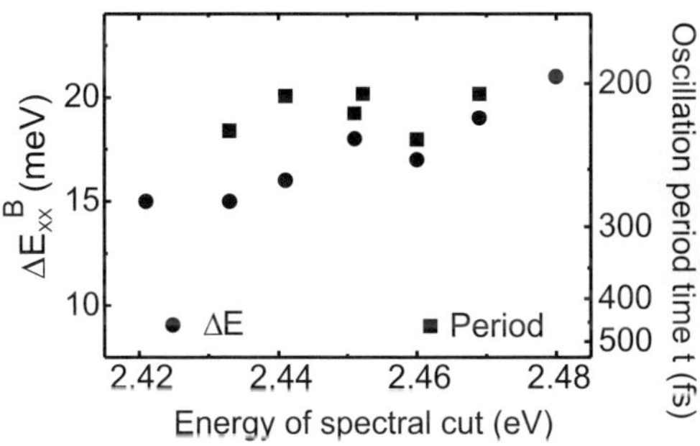

Fig. 9. Biexciton binding energy $\Delta E^{\mathrm{B}}_{\mathrm{XX}}$ determined from both the oscillation period t for different spectral cuts within the FWM spectrum shown in Fig. 8 and the energy difference ΔE between the peak positions of the exciton and biexciton observed in the FWM spectrum [34, 66].

PL signals is obtained (see Fig. 9). From that FWM experiment biexcitonic binding energies $\Delta E^{\mathrm{B}}_{\mathrm{XX}}$ between 15 and 24 meV are derived, depending on the strength of localization due to the composition fluctuations.

In general, biexciton binding energies between 15 meV and 25 meV are measured for self-assembled CdSe islands of average size of 1.5 to 2 nm height in growth direction and 5 to 10 nm lateral extension [34, 35, 66–68] while for flat, more 2D-like islands a biexciton binding energy of 11.5 meV is reported in

[69] and of 10 meV for localized biexcitons in ternary ZnCdSe/ZnSe quantum wells [70]. The experimentally found values for $\Delta E^{\mathrm{B}}_{\mathrm{XX}}$ in CdSe/ZnSe-based self-organized quantum dots thus correspond to a ratio $\Delta E^{\mathrm{B}}_{\mathrm{XX}}/E_{\mathrm{Ryd}} \geq 1$, i.e. $\Delta E^{\mathrm{B}}_{\mathrm{XX}}$ can become larger than the bulk exciton Rydberg energy which amounts to $E_{\mathrm{Ryd}} = 15\,\mathrm{meV}$ in CdSe.

3.2 Biexciton Binding Energies Determined by Two-Photon Absorption

A very large value of $\Delta E^{\mathrm{B}}_{\mathrm{XX}}$ and a ratio of $\Delta E^{\mathrm{B}}_{\mathrm{XX}}/E_{\mathrm{Ryd}} \approx 1.4$ are found in CdS/ZnS-based quantum structures. Within the group of II-VI semiconductors, CdS has comparatively high bulk exciton and biexciton binding energies of 28 meV and 5 meV, respectively [8]. The electron–hole mass ratio is $m_{\mathrm{e}}/m_{\mathrm{h}} \approx 0.14$ [4] and for the ratio of the dielectric constants holds $\epsilon_{\mathrm{CdS}}/\epsilon_{\mathrm{ZnS}} \approx 1$. For a strained CdS/ZnS quantum structure, the confinement potentials are large (1.1 eV for the electron and 260 meV for the hole with $E^{\mathrm{CdS}}_{\mathrm{G}} = 2.48\,\mathrm{eV}$ (cubic modification) and $E^{\mathrm{ZnS}}_{\mathrm{G}} = 3.84\,\mathrm{eV}$) [71]. These parameters make a CdS-based quantum structure to a good candidate for large biexciton binding energies of a few tens of meV, in particular for quantum confinement in all three dimensions. In [37, 38] the biexciton binding energy is investigated by polarization-dependent two-photon absorption experiments (TPA) in epitaxially grown CdS/ZnS quantum structures.

Figure 10 shows the photoluminescence spectrum (PL) after two-photon excitation. For co-circularly polarized excitation, TPA in the biexciton ground state is forbidden while for linearly polarized excitation, the biexciton ground state can be directly populated by resonantly enhanced TPA. The radiative decay of the biexciton proceeds then via the one-photon transition between the biexciton and the exciton state (low energy peak E_{XX}, see scheme in the inset), followed by the transition between the exciton and the ground state (high energy peak E_{X}). When using circularly polarized light the emission peaks E_{XX} and E_{X} disappear because of the selection rules for TPA (see Fig. 1). The difference spectrum between the PL spectra measured for excitation with linearly and with circularly polarized light is plotted in Fig. 10. According to the scheme in the inset, the energy difference between the peak maximum E_{X} and the excitation energy E_{laser} corresponds to half of the value of the biexciton binding energy. The observation of two PL bands situated in energy 15 to 20 meV below and above the two-photon excitation energy indicates huge biexciton binding energies in the CdS/ZnS structures under study. The polarization-dependent two-photon absorption experiments on non-mesa-etched CdS/ZnS structures indicated biexciton binding energies between 26 and 38 meV, i.e. between 1.0 to 1.4 E_{Ryd} of the bulk exciton Rydberg energy [37, 38].

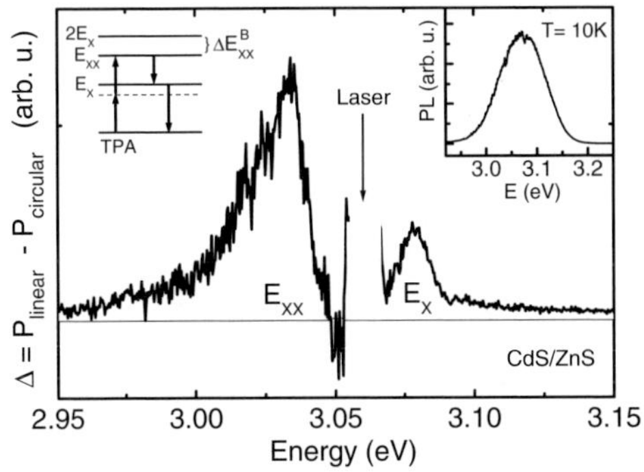

Fig. 10. Determination of the biexciton binding energy $\Delta E^{\mathrm{B}}_{\mathrm{XX}}$ from photoluminescence (PL) after two-photon absorption (TPA) in epitaxially grown CdS/ZnS quantum structures [37] ($T = 10\,\mathrm{K}$, laser energy 3.06 eV, pump intensity 43 kW/cm^2. The sample consists of 2.7 ML CdS grown on a 100 nm ZnS buffer and capped with 50 nm ZnS.). According to the sketch of the radiative recombination process in the inset, the biexciton binding energy can be derived by calculating the difference spectrum Δ for excitation with linearly and circularly polarized light. The energy difference between the peak maximum E_{X} and the excitation energy E_{laser} corresponds to half of the value of the biexciton binding energy. The inset shows the PL spectrum for non-resonant excitation at 3.5 eV.

3.3 Biexciton Binding Energies Determined by Single-Dot Spectroscopy

Finally, we compare experimental data obtained in ensemble measurements with those from single-dot spectroscopy. Optical spectroscopy on single, zero-dimensional excitons is presently a fascinating new field because of the possible observation of discrete quantum states without the effect of inhomogeneous broadening. These measurements are feasible by the use of new experimental techniques with high spatial resolution and high sensitivity, and allow the study of the electronic properties of individual quantum dots.

From the CdSe/ZnSe structures studied in FWM, optically active mesa structures are prepared [72] and single excitons could be studied. Because the average dot densities in the epitaxially grown structures are 10^{10} to $10^{11}/\mathrm{cm}^2$, mesa sizes below $(100\,\mathrm{nm})^2$ contain only a few dots. Single-dot spectroscopy in systems of semiconductor quantum dots often relies on the ability to isolate the spectrum of one dot from a multi-dot spectrum, even for a small mesa structure. Transitions belonging to the same quantum dot can be identified by the long-term time-evolution of their spectrum (jitter analysis and its correction), their intensity-dependence and polarization splittings [73, 74].

In [73] a systematic study of the biexciton (and trion (see below)) binding energy as well as of the exciton exchange splitting energy is given for CdSe/ZnCdSe single quantum dots grown by MEE. Figure 11 shows jitter-corrected time-integrated photoluminescence (PL) spectra for the two linear

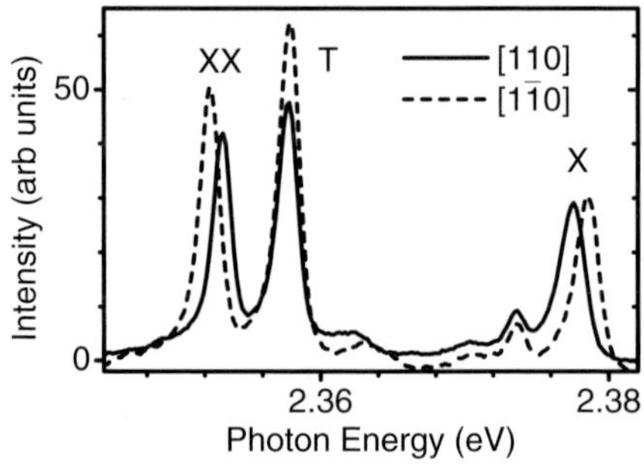

Fig. 11. Jitter-corrected time-integrated photoluminescence (PL) spectra for the two linear polarizations along [110] and [1$\bar{1}$0] measured at a single mesa-structure $(120\,\mathrm{nm})^2$ in epitaxially grown CdSe/ZnSe samples. The three peaks are assigned to exciton (X), trion (T), and biexciton (XX) transitions of a single quantum dot [73]. The experimental parameters are: Excitation with frequency-doubled, mode-locked Ti-sapphire laser, excitation intensity $I_0 = 250\,\mu\mathrm{W}$ focused into a 50-μm-diam. excitation spot, collection of PL with a 0.4-NA-long working distance objective lens, a sample consisting of 3 ML CdSe grown between 25 nm and 50 nm ZnSe layers, $T = 13\,\mathrm{K}$.

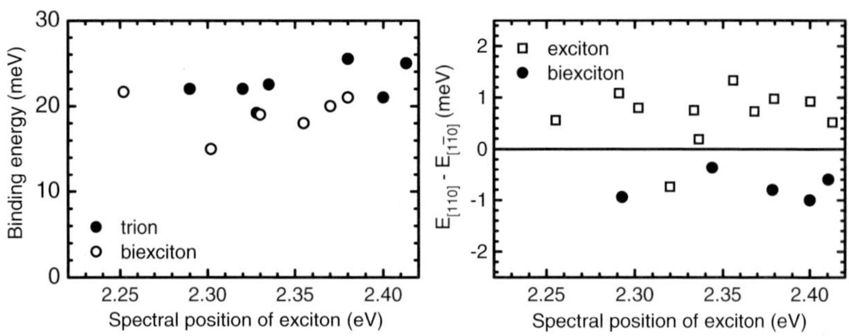

Fig. 12. Binding energies and exchange splitting energies measured for single CdSe/ZnSe quantum dots by polarization-sensitive micro-photoluminescence. Left: Binding energy of trions and biexcitons vs. the spectral position of the transition. Right: Exchange splitting Δ_1 vs. the spectral position of the transition observed for biexcitons and excitons [73].

polarizations along [110] and [1$\bar{1}$0] measured at a single mesa-structure of $(120\,\mathrm{nm})^2$. The three peaks are assigned to exciton (X), trion (T), and biexciton (XX) transitions of a single quantum dot by applying a spectral jitter correlation technique and by their polarization properties and excitation density dependencies. The exchange splitting $\Delta_1 = E_{[110]} - E_{[1\bar{1}0]}$ of the bright exciton has been inferred from the peak fine structure observed when detecting the photoluminescence polarized along the [110] and [1$\bar{1}$0] crystal directions. Fig. 12 shows the exchange splitting Δ_1 versus the spectral position of the transition observed for biexcitons and excitons of MEE-grown CdSe/ZnCdSe

QD at $T = 13\,\mathrm{K}$. The exciton splitting has dominantly a positive sign, so that the exciton anisotropy is not random, but its size is preferentially larger in the $[1\bar{1}0]$ crystal direction. The peak splitting observed for the biexciton ranges between 0.2 meV and 1.2 meV. Because the biexciton has no spin degeneracy, the biexciton to exciton transition shows the (inverted) fine structure of the exciton transition as can be seen in Fig. 12 (right). The observed polarization splitting of exciton and biexciton is with the experimental accuracy equal as expected from theory.

The biexciton binding energy is derived by calculating the energy difference of the biexciton peak XX with respect to the energy of the bright exciton peak X, i.e., we neglect the correction due to polarization splitting introduced in [23] and discussed in (5). Since the observed polarization splitting is smaller than the experimentally obtained standard deviation of the biexciton binding energy itself, this approximation is justified. The biexciton binding energy vs. the spectral position of the exciton is shown in Fig. 12 (left). Biexciton binding energies between 19 and 26 meV are found, confirming the data obtained from measurements on ensembles (see Fig. 9).

4 Trion Properties in Quantum Dots

The trion is an interesting elementary excitation because its radiative decay can be used to monitor properties of single electrons or holes which are usually optically inaccessible. It is expected that the number of charge carriers within a quantum dot will affect both the lifetime of the state and its other spectroscopic properties such as exchange splitting and binding energy. The existence of trions was first reported almost ten years ago [39–41] and predominantly studied in quantum well structures (see e.g. [42–44] and Refs. therein). Only recently trion states were studied in self-assembled quantum dots using magneto-photoluminescence and micro-photoluminescence experiments [25, 38, 45–50, 52, 53, 73]. The observation of charged exciton states shows that a surplus of negative charge is present in the samples which may be attributed to a n-type background doping almost always present in II-VI materials. Figure 11 shows the time-integrated spectra of exciton (X), biexciton (XX) and trion (T) at a low excitation intensity for a single CdSe/ZnCdSe quantum dot. To within the experimental resolution (100 μeV) no polarization splitting is found for the trion. This fact, and the nonlinear increase of peak intensity vs. pump power, is used to distinguish trion peaks from exciton and biexciton peaks. For these MEE-grown CdSe/ZnCdSe samples, the trion emission is always observed at a higher energy than that of the corresponding biexciton. As can be seen in Fig. 12, trion binding energies between 15 and 22 meV are found. Slightly smaller trion binding energies of 10 meV are observed for CdSe quantum dots embedded in a ZnSSe matrix grown by low-pressure MOCVD [53]. The observed variations in the trion binding energy is attributed to a variation of material composition, most likely a strong fluctuation of the Zn content in the quantum dots.

5 Dynamics of Single Excitons, Biexcitons, and Trions

5.1 Dynamics at Low-Excitation Density

The ability to identify single exciton, biexciton and trion transitions by making use of polarization spectroscopy and jitter analysis allowed us to identify the dynamical properties of exciton, multiexciton and trion states. In Fig. 13 we illustrate the exciton (X) and trion (T) dynamics of a single CdSe/ZnSe quantum dot after pulsed excitation at low temperature and low excitation density. With Fig. 14 we discuss the changes in the single dot dynamics when applying higher pump power.

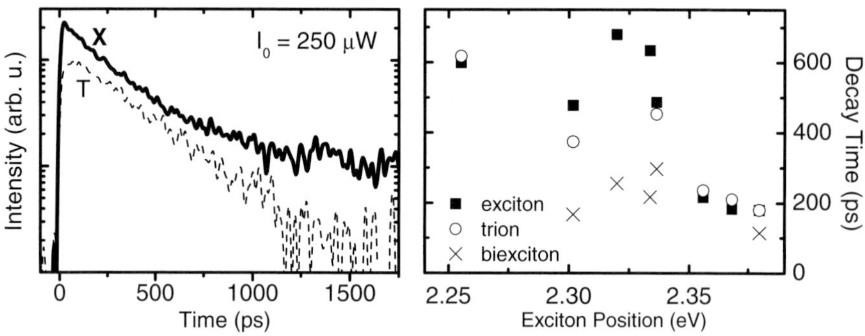

Fig. 13. Exciton (X) and trion (T) dynamics of a single CdSe/ZnSe quantum dot after pulsed excitation at low temperature [73]. The time-resolved data were collected by a streak camera with 4 ps time resolution. All other experimental parameters are identical to those in Fig. 11. Right: Overview about measured PL intensity decay times for exciton, biexciton, and trion transitions in different quantum dots vs. the corresponding exciton transition energy.

The recombination dynamics of the single dot emission is governed by the exchange interaction which splits the lowest electron-hole pair state into dark states and bright states (see Fig. 2). The repopulation of the bright state from the dark state results in the observation of a so-called "effective" radiative decay time which is temperature-dependent and reflects the interplay between bright and dark state population. At the lowest excitation intensity (Fig. 13, left) the exciton shows a biexponential decay, with an initial fast decay with a time constant of 240 ps followed by a longer decay with a time constant of 2.5 ± 0.5 ns. We attribute this long-time component to a refilling of the optically active state from the dark exciton state. The trion does not possess

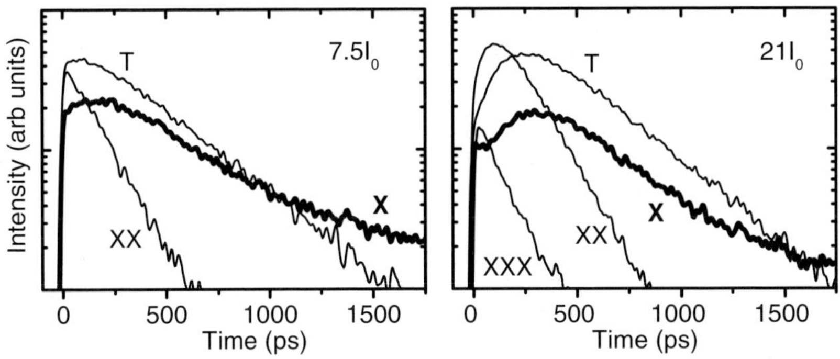

Fig. 14. Time-resolved photoluminescence intensity (logarithmic scale) of quantum dot emission after pulsed excitation with high pump intensity [73, 74]. The peaks are assigned to exciton (X), trion (T), biexciton (XX), and multiexciton (XXX) transitions of a single quantum dot. The experimental parameters are identical to those in Figs. 11 and 13.

such a dark ground state and, consequently, a single exponential decay with a time constant of 320 ps is observed. In addition, it shows an initial refilling behavior (of between 20 and 50 ps lifetime) followed by the monoexponential decay of similar lifetime to that of the exciton. We attribute this refilling to a spin-relaxation process in which a carrier is Pauli-blocked from relaxing into the ground state whence it can recombine. A spin-flip lifts the blocking and allows the carrier to relax and optical decay subsequently occurs. The measured rise time is therefore identified with a spin relaxation time. The trion dynamics itself monitors the single electron properties, i.e. by evaluating the spin-flip time from the rise time and the combined lifetime from the up to 600 ps monoexponential decay time.

An overview about the measured decay times for the different transitions in the investigated single dot ensemble is given in the right part of Fig. 13. It can be seen that in general the trion has a lifetime comparable to that of the exciton (while the biexciton has a lifetime which is shorter than the exciton, typically by a factor of 2 (see below)). The effective radiative decay time derived from a temperature-dependent analysis of the PL decay are in the range between 200 ps and 600 ps for both exciton and trion.

5.2 Dynamics at High Excitation Density

Figure 14 shows the photoluminescence decay curves for the exciton, trion, and biexciton at different excitation intensities. At higher intensities the biexciton (XX) appears, showing a monoexponential decay characteristics with a decay constant about half that of the exciton (X). In most cases we found that the ratio of the exciton decay time to the biexciton decay time was approximately 2:1, in contrast to some previous results [67] but in agreement with ensemble data [34]. For higher intensities, the exciton now shows an initial refilling which can well be fitted by setting the time constant for this refilling to the decay time of the biexciton. At the highest excitations we see that even the biexciton shows some refilling by decays of multiexcitonic states (XXX) with an even greater number of excitons. We have been unable to isolate spectral peaks corresponding to multiexciton transitions and further examination of multiexciton states is required in future experiments.

6 Summary and Outlook

In both investigated systems, II-VI nanocrystals in glass and epitaxially grown II-VI based self-assembled islands, we observe $\Delta E_{\mathrm{XX}}^{\mathrm{B}}$ values in the order of the thermal energy at room temperature, which makes biexcitons in quantum dots stable even at elevated temperatures. The fact that the biexciton binding energies have so similar values in both systems is surprising since the confinement geometries and the matrix material are completely different. To clarify why, and to find out why no clear trion features are observed in nanocrystals until now are interesting questions for future research. Our experimental data show that biexciton binding energies reach the order of the LO-phonon energy in both CdSe- and CdS-based systems which is not observed in III-V based quantum dot structures. While the current theory of excitons in quantum dots involves more and more polaron properties, e.g. to explain homogeneous linewidths and intensities of LO-phonon replica, the present theory for biexciton states does not. Intuitively we would expect polaron effects also in multiexciton systems which might be another interesting problem for future studies. The large biexciton binding energy leads to an optical gain strongly influenced by Coulomb interaction. Any strong optical pumping will result in the formation of biexcitons simultaneously with the population and filling of higher excited exciton states. Thus, an exact theoretical description of all gain mechanism has to take into account Coulomb correlation effects. Recently, multiexciton effects [75] and charged excitons [76] are likewise addressed for wet chemically grown nanocrystals. Since this class of quantum dots possess much more flexibility to engineer surface properties, the influence of dielectric confinement and barrier height on biexciton and trion properties might be addressed more in future studies of multiexciton states in quantum dots aiming at optimizing their binding energy, radiative lifetimes, and homogeneous lineshapes.

Acknowledgments

The author wishes to thank H. Giessen, W. Langbein, O. Wind, F. Gindele, K. Hild, B. Patton, for fruitful cooperations and many valuable discussions. We acknowledge the group of D. Hommel in Bremen for growing the CdSe/ZnSe samples, the group of Prof. Klingshirn in Karlsruhe, in particular M. Hetterich and M. Grün, for growing the CdS/ZnS samples, and the group of A. Forchel in Würzburg for patterning the mesa structures.

Special thanks go to C. Klingshirn who introduced me to the field of biexciton physics. His continuous interest, encouraging discussions and suggestions, combined sometimes with some doubts, have stimulated many of the here presented research projects.

References

1. W.F. Brinkmann, T.M. Rice, B. Bell: Phys. Rev. B **8**, 1570 (1972)
2. O. Akimoto, E. Hanamura: Solid State Comm. **10**, 253 (1973)
3. J.M. Hvam: Solid State Comm. **12**, 95 (1973); phys. stat. sol. (b) **63**, 511 (1974)
4. C. Klingshirn, H. Haug: Physics Reports **70**, 315 (1981); phys. stat. sol. (b) 71, 547 (1975)
5. F. Henneberger, K. Henneberger, J. Voigt: phys. stat. sol. (b) **79**, K 81 (1977); F. Henneberger, J. Voigt: phys. stat. sol. (b) **76**, 313 (1976)
6. M. Ueta, H. Kazanaki, K. Kobayashi, Y. Toyozawa, E. Hanamura: *Excitonic Processes in Solids*, Springer Ser. Solid State Sc. **60** (Springer, Berlin Heidelberg 1986)
7. C. Dörnfeld, J.M. Hvam: IEEE J. Quantum Electron. **25**, 904 (1989)
8. Landolt–Börnstein: *Physics of II-VI and I-VII Compounds*, vol **17b**, New Series, ed by O. Madelung (Springer, Berlin Heidelberg 1982)
9. C.F. Klingshirn: *Semiconductor Optics*, 2nd corr. printing (Springer, Berlin Heidelberg New York 1997)
10. Y.Z. Hu, M. Lindberg, S.W. Koch: Phys. Rev. B **42**, 1713 (1990)
11. Y.Z. Hu, S.W. Koch, M. Lindberg, N. Peyghambarian, E.L. Pollock, F.F. Abraham: Phys. Rev. Lett. **64** 1805 (1990)
12. S.H. Park, R.A. Morgan, Y.Z. Hu, M. Lindberg, S.W. Koch, N. Peyghambarian: J. Opt. Soc. Am. **7**, 2097 (1990)
13. L. Banyai, S.W. Koch: *Semiconductor Quantum Dots*, Series on Atomic, Molecular and Optical Physics vol 2, (World Scientific, Singapore 1993)
14. L. Banyai, P. Gilliot, Y.Z. Hu, S.W. Koch: Phys. Rev. B **45**, 14136 (1992)
15. T. Takagahara: Phys. Rev. B **39**, 10206 (1989)
16. Al.L. Efros, A.V. Rodina: Solid State Comm. **72**, 645 (1989)
17. E.L. Pollock, S.W. Koch: J. Chem. Phys. **94**, 6766 (1991)
18. O. Stier, R. Heitz, A. Schliwa, D. Bimberg: phys. stat. sol. (a) **190**, 477 (2002)
19. A. Franceschetti, A. Zunger: Phys. Rev. B **62**, R16287 (2000)
20. A. Franceschetti, A. Williamson, A. Zunger: J. Phys. Chem. B **104**, 3398 (2000)
21. J. Shumway, A. Franceschetti, A. Zunger: Phys. Rev. B **63**, 155316 (2001)
22. L.W. Wang: J. Phys. Chem. B **105**, 2360 (2001)
23. V.D. Kulakovskii, G. Bacher, R. Weigand, T. Kümmell, A. Forchel, E. Borovitskaya, K. Leonardi, D. Hommel: Phys. Rev. Lett. **82**, 1780 (1990)
24. L. Besombes, K. Kheng, D. Martrou: Phys. Rev. Lett. **85**, 425 (2000)
25. L. Besombes, K. Kheng, L. Marsal, H. Mariette: Phys. Rev. B **65**, 121314(R) (2002)
26. F. Bassani, J. Forney, A. Quattropani: phys. stat. sol. (b) **65**, 591 (1974)

27. A. Uhrig, L. Banyai, Y.Z. Hu, S.W. Koch, C. Klingshirn, N. Neuroth: Z. Phys. B – Condens. Matter **81**, 385 (1990)
28. N. Peyghambarian, B. Fluegel, D. Hulin, A. Migus, M. Joffre, A. Antonetti, S.W. Koch, M. Lindberg: IEEE J. Quant. Electron. **25**, 2516 (1990)
29. J.Y. Bigot, A. Daunois, J. Oberle, J.C. Merle: Phys. Rev. Lett. **71**, 1820 (1993)
30. K.I. Kang, A.D. Kepner, S.V. Gaponenko, S.W. Koch, Y.Z. Hu, N. Peyghambarian: Phys. Rev. B **48**, 15449 (1993)
31. Y. Masumoto, S. Okamoto, S. Katayanagi: Phys. Rev. B **50**, 18658 (1994)
32. U. Woggon, O. Wind, W. Langbein, O. Gogolin, C. Klingshirn: J. Lum. **59**, 135 (1994)
33. H. Spöcker, M. Portune, U. Woggon: Optics Lett. **23**, 427 (1998)
34. F. Gindele, U. Woggon, W. Langbein, J.M. Hvam, K. Leonardi, D. Hommel, H. Selke: Phys. Rev. B **60**, 8773 (1999)
35. F. Gindele, K. Hild, W. Langbein, U. Woggon: Phys. Rev. B **60**, R21570 (1999)
36. Young-Nam Hwang, Koo-Chul Je, Dongho Kim, Seung-Han Park: Phys. Rev. B **64**, 041305 (2001)
37. U. Woggon, K. Hild, F. Gindele, W. Langbein, M. Hetterich, M. Grün, C. Klingshirn: Phys. Rev. B **61**, 12632 (2000)
38. K. Hild, D. Miller, W. Langbein, U. Woggon, M. Hetterich, C. Klingshirn: phys. stat. sol. (b) **224**, 379 (2001).
39. K. Kheng, R.T. Cox, Y. Merle d'Aubigné, F. Bassani, K. Saminadayar, S. Tatarenko: Phys. Rev. Lett. **71**, 1752 (1993)
40. G. Finkelstein, H. Shtrikman, I. Bar-Joseph: Phys. Rev. Lett. **74**, 976 (1995)
41. A.J. Shields, M. Pepper, M.Y. Simmons, D.A. Ritchie: Phys. Rev. B **52**, 7841 (1995)
42. E. Vanelle, M. Paillard, X. Marie, T. Amand, P. Gilliot, D. Brinkmann, R. Levy, J. Cibert, S. Tatarenko: Phys. Rev. B **62**, 2696 (2000)
43. A. Esser, E. Runge, R. Zimmermann, W. Langbein: Phys. Rev. B **62**, 8232 (2000).
44. G.V. Astakhov, D.R. Yakovlev, V.P. Kochereshko, W. Ossau, W. Faschinger, J. Puls, F. Henneberger, S.A. Crooker, Q. McCulloch, D. Wolverson, N.A. Gippius, A. Waag: Phys. Rev. B **65**, 165335 (2002)
45. T. Taliercio, P. Lefebvre, V. Calvo, D. Scalbert, N. Magnea, H. Mathieu, J. Allegre: Phys. Rev. B **58**, 15408 (1998)
46. J.J. Finley, A.D. Ashmore, A. Lemaitre, D.J. Mowbray, M.S. Skolnick, I.E. Itskevich, P.A. Maksym, M. Hopkinson, T.F. Krauss: Phys. Rev. B **63**, 073307 (2001)
47. V. Türck, S. Rodt, R. Heitz, O. Stier, M. Strassburg, U.W. Pohl, D. Bimberg: phys. stat. sol. (b) **224**, 217 (2001)
48. M. Bayer, G. Ortner, O. Stern, A. Kuther, A.A. Gorbunov, A. Forchel, P. Hawrylak, S. Fafard, K. Hinzer, T.L. Reinecke, S.N. Walck, J.P. Reithmaier, F. Klopf, F. Schäfer: Phys. Rev. B **65**, 195315 (2002)
49. J.G. Tischler, A.S. Bracker, D. Gammon, D. Park: Phys. Rev. B **66**, 081310(R) (2002)
50. I.A. Akimov, A. Hundt, T. Flissikowski, F. Henneberger: Appl. Phys. Lett. **81**, 4730 (2002)
51. A. Wojs, P. Hawrylak: Phys. Rev. B **55**, 13066 (1997)
52. J. Puls, M. Rabe, H.J. Wünsche, F. Henneberger: Phys. Rev. B **60**, R16303 (1999)

53. S. Rodt, A. Schliwa, R. Heitz, V. Türck, O. Stier, R.L. Sellin, M. Strassburg, U.W. Pohl, D. Bimberg: phys. stat. sol. (b) **234**, 354 (2002)
54. U. Woggon: *Optical Properties of Semiconductor Quantum Dots*, Springer Tract in Modern Physics **136** (Springer, Berlin Heidelberg New York 1997)
55. S.V. Gaponenko: *Optical Properties of Semiconductor Nanocrystals* (Cambridge University Press, Cambridge 1998)
56. U. Woggon, H. Giessen, F. Gindele, O. Wind, B. Fluegel, N. Peyghambarian: Phys. Rev. B **54**, 17681 (1996)
57. U. Woggon, O. Wind, W. Langbein, C. Klingshirn: Jpn. J. Appl. Phys. **34**, 232 (1995)
58. U. Woggon, O. Wind, F. Gindele, E. Tsitsishvili, M. Müller: J. Luminescence **70**, 269 (1996)
59. H. Giessen, U. Woggon, B. Fluegel, G. Mohs, Y.Z. Hu, S.W. Koch, N. Peyghambarian: Optics Letters **21**, 1043 (1996)
60. H. Giessen, B. Fluegel, G. Mohs, Y.Z. Hu, N. Peyghambarian, U. Woggon, C. Klingshirn, P. Thomas, S.W. Koch: J. Opt. Soc. Am. B **13**, 1039 (1996)
61. H. Giessen, U. Woggon, B. Fluegel, G. Mohs, Y.Z. Hu, S.W. Koch, N. Peyghambarian: Chemical Physics **210**, 71 (1996)
62. D.J. Norris, M.G. Bawendi: Phys. Rev. B. **53**, 16338 (1996)
63. H. Giessen: *Ultrafast carrier and gain dynamics in strongly confined semiconductor quantum dots*, PhD thesis, The University of Arizona, Tucson (1995)
64. V.D. Kulakovskii, K. Babocsi, M. Schmitt, N.A. Gippius, W. Kiefer: Phys. Rev. B **67**, 113303 (2003)
65. According to Ref. T.F. Albrecht, K. Bott, T. Meier, M. Koch, S.T. Cundiff, J. Feldmann, W. Stolz, P. Thomas, S.W. Koch, E.O. Göbel: Phys. Rev. B **54**, 4436 (1996), the biexcitonic beating in strongly inhomogeneously broadened systems starts with a minimum (maximum) at zero-delay for colinear (cross-linear) polarization.
66. F. Gindele: *Kohärente und inkohärente Dynamik in II-VI Halbleiterstrukturen verschiedener Dimensionalität*, Dissertation, Universität Dortmund (1999)
67. G. Bacher, R. Weigand, J. Seufert, V.D. Kulakovskii, N.A. Gippius, A. Forchel, K. Leonardi, D. Hommel: Phys. Rev. Lett. **83**, 4417 (1999)
68. M. Lowisch, M. Rabe, F. Kreller, F. Henneberger: Appl. Phys. Lett. **74**, 2489 (1999)
69. H.P. Wagner, H.P. Tranitz, R. Schuster, R. Engelhardt, U.W. Pohl, D. Bimberg: phys. stat. sol. (b) **224**, 195 (2001)
70. F. Kreller, M. Lowisch, J. Puls, F. Henneberger: Phys. Rev. Lett. **75**, 2420 (1995)
71. M. Hetterich, Ch. Märkle, A. Dinger, M. Grün, C. Klingshirn: Phys. Rev. B **59**, 10268 (1999)
72. T. Kümmell, R. Weigand, G. Bacher, A. Forchel, K. Leonardi, D. Hommel, H. Selke: Appl. Phys. Lett. **73**, 3105 (1998).
73. B. Patton, W. Langbein, U. Woggon: Phys. Rev. B **68**, 125316 (2003)
74. B. Patton, W. Langbein, U. Woggon: Proc. ICPS 26, Edinburgh, Scotland, July 2002
75. A.A. Mikhailovsky, A.V. Malko, J.A. Hollingsworth, M.G. Bawendi, V.I. Klimov: Appl. Phys. Lett. **80**, 2380 (2002)
76. K.T. Shimizu, W.K. Woo, B.R. Fisher, H.J. Eisler, M.G. Bawendi: Phys. Rev. Lett. **89**, 117401 (2002)

Dynamics of Excitons and Exciton Complexes in Wide-Gap Semiconductors

Jürgen Gutowski, Hans-Georg Breunig, and Tobias Voss

1 Introduction

Optical properties of direct-gap semiconductors are, at least for moderate excitation densities and temperatures, strongly governed by excitonic processes, i.e., in the most simple case by the formation and decay of the two-particle system exciton composed of each one conduction-band electron and valence-band hole bound by their mutual Coulomb attraction. However, even unbound electrons and holes in the bands experience Coulomb correlation leading to the so-called Coulomb enhancement of the band-edge absorption exceeding the simple square-root absorption deduced from the treatment of optical excitation in the single-particle approach for which Coulomb interaction of excitations in the bands is neglected.

The role of excitonic correlation is, in general, increasing with increasing semiconductor band gap and for reduction of the material dimension, i.e., for quantum wells, quantum wires and, eventually, quantum dot systems. The first aspect results in the dominance of excitonic features for wide-gap materials such as the II-VI and the III-nitride semiconductors. If the second aspect, the reduction of dimension, is particularly applied to those materials, exciton binding energies easily exceed the materials' phonon energies and/or the room-temperature thermal energy so that excitons continue to dominate their optical properties even beyond room temperature and for elevated excitation densities.

These general trends also hold for more complex excitonic systems. The most simple exciton complex is the trion being composed of either two electrons and one hole (negatively charged trion X^-) or one electron and two holes (positively charged trion X^+) whose existence, however, requires the excess of one kind of these carriers. Furthermore, trions are partly not stable in these materials unless additional conditions are given as is the application of high magnetic fields. Probably more important are the four-particle biexcitons having been observed in practically all wide-gap semiconductors and approaching rather high binding energies (up to several ten meV) under reduction of material dimension. Excitonic complexes involving more than four particles (so-called multiexcitons) have been observed in some wide-gap materials but remained of less importance for the general features in the optical spectra until recently their relevance in quantum dots became evident[1].

Excitonic systems in quantum dots are addressed at least in the two contributions by U. Woggon [1] and J. Hvam [2] in this book. In the present contribution, we will therefore concentrate on the role of excitonic excitations for the dynamics in two-dimensional (and a few one-dimensional) wide-gap semiconductor systems as can be investigated by different methods used in coherent optics. In Sect. 2, just some basics of the dynamics of excitonic systems in lower-dimensional wide-gap materials are sketched in so far they are necessary for the understanding of Sect. 3 dealing with the main topic of this article.

2 Dynamics of Excitons, Trions and Biexcitons in Low-Dimensional Wide-Gap Semiconductors

The principle description of excitons and biexcitons in three- and lower-dimensional wide-gap semiconductors can be found in a manifold of review papers and textbooks (e.g., [3, 4] just to name a few written by C. Klingshirn). For the understanding of the (coherent) dynamics of excitons described in the following chapters, it is just worth mentioning that the increase of the exciton and biexciton binding energy is not only a function of the pure size of the confined structures in a sample but also of the localization of the excitonic systems in potential fluctuations introduced by well or wire width and/or chemical composition variations on the nm scale. Some preliminary studies on exciton recombination in II-VI quantum wells pointed already to the strong impact of localization to the measured decay times as well as to the dominant radiative decay of localized excitons at least at low temperatures but involved some unclear and deviating cases of exponential and non-exponential decay characteristics [5–7]. Neukirch et al. [8] showed that for systems with a long-range potential fluctuation as are ultrathin binary CdSe/ZnSe quantum wells the dynamics of excitons has to be described by taking into account locally completely trapped and non-trapped excitons at any energy. This yielded a non-exponential decay at higher energies within the broadened exciton luminescence band but eventually purely exponential decay at lowest energies

[1] In quantum dots, the terminus "exciton" is not fully accepted in the scientific community since the excitation of pairs of electrons and holes in such systems with diameters of about or less than one three-dimensional exciton Bohr radius is described, e.g., in the so-called shell models of electron and hole energy levels where Coulomb interaction has to be taken into account but not necessarily leads to Coulomb binding. Thus, four-particle systems may possess transition energies lying beyond those of two-particle excitations resulting in no pure biexcitonic binding, however, the escape of carriers is nevertheless prevented by the confinement potential of the dot. For simplicity, Coulomb-correlated electron-hole systems in dots are addressed as excitonic in this paper.

corresponding to excitons fully trapped in potential minima. For short-range fluctuations, however, as found in ternary ZnCdSe/ZnSe wells, this does not apply since the relaxation probability of the excitons then only depends on their absolute energy [9]. For the latter kind of wells, it was shown later experimentally and theoretically that even long time after excitation excitons in strongly localizing systems never reach global thermal equilibrium distribution but remain distributed according to the potential minima topology yielding a so-called topological lineshape of the exciton emission band [10, 11]. The same samples exhibited, at least at low temperatures, a rather large radiative decay rate pointing to strong 2D confinement at high structural quality [10]. Recently, first picosecond time-resolved studies on InGaN/GaN quantum-well systems [12] revealed the dominant role of potential fluctuations most probably due to In composition fluctuations. However, in thin InGaN wells the local potential minima were found to be that close that excitons remain mobile even at lowest energies.

Transport properties of excitons in wide-gap quantum wells have been investigated by using optical methods by a number of authors, see, e.g., [13, 14] (exciton and exciton-complex diffusion often investigated in transient-grating experiments shall not be addressed in this paper). Excitonic transport in quantum wells sensitively depends on the exciton velocity which has, in the weak-confinement limit (well widths from few to some ten excitonic Bohr radii) to be described in the exciton polariton picture [13]. In this case, the polariton group velocity is determined by the center-of-mass quantization $\boldsymbol{k}_{\mathrm{z}} = p/L_{\mathrm{z}}$ superimposed to the polariton dispersion curves. This results in a dramatic increase of the transport coefficient when approaching the bottleneck near-photonic part of the polariton dispersion for increasing well thickness which could be confirmed experimentally by delay-time dependent transient-grating experiments (Fig. 1 [13]). The non-diffusive nature of exciton transport and the strong coupling to phonons in ZnSe quantum wells has been proven by recording spatial profiles of the heavy-hole (hh) exciton luminescence as a function of the excess energies of the excited non-thermal exciton population, e.g., [14]. Details of hot exciton analysis are addressed in another contribution to this book [15].

Some studies dealt with the interesting question of exciton tunnelling between adjacent wide-gap asymmetric quantum wells with barriers guaranteeing complete decoupling of the electronic resonances. Anyway, the tunnelling has been found to be rather fast, between a few and a few tens of picoseconds. For CdTe/CdZnTe asymmetric wells the experimental results pointed to complete exciton tunnelling if the energy difference of the exciton levels in the adjacent wells exceeds the LO phonon energy [16, 17]. Measurements and calculations for ZnCdSe/ZnSe asymmetric wells yielded that, in spite of the energetically allowed complete LO-phonon assisted exciton tunnelling, efficient tunnelling takes place as a two-step process where first the electron and after that the hole transmits the barrier [18].

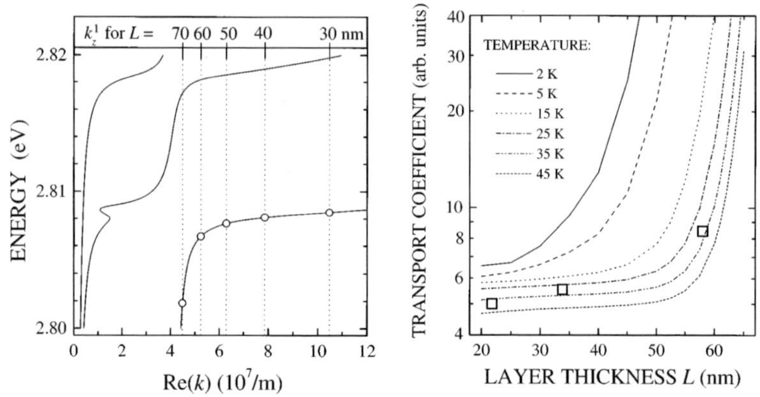

Fig. 1. (a) Polariton dispersion of compressively strained ZnSe (two-oscillator polariton model with heavy-hole exciton resonance at 2.8081 eV, light-hole exciton resonance at 2.8181 eV) for different well (or layer) thicknesses between 30 and 70 nm (corresponding to 7 to 18 exciton Bohr radii). (b) Mean transport coefficient versus layer thickness for different temperatures. Open squares are deduced from transient light-induced grating experiments. From [13].

Rather few studies have been concerned with exciton dynamics in wide-gap semiconductor quantum wires since, in contrast to the growth of quantum dots, not too many systematic and successful experiments have been performed to achieve wires on base of these materials. In ridge quantum wires obtained on prepatterned GaAs substrates, an increase of the exciton lifetime due to the reduced exciton transition probability compared to the 2D well excitons could be clearly proven by time-resolved photoluminescence measurements [19].

For biexcitons (XX), Yamada et al. [20] reported a temporal luminescence behavior for which, at highest applied densities, not only a fastening-up of the XX luminescence onset but also the occurrence of a very fast (less than 10 ps) component on the usually slower decay transient was observed. The latter yielded time constants between 150 and 210 ps in ZnCdSe/ZnSe quantum wells. They explained the shortening of the rise time by the enhanced biexciton formation rate at increased exciton densities and by the rising direct biexciton formation from the pump light. The fast decay component was attributed to the radiative decay while the longer one was assigned to the re-formation of XX from the longer-living excitons. Puls et al. [21] contradicted this interpretation by stating that neither theory nor their experiments resulted in such a fast component. They could show that the XX binding energy in ZnCdSe/ZnSe quantum wells slightly increases with increasing exciton localization energy, however, the biexciton localization energy shows a strong shift by approximately twice the latter one. In the case of strong biexciton localization, the decays of excitons and biexcitons are monoexponential and exhibit basically the same or very similar time constants due to the fact that the exciton localization energy is comparable with the biexciton binding energy. In systems with less strong localization, excitons and biexcitons are better described as staying in chemical equilibrium concerning decay of biexcitons into each one photon and exciton and their re-creation from free excitons so that the exciton and biexciton decay are described by a system of rate equations [22, 23] (also in these studies, no fast component of the XX decay was observed). This results in a biexciton decay twice as fast as that of the exciton which was excellently confirmed by measurements in a weakly localizing ZnSe/$ZnS_{0.06}Se_{0.94}$ quantum-well system (Fig. 2).

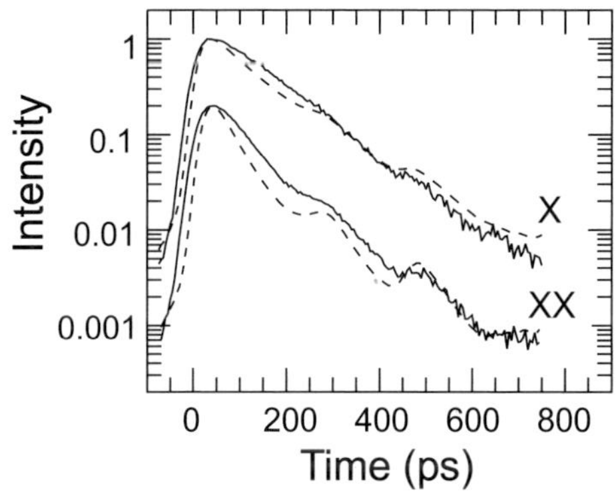

Fig. 2. Decay transients of the X_{hh} heavy-hole and the XX biexciton (measured at the band peaks) in a 15 nm $Zn_{0.92}Cd_{0.08}Se$/ZnSe quantum well yielding decay time constants of $t_{hh} = 85$ ps and $t_{XX} = 43$ ps. From [22].

Another study of Yamada et al. [24] resulted in the observation of a strongly superlinear intensity behavior of XX with a fast recovery time of 30 ps while the exciton showed a sublinear intensity behavior with 133 ps decay time, the first having been attributed to stimulated emission of XX at two-pulse excitation. This indication of possible biexciton lasing was in accordance with the clear identification of XX lasing at least at low temperatures up to 100–160 K in ternary [25, 26] and binary wells [27–29]. However, the discussion of details concerning biexcitonic gain goes beyond the scope of this paper.

Especially through the introduction of doped layers in semiconductor low-dimensional structures being necessary to realize current- or voltage-driven devices as light emitters or detectors, charged excitons, i.e., positively (X^+) or negatively (X^-) charged trions, came increasingly into play which do hardly show up at least in undoped bulk semiconductors. For a recent review on trion research in direct-gap semiconductors, the reader is referred to the proceedings of the Mini-Conference on Trion Physics, Berlin 2001 [30]. Here, we concentrate on some dynamics of trions found recently in II-VI layer structures. These structures are especially suited for the study of trion physics due to the relatively large excitonic binding energies since trions usually are rather weakly bound by energies of just a few per cent of the free-exciton binding energies, the trion triplet states even being often unstable unless strong magnetic fields are applied.

The use of CdTe based quantum wells allows for the study of systems with small (pure CdTe wells) or enhanced localization (ternary wells), the latter accompanied by a giant Zeeman effect supporting the occurrence of trion transitions if a magnetic compound as Mn is used to realize the ternary well (e.g., $Cd_{1-y}Mn_yTe/Cd_{1-x}Mg_xTe$ [31–33]). For such structures it was shown that under nonresonant excitation X and X^- exhibit different rise times since the latter can only be formed if an X finds an electron. However, alike long-term decay times in PL were reported and explained by the fact that on a short time-scale an equilibrium between excitons and trions is established. These decay times strongly decrease even below that free-exciton lifetime known and observed in samples without remarkable trion contributions. Definitely, increasing electron density enhances the trion formation probability. Further, if the trion transition is resonantly excited without preceding X formation, one gets direct access to the radiative trion lifetime at least for very low temperatures where non-radiative processes and thermalization are negligible (Fig. 3). The trion lifetime turned out to be much shorter than that of free excitons in the same material. This short decay time is interpreted as being due to the fact that the electron remaining after an electron-hole-recombination within the trion may carry all $\boldsymbol{k}$ with it so that here recombination is not restricted to small $\boldsymbol{k}$ values within the light cone which is a necessary precondition for radiative decay of uncharged excitons [31–33].

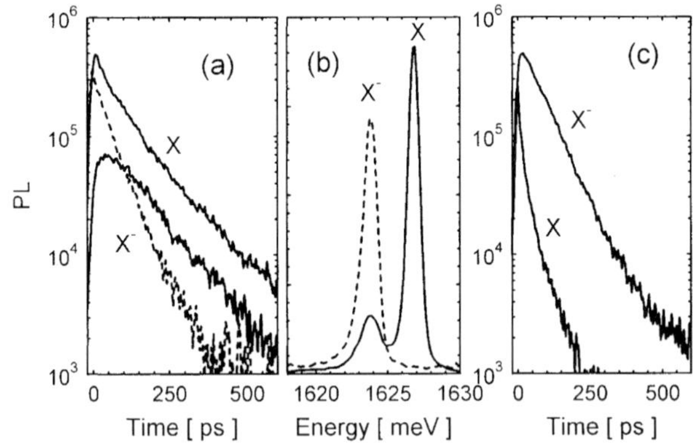

Fig. 3. (a) X and trion X^- decay transients for nonresonant (solid lines) and resonant X^- excitation (dashed line) in a pure 8 nm $CdTe/Cd_{1-x}Mg_xTe$ one-side modulation-doped quantum well at an electron concentration of $1 \times 10^{10}\,cm^{-2}$, (c) nonresonant excitation at $5 \times 10^{10}\,cm^{-2}$, T=2 K. (b) Time integrated PL for the two cases described in (a). From [22].

For a $Cd_{1-y}Mn_yTe/Cd_{1-x}Mg_xTe$ well, the authors could show that the decay rate is reduced and the emission band shifted to the red if the resonant excitation energy is tuned to lower values in the range of the original emission band as obtained for nonresonant excitation. This clearly indicated the role of localization also for trions in this ternary well having not been observed in the above-mentioned binary-well structure.

In a further study on the same kind of samples, the same group of authors [34] investigated the role of spin-flip processes in the trion luminescence dynamics. For this purpose, they have chosen the detection polarization cross-linearly and cross-circularly with respect to the low-density excitation energetically well below the free exciton X. For the first polarization configuration the trion PL onset was found to be almost instantaneously. The latter configuration requires a spin-flip being expressed by an onset delay yielding a spin-flip time of about 30 ps for resonant (cold) trion formation but being reduced to about 10 ps for 2 meV excess energy ("hot" trion formation). The latter short time, however, just gives an upper bound of the spin relaxation time in a hot trion distribution since it includes also the necessary trion relaxation time.

Trionic gain and lasing was found in δ-doped ZnSe/ZnBeMgSe single-quantum-well samples [35, 36]. The peculiarity of trionic gain is based on the fact that the excitonic recombination process of the fermion-like trion leaving an electron behind does not require degeneracy, and that the rather different trion and electron masses ($m_t/m_e \sim 10$ in ZnSe) allows for lasing of the trion-electron transition for larger $\boldsymbol{k}$ values for which $n_t(\boldsymbol{k}) > n_e(\boldsymbol{k})$ can hold although general inversion $n_t > n_e$ is usually not given. Based on polarization dependent measurements, trionic gain was stated proven with modal gain values up to $100\,\mathrm{cm}^{-1}$ (QW gain up to $10^3\,\mathrm{cm}^{-1}$) at $T = 2\,\mathrm{K}$ [32].

3 Coherent Optics and Higher Coulomb Correlations in the Exciton Range of Wide-Gap Semiconductors

If coherent optical radiation is incident on a material it induces a coherent polarization, i.e., this polarization has initially a well-defined spatial and temporal phase relation (coherence), it carries the phase information of the incident radiation. The coherence is subsequently destroyed by scattering and other dephasing processes within the material. The decay time of the coherence, i.e., the dephasing time, of optically induced excitations in bulk semiconductors and quantum-confined semiconductor structures typically lies in the range of femtoseconds at room and picoseconds at liquid helium temperatures. The study of elementary coherent processes in semiconductors therefore requires the use of mode-locked pico- and femtosecond lasers.

The analysis of the coherent dynamics of optical excitations in semiconductors offers the possibility to investigate fundamental quantum mechanical interactions. In particular the coherent dynamics of excitations in one-dimensional quantum wires (e.g., [37]) and especially in two-dimensional quantum wells [38, 39] have attracted a lot of interest in the last years. In these structures the enhancement of the Coulomb interaction due to the lowering of the dimensionality leads to very distinct exciton as well as higher Coulomb correlation effects which influence the optical response in II-VI materials even stronger than in conventional III-V semiconductors. The coherent dynamics of excitons mainly in wide-gap semiconductors and basic measurement techniques are discussed in the next two sections.

The phase memory of semiconductors also enables another interaction: An excitation of the semiconductor system can interfere with a further coherent pulse even after the initial pulse has passed through the system. Since the relative phase between two femtosecond pulses can be accurately controlled up to a fraction of the period of one optical cycle it is possible to "coherently control" the excitation in the material. This provides the opportunity to prepare desired quantum mechanical states or to coherently control a variety of properties of the semiconductor. Examples for the coherent control of excitonic excitations in II-VI quantum structures are discussed in Sect. 3.3.

3.1 Coherent Dynamics of Excitons and Exciton-Polaritons in Transmission Experiments

The conceptually easiest way to analyze the coherent dynamics of excitons is to send an ultrashort pulse through an appropriate sample and to analyze its resulting shape after transmission. If an excitonic resonance lies within the spectral range of the pulse the corresponding polarization will be excited inside the sample which re-emits the absorbed energy usually for several picoseconds after the pulse has passed through the sample. This will result in a signal with an intensity I being proportional to the square of the coherent polarization P, $I \propto |P|^2$.

In order to get some information on the dynamics of the excitonic polarization in pulse-transmission experiments it is necessary to analyze the signal real-time resolved. The time resolution must be better than the dephasing time of the induced polarization which is usually in the order of picoseconds. One possibility to achieve a high temporal resolution even for weak signals is the so-called up-conversion (UPC) method which makes use of a cross-correlation technique (see, e.g., [38, 40]). The residual output of a short-pulse laser (reference pulse, photon energy ω_r) is overlapped spatially and temporally with the signal from the polarization (photon energy ω_s) in a beta-barium-borate (BBO) crystal. For an appropriate orientation of the crystal, energy and momentum conservation will allow for the generation of photons with the sum frequency (photon energy $\omega_r + \omega_s$) which can be spectrally filtered and finally recorded by a photo multiplier. The intensity of the up-converted light is measured as a function of the time delay t_{upc} between the signal and the reference pulse which does not directly give the real-time trace of the signal but rather a convolution of the signal and the reference pulse. However, if the temporal width Δt_{ref} of the reference pulse is much smaller than the time scale on which structures in the signal shall be analyzed, the convolution only leads to a smoothing of the signal on the time scale of Δt_{ref}. In most experiments which will be discussed in this section this precondition is fulfilled, therefore the UPC signal can be regarded as the real-time signal in a good approximation.

First pulse transmission experiments in the spectral range of excitonic resonances in wide-gap semiconductors were carried out by Fröhlich et al. [41] on Cu_2O (see Fig. 4a). They found that the propagation of a picosecond pulse through a $\approx 1\,\mathrm{mm}$ thick crystal led to a strong distortion of its temporal shape. In time-resolved measurements the leading pulse peak was followed by a trailing edge which lasted for a few nanoseconds and showed strong oscillations with increasing periods. The authors were able to explain their results on the basis of an exciton-polariton theory. The concept of the exciton-polariton is extensively treated in many textbooks (see, e.g., [4, 42]); its basic idea is the introduction of the polariton as a new quasi-particle which is created by the strong interaction of the light field and the excitonic polarization of the semiconductor which cannot be treated just as a perturbation in a proper theoretical way. In particular, a splitting of the energy dispersion relation into an upper and lower polariton branch results. A short laser pulse will usually excite several polariton modes on both branches which propagate through the sample and give rise to a complex quantum-beat structure in the transmitted signal.

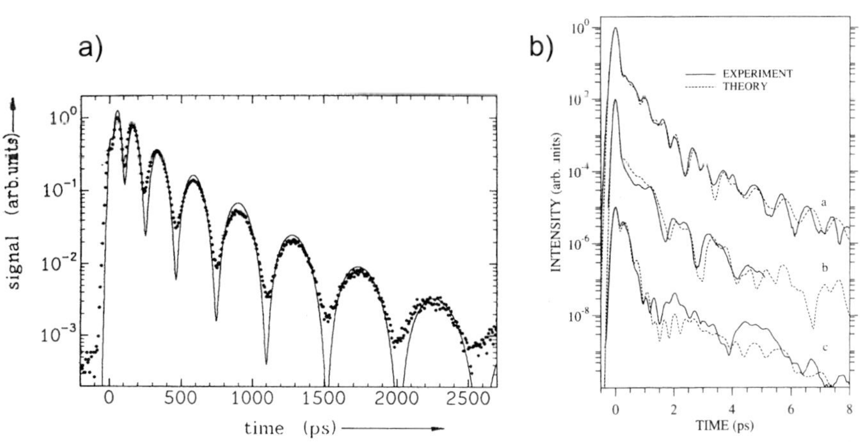

Fig. 4. Pulse-transmission experiments. (a) Transmission of a picosecond pulse through bulk Cu_2O. From [41]. (b) Transmission of femtosecond pulses through thin ZnSe layers. From [43].

Many similar experiments on other wide-gap II-VI bulk materials followed. Among other topics some authors focused on polariton propagation effects of linear and of nonlinear polarizations [43–46] (in [44] four-wave-mixing polarizations). In [46] a "coherent dip" was found, i.e., a distinct drop in the intensity near the position of the maximum of the transmitted laser pulse itself. This dip could be theoretically modelled by use of an harmonic-oscillator model taking into account the nonzero temporal width of the laser pulse and the nonzero thickness of the sample.

Epitaxially grown semiconductor heterostructures such as multiple quantum wells were also taken as samples for pulse transmission experiments (for work done on the GaAs system see, e.g., [47–49]) where despite of the lowered thickness of the samples still a significant distortion of femtosecond pulses after transmission was found. The distortion results from the interference between the laser field and the re-radiated field of the induced dipoles and was shown to strongly depend on many parameters such as sample thickness, temperature, excitation density, spectral position of the excitation pulse, and inhomogeneous broadening of the excited resonances. Also the influence of disorder and interface roughness which lead to a nonlocal susceptibility in the sample was investigated both experimentally and theoretically.

On the II-VI material system Neukirch et al. [43] showed in pulse-transmission experiments that in thin ZnSe layers with thicknesses between 23 and 100 nm embedded in ZnSSe cladding layers, the quantization of the polariton wave vector addressed as the "center-of-mass quantization" together with the spatial dispersion produces a variety of allowed propagating polariton modes. This also leads to a very pronounced and complex beating behavior (see Fig. 4b) which could be explained by a model based on a harmonic oscillator theory using phenomenologically introduced additional boundary conditions [43].

Besides signatures due to the influence of polariton modes the study of femtosecond pulse-transients can even reveal effects related to the excitation of a bound-biexciton state. Fig. 5a shows typical real-time transients of 120 fs pulses on a logarithmic scale at low excitation intensities (pulse energy $\approx$ 0.1 μJ). The center wavelength of the pulses was tuned to the heavy-hole excitonic resonance of the ZnSe quantum well at about 2.81 eV. The dotted line in Fig. 5a shows a cross-correlation measurement of the blue laser pulse without a sample which results in a symmetric, nearly Gaussian shaped signal around $t_{\mathrm{upc}} = 0$ ps. The dotted transient indicates the strongly asymmetric shape of a pulse that has been transmitted through a 10 nm quantum well. The signal which is generated by the coherent polarization of the sample can be easily identified as the difference between the dotted and the dashed transients. Because of the relatively small spectral width of the pulse only the hh excitonic resonance was excited which results in a simple exponential decay of the signal due to the dephasing of the excitonic polarization. From the decay-time constant of the signal the dephasing time of the polarization,

$T_2 = 1.9\,\text{ps}$, can be extracted. The solid line corresponds to the real-time transient of a pulse after transmission through a 25 nm sample. As described above in such thicker samples different center-of-mass quantized polariton modes are usually excited simultaneously by the 120 fs pulse which results in a strong and complex beating structure in the tail of the transient.

If in measurements using the 10 nm sample the excitation density is strongly increased (pulse energy $\approx$ 18 pJ, diameter of the laser spot $\approx$ 100 µm) a weak but clearly resolved beat structure will occur on the tail of the signal (Fig. 5b). This beat structure disappears if the polarization state of the femtosecond pulse is changed from linear (x) to circular (σ^+). This polarization sensitivity is a clear signature of the excitation of the bound-biexciton state in the quantum well. In analogy to the hydrogen molecule bound biexcitons are in their electronic ground state composed of two excitons with opposite spin orientations. Therefore their excitation can be prevented by creating only excitons with exclusively one spin orientation as it is the case if a single circularly polarized excitation pulse is used.

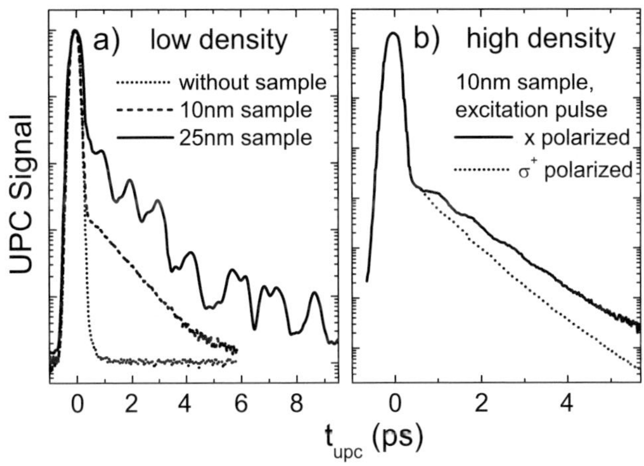

Fig. 5. Real time measurements of femtosecond pulses. (a) Low density references without sample, after transmission through a 10 nm ZnSe/ZnSSe sample (one excitonic resonance excited) and after transmission through a 25 nm sample (several exciton-polariton modes excited). (b) High-density measurements at the 10 nm sample with a linearly polarized pulse where exciton-biexciton beats are visible on the transient, and with a circularly polarized pulse for which the transition towards a bound-biexciton state is forbidden and the beats are absent.

3.2 Nonlinear Optical Spectroscopy of the Coherent Dynamics

The most frequently used nonlinear optical spectroscopy techniques to investigate the coherent regime are pump-probe and coherent wave-mixing spectroscopy. The basic idea is to excite the sample by (usually) two ultrashort laser pulses separated by a time delay. The first pulse creates a nonequilibrium excitation in the sample that is probed by the subsequent pulse. By varying the time delay between the pulses it is possible to directly follow the dynamics of the (coherent) excitations.

In pump-probe experiments small changes in the transmission of a sample induced by a pump pulse are detected by a weaker probe pulse. The effects of exciton-polariton modes on pulse transient measurements exemplarily discussed mainly for the linear regime in the previous section can also be studied with the pump-probe technique in particular suitable for the study of nonlinear phenomena. This way, Neukirch and Wundke [45] analyzed the polariton propagation at high densities. They found that the transmitted intensity of the weak probe pulse is increasingly shifted towards the leading maximum of the pulse, together with a strong decrease of the polariton beat structure, if an additional strong pump pulse excites an incoherent carrier distribution several picoseconds before the probe arrives. In a comparison of their results with a semiclassical polariton model which still used a single, however nonlinear damping constant for the exciton oscillator the authors found a good agreement. As a consequence, the consideration of higher quantized polariton modes in structures with dimensions of several excitonic Bohr radii is of crucial importance for the understanding of nonlinear optical phenomena in such semiconductor nanostructures.

In wave-mixing measurements the nonlinear interference of excitations of two ultrashort pulses leads to the emission of signals in background-free directions which are determined by the conservation of photon momentum. For this technique, ultrashort pulses from a mode-locked laser are divided into two parts so to realize the common two-pulse geometry (for certain applications a three-pulse geometry is used, this will not be treated here). The pulses arriving from directions $\boldsymbol{k}_1$ and $\boldsymbol{k}_2$ are spatially overlapped on the sample (Fig. 6). The coherent polarizations created by the two pulses form a polarization grating in the sample giving rise to the creation of various signals radiating in different directions due to the phenomenon of self-diffraction at this grating. Thus, these signals are directly related to the coherent nonlinear polarization. The easiest and most commonly used method is the detection of one of the wave-mixing signals with a slow detector as a function of the delay between the pulses (time-integrated signal). Additionally, for each delay the signal can be time-resolved by use of the up-conversion technique described in 3.1 or it can be frequency resolved by a spectrometer. The radiation in the direction $\boldsymbol{k}_{\mathrm{signal}} = 2\boldsymbol{k}_2 - \boldsymbol{k}_1$ is one of the "four-wave-mixing signals" corresponding to a first-order diffraction (a second signal of that kind is emitted in $2\boldsymbol{k}_1 - \boldsymbol{k}_2$). The term reflects the fact that in lowest order three incident photons and one signal photon are involved in the wave-mixing process. However, also signals due to higher-order diffraction are emitted which are more difficult to be interpreted and described theoretically but sometimes necessary to be investigated for clearing up the influence of high-order correlations. The

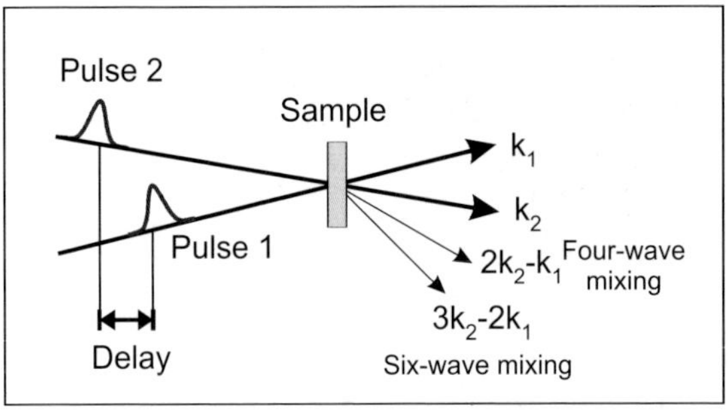

Fig. 6. Scheme of a wave-mixing experiment with two excitation pulses.

second-order signal is the so-called six-wave-mixing signal, since five photons plus one signal photon are necessary for its creation. An example is discussed at the end of this section.

In particular, the biexcitonic correlations affecting the optical response in the coherent regime can be identified in polarization-selective four-wave-mixing experiments (demonstrated for II-VI quantum wells (ZnSe/ZnSSe) by, e.g., [50, 51], and just recently also for GaN [52]). Exemplarily, Fig. 7 shows time-integrated four-wave-mixing signals emitted in the direction $2\boldsymbol{k}_2 - \boldsymbol{k}_1$. The measurements were performed on a 10 nm ZnSe single quantum well in transmission geometry at a temperature of 4 K. The pulses were spectrally peaked at the exciton resonance of 2.816 eV, i.e., 30 meV below the bandgap energy. In that energy range the spectral width of the 110 fs pulses is about 15 meV, so that exclusively the heavy-hole exciton-biexciton system was excited.

The left column in Fig. 7 shows contour maps of four-wave-mixing signals as a function of energy and delay time. The intensity of the signals is decoded in a grey scale. In the linear polarization configurations (top and middle row) two contributions at the spectral positions of the exciton energy (X) and the exciton-biexciton energy (XX) are visible. The signals exhibit modulations which are differently pronounced at X and XX. In the co-linear configuration the signal at X is much less modulated than that at XX. The strength of the modulations is reversed for the cross-linear polarization configuration. This can be clearly seen in the right column of Fig. 7 showing cuts along the delay-time axis at the energy positions of the exciton and exciton-biexciton transitions. The modulations are caused by exciton-biexciton quantum beats resulting from the interference of the excitonic and biexcitonic wave functions. These beats therefore have to be distinguished from polarization beats due to interference of emitted photons after the excitation of several distinct exciton resonances [53].

In the co-circular polarization configuration (bottom row) no signal at XX is observed (except for a very small residual contribution resulting from not perfectly co-circularly polarized beams). The cut at X shows an exponential

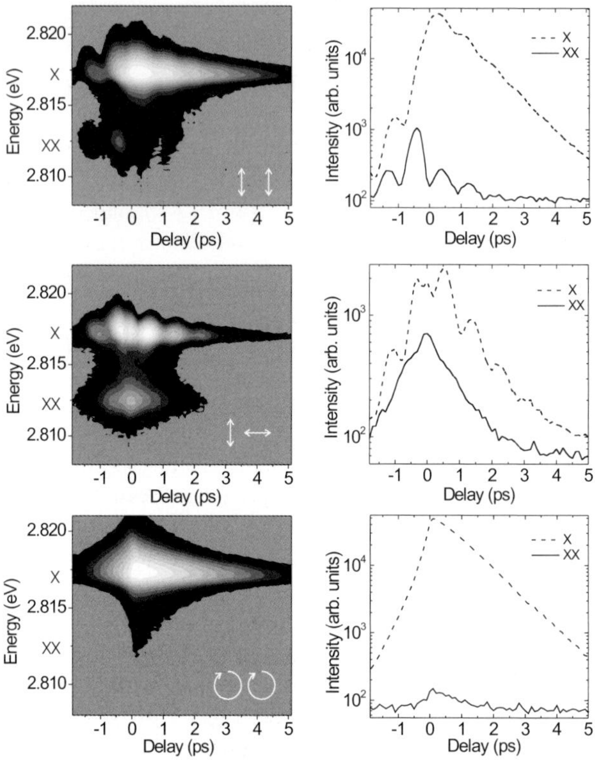

Fig. 7. Time-integrated four-wave-mixing signals from a 10 nm ZnSe quantum well at 4 K. The measurements with linearly polarized excitation pulses (top and middle row) show exciton-biexciton quantum beats which are absent in the co-circularly configuration (bottom row). The arrows indicate the polarization state of the $\boldsymbol{k}_1$ and $\boldsymbol{k}_2$ pulses, respectively. From [54].

rise and decay. This behavior is in agreement with the polarization selection rules for the excitation of the biexcitonic resonance [55] and confirms the latter as the origin of the signals at XX.

By comparing experimental results with theoretical calculations contributions of third and fifth order in the excitation fields were identified to cause the observed beat structure [56–58]. In calculations limited to contributions of third order no beats appear for positive delay times what corresponds to the case of the pulse from direction $\boldsymbol{k}_1$ to arrive first. This can be understood from the fact that the $\boldsymbol{k}_1$ pulse contributes linearly to the signal in third order for positive delays and, therefore, induces exclusively excitonic but no biexcitonic polarization. The excitonic polarization immediately starts to decay thus determining the strength of the wave-mixing signal which is generated on arrival of the second pulse. Since only excitonic polarization is generated by the $\boldsymbol{k}_1$ pulse no exciton-biexciton beats occur in the wave-mixing signal for positive delay times.

The situation differs for negative delay times which correspond to the arrival of the $\boldsymbol{k}_2$ before the $\boldsymbol{k}_1$ pulse. Since the $\boldsymbol{k}_2$ pulse contributes quadratically to the signal it can excite exciton and biexciton states (the latter only if the $\boldsymbol{k}_2$ pulse is linearly polarized). If indeed several states are excited the dynamics of the wave packet is mapped onto the four-wave-mixing signal. The arrival of the $\boldsymbol{k}_1$ pulse thus leads to the occurrence of a beat structure on the signal as a function of delay time. The frequency of the beats is determined by the inverse of the energy difference of the states excited. Hence, the beat structure modulating the signal in Fig. 7 at positive delay times results from the nonlinear interaction of higher than third-order contributions where both the $\boldsymbol{k}_1$ and $\boldsymbol{k}_2$ pulse excite one- and two-exciton states.

The previous arguments are only valid for systems with purely homogenously broadened resonances. A strong inhomogeneous broadening of the resonances can lead to the occurrence of beats even within a description limited to third-order contributions [56]. However, this is only possible if the energy width of the homogenous broadening exceeds the binding energy of the biexcitons which was not the case in the measurements presented in Fig. 7.

To investigate in particular the influence of high-order correlations the wave-mixing signal emitted in the second order of diffraction, the six-wave-mixing signal, is more suitable than the four-wave-mixing signal. In the six-wave-mixing signal the effect of higher than four-particle correlations can be identified although lower-order contributions also significantly contribute to the signal. By comparing experiments with theoretical results obtained from microscopic calculations based on the dynamics-controlled truncation method [59, 60] Bolton et al. could show the specific influence of six-particle correlations in experiments performed on a ZnSe quantum well [61, 62]. Fig. 8 shows the contour diagram of the six-wave-mixing signal emitted in the direction $3\boldsymbol{k}_2 - 2\boldsymbol{k}_1$. The exciting laser pulses were cross-linearly polarized since in this configuration the signals at X and XX are of comparable intensity. The contour diagram shows the signals as functions of energy and delay time. The grey scale covers three orders of magnitude. Signal contributions at the spectral position of the exciton transition at 2.825 eV and of the exciton-biexciton transitions at 2.817 eV are clearly visible. Although both resonances are excited the signals exhibit no beats as a function of delay time.

For the model a microscopic description of the correlation dynamics was used which could be evaluated at different levels of sophistication. In Fig. 8b only two- and four-particle correlations in the purely coherent emission (one-photon and two-photon coherences) were taken into account. In Fig. 8c additionally incoherent exciton densities were taken into consideration, in Fig. 8d also six-point correlations describing transitions from incoherent exciton densities and two-pair states were included in the model. The plots in Figs. 8b–d were obtained by solving numerically the equations of motion of the two-,

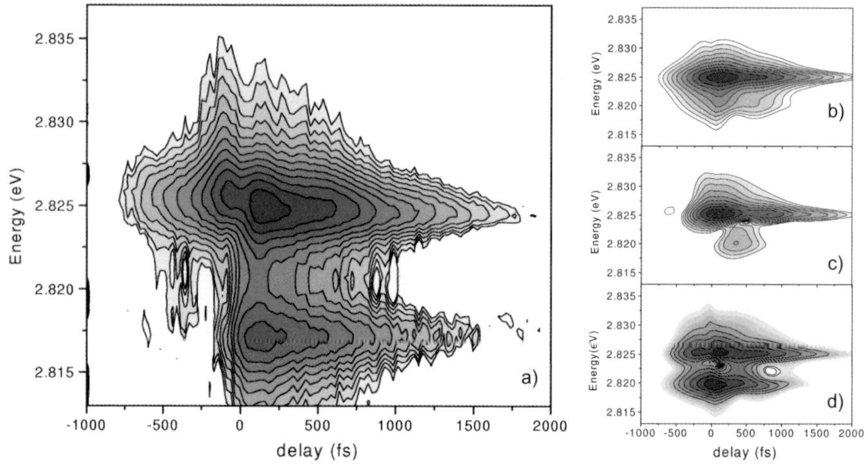

Fig. 8. Contour diagram of a six-wave-mixing signal from a ZnSe single-quantum well. (a) Experimental data with signal contributions at the exciton and exciton-biexciton transitions. The grey scale represents logarithmically three orders of magnitude. (b–d) Results of a theoretical modelling of the signal on three different levels as discussed in the text. From [61].

four- and six-particle correlation functions. Figs. 8b and c show large discrepancies between theory and experiment. However, good agreement is obtained by including the three-pair correlations as done in Fig. 8d clearly demonstrating the necessity to consider correlations up to sixth order in the description of six-wave-mixing experiments. Thus, this provides a convincing example of the effect of high-order correlations which are otherwise rarely directly accessible in experiments.

3.3 Coherent Control of Excitonic Excitations

In recent years the optical coherent-control technique has attracted much interest in many different fields of physics and chemistry since it allows not only to observe coherent processes but even to actively manipulate and direct them in a desired way. The basic idea is to excite one or more resonances with ultrashort laser pulses such that the relative phase of the excitations favors the transition from the initial to a desired final state due to constructive interference. All pathways to undesired final states, however, ideally should cancel each other due to destructive interference of the excitations. Such a coherent control with femtosecond laser pulses can be achieved by applying a special pulse-shaping technique which directly manipulates the different spectral components of the excitation pulses with regard to their amplitudes and relative phases. In most cases an active feedback loop is added to the experimental setup to adjust the pulse shaping such that the ratio of the excitations in the desired final state to those in "wrong" states is maximized. A review of this technique which is especially used in atomic and chemical physics can be found in [63].

For the coherent control of excitonic polarizations and populations in semiconductor nanostructures, however, a different and somewhat more direct approach has been used: by use of an actively stabilized Michelson interferometer a collinearly propagating pair of phase-locked laser pulses is generated. The term "phase-locked" expresses the fact that the relative phase of these two pulses can be tuned and stabilized by controlling the relative length of the two arms of the interferometer with an accuracy of just a fraction of the central wavelength of the laser pulses. The first pulse is used to excite a coherent polarization inside the sample. Upon its arrival at the sample the second pulse will interact with the remaining coherent polarization and depending on their relative phase enhance or diminish this polarization. If the relative phase of the two laser pulses is continuously varied this will result in a successive change between constructive and destructive interference of the induced polarizations. Consequently, the signal from the effectively resulting polarization will show a cosine-like oscillation as a function of the relative phase, i.e., the inter-pulse delay.

The pioneering work in coherent control of optical excitations in semiconductors was done by Heberle, Baumberg, and coworkers [64–66] who demonstrated the coherent enhancement of excitonic polarization and population in GaAs quantum wells. Other groups later showed that this coherent control of excitons in semiconductors can also be achieved by applying the pulse-shaping technique [67, 68]. In [69], Yee et al. especially focused on the coherent control of excitonic polarization in pulse-transmission experiments where they analyzed the interaction between the coherent excitonic polarization and the controlling phase-locked pulses. They were able to show that an energy exchange between the excited exciton system and the second pulse occurs which crucially depends on the relative phase of the phase-locked pulses as well as on the ratio of their intensities.

For an experimental study of the influence of Coulomb correlations and high-density effects in coherent-control experiments wide-gap semiconductors offer an ideal model system. Because of the large exciton and biexciton binding energies and strong Coulomb correlation present in those materials it is possible to analyze if and how these correlations as well as high-density effects can be subject to coherent manipulation on ultrashort timescales.

In order to compare the coherent control of excitonic polarization at low and high densities real-time transients obtained from transmission of the phase-locked pulse pair are recorded for different inter-pulse delay times t_{int}. For low excitation density the polarization created by the second pulse can indeed distinctly enhance or diminish the coherent polarization induced by the first pulse as can be seen in Fig. 9a which shows results obtained from a 10 nm ZnSe/ZnSSe-quantum-well sample. The solid transient was recorded with a single pulse as a reference. The second pulse was applied to the sample at a time around $t_{int} \approx -500$ fs. The dashed line corresponds to a situation where destructive interference of the induced polarizations occurs. If t_{int} is increased by just a small amount of $\Delta t_{int} = 0.6$ fs the polarizations interfere constructively which results in a stronger signal in the tail of the transient (dotted curve). However, the dephasing times of the polarization remain the same for all three cases. As a consequence, a cosine-line coherent switching signal can be measured if the up-conversion time t_{upc}, i.e., the delay between signal and the reference pulse (see 3.1), is fixed at +1 ps and t_{int} is continuously varied by small amounts Δt_{int} around a central delay time $t^0_{int} \approx -500$ fs at a low excitation density (0.1 pJ/pulse). This case is exemplarily depicted as dotted line in Fig. 9b. The signal shows the successive change between constructive and destructive interference of the induced polarizations and illustrates why the optical coherent control technique is expected to provide a possibility for ultra-high-speed all-optical switching.

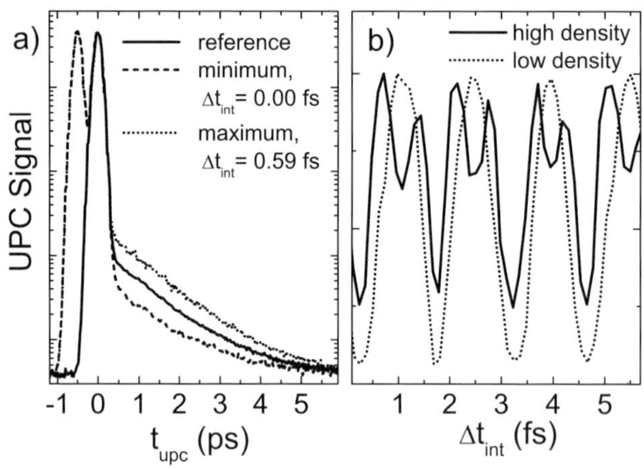

Fig. 9. (a) Coherent enhancement and destruction of polarization by a phase-locked pulse pair. (b) Coherent switching signal at low density and high density. In the latter configuration "dips" occur at the positions of constructive interference.

If the excitation density is increased by a factor of ≈ 100 distinct changes in the coherent switching signal occur (solid line in Fig. 9b). At the positions of constructive interference additional "dips" (local minima) show up and give rise to a fine structure in the cosine-like signal. An explanation for the occurrence of this additional fine structure can be obtained by again analyzing the corresponding real-time transients caused by the phase-locked pulse pair. Three transients that were obtained under similar conditions as the solid-line switching signal in Fig. 9b are shown in Fig. 10a. It can be clearly seen that under the high excitation density applied in these experiments the dephasing time of the polarization is not constant but shows a strong dependence on the relative phase of the phase-locked pulses. More precisely, the transient at a delay time corresponding to constructive interference (Fig. 10a, solid line) shows a much faster dephasing than the transients at the delay times corresponding to destructive interference (dotted line) or an intermediate position (dashed line). A quantitative analysis of the dephasing times yields a continuous change between 1.1 and 1.8 ps if $t_{\rm int}$ is slightly varied around $t^0_{\rm int} \approx -350$ fs. A simple model for the coherent manipulation of the dephasing time is based on exciton-exciton scattering. Since for constructive interference more excitonic polarization and finally also population is excited these scattering processes become more important and significantly increase the dephasing rate of the coherent polarization. For destructive interference or an intermediate case the exciton-exciton scattering is less pronounced and does less significantly decrease the dephasing time.

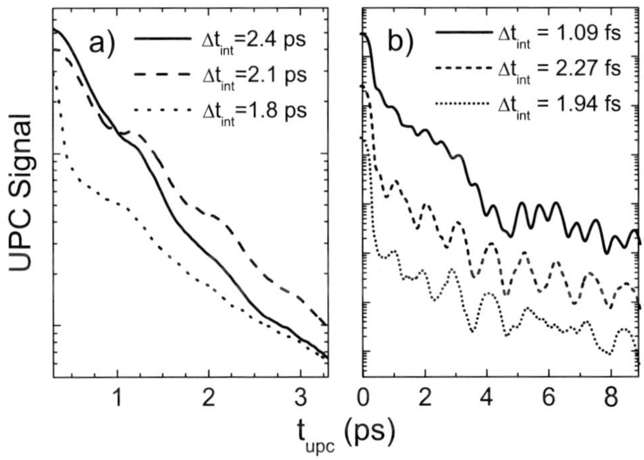

Fig. 10. Coherent-control measurements with (a) biexcitons (10 nm ZnSe/ZnSSe-quantum-well sample) and (b) polaritons (25 nm ZnSe/ZnSSe-quantum well sample).

Furthermore, the results in Fig. 10a show that not only the dephasing time of the excitonic polarization can be coherently manipulated at high excitation densities. Also the amplitude and relative phase of the exciton-biexciton beats clearly changes as $t_{\rm int}$ is varied. Remarkably, for the case of constructive interference (solid line) the amplitude of the beats is not maximized but is even reduced compared to the intermediate case (dashed line). This behavior can be understood by considering that the chosen value $t^0_{\rm int} \approx -350$ fs corresponds to half a period of the exciton-biexciton beats. This implies that with this special value of $t_{\rm int}$ constructive interference for the exciton transition will occur simultaneously with destructive interference for the exciton-biexciton transition. Therefore, the transients for constructive interference of the excitonic polarization should show a less pronounced beating than those for an intermediate position. This is exactly what can be observed in Fig. 10a.

Experiments that are currently in progress deal with the coherent manipulation of different simultaneously excited polariton modes in ZnSe nanostructures. In Fig. 10b coherent-control measurements are exemplarily shown (vertically displaced) that were obtained at low excitation densities with a 25 nm ZnSe/ZnSSe sample. The signal was recorded for three different but fixed values of Δt_{int} around $t^0_{\mathrm{int}} \approx -450$ fs as a function of t_{upc}. The transients for $\Delta t_{\mathrm{int}} = 1.09$ fs and $\Delta t_{\mathrm{int}} = 2.27$ fs show a significantly simpler beat structure than the transient that was measured with one single pulse (compare to Fig. 5). Especially the transient for $\Delta t_{\mathrm{int}} = 2.27$ fs exhibits a beat structure with a nearly constant period $T_{\mathrm{beat}} = 1.1$ ps which can be attributed to the coherent interplay of the polariton resonances m=1 and m=3 (center-of-mass quantization, see Sect. 3.1) in the sample. The measurement clearly shows first evidence that the contributions from different polariton modes can be separately enhanced or suppressed by use of coherent control at low excitation densities.

Not only the transient signals in pulse-transmission experiments but also the four-wave-mixing technique can be used to monitor the coherent control of the polarization by a pair of phase-locked pulses. In fact this method is well suited for this purpose since it is very sensitive and can be applied, due to background-free signals, in measurements with lowest intensity, or at low to medium intensities to study particularly the influence of higher-order contributions to the signal [70–72].

In the following the effect of a phase-locked pulse pair on the excitonic and biexcitonic polarization is directly demonstrated in spectrally resolved measurements. The contour diagrams in Fig. 11 show the four-wave-mixing signal as a function of energy and inter-pulse delay t_{int} of the phase locked-pulse pair. The latter was applied from direction $\boldsymbol{k}_1$ and an additional single pulse from direction $\boldsymbol{k}_2$. The signal emitted in the direction $2\boldsymbol{k}_1 - \boldsymbol{k}_2$ was detected. In this configuration the phase-locked pulse pair contributed in third order of the electrical field quadratically to the signal. To simplify the interpretation of the result the single $\boldsymbol{k}_2$ pulse was applied to the sample 0.4 ps before the pulse pair to avoid further complicating interactions of the induced polarizations [73] so that the effect of the phase-locked pulses is directly mapped onto the signal. To investigate simultaneously the signals at the exciton and biexciton positions the cross-linear polarization configuration is favorable, since those signals are of comparable intensity in this configuration.

The delay time of the single $\boldsymbol{k}_2$ pulse was kept fixed with respect to one of the phase-locked pulses during the measurement. In the contour diagrams in Figs. 11a, c signals are visible at the spectral positions of the exciton and the exciton-biexciton transition. The successive change of constructive and destructive interference with changing inter-pulse delay leads to oscillations which are more clearly visible in the cuts at X and XX along the axis of the

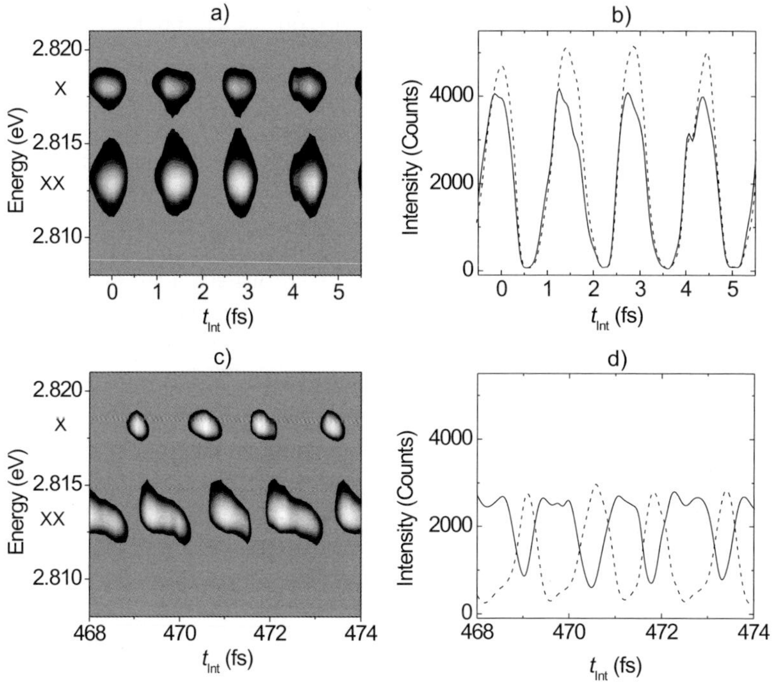

Fig. 11. (a), (c) Contour diagrams of coherent control FWM signals from a 10 nm ZnSe/ZnSSe-quantum well for two different ranges of the inter-pulse delay. The diagrams show the signal as a function of energy and inter-pulse delay. The spectral position of the exciton transition (exciton-biexciton transition) is labelled X (XX). (b), (d) Cuts from (a) and (c) along the inter-pulse-delay axis at X and XX, respectively.

inter-pulse delay time as shown in Figs. 11b, d. The period of the oscillations corresponds to that associated with the energy of the resonances via h/E_{res}. Since the energy difference of the exciton and exciton-biexciton transition is very small compared to the exciton energy no phase difference between the oscillations is visible for inter-pulse delay times close to zero (Figs. 11a, b). However, for larger delay times (Figs. 11c, d) the conditions for destructive and constructive interference are fulfilled for different inter-pulse delay times and a phase difference between the oscillations occurs, indicating the coherent control of the signals at the exciton and exciton-biexciton transition [74].

Outlook

Although much work on the dynamics of exciton-like excitations in wide-gap semiconductors has been performed just in the last decade this will remain a lively field of research. In particular the improvements of the very special III-nitrides, the capability of controlling quantum dot growth in all wide-gap materials, and modern questions of quantum optics will open a large number of new aspects concerning the dynamics of these materials already in the near future.

Acknowledgments

This contribution is dedicated to C. Klingshirn on the occasion of his 60th birthday. Although it is a (limited) review, the topics have been selected along the contributions of the Bremen group to the field during the last 12 years. Thus, the authors would explicitly like to thank the group of D. Hommel (Bremen) and W. Faschinger (Würzburg, deceased 2001) for providing excellent low-dimensional, wide-gap semiconductor samples over a decade of work, the theorists V.M. Axt (Aachen, Münster), A. Stahl (Aachen) and coworkers as well as the group of F. Jahnke (Bremen) for intense collaboration, and the Deutsche Forschungsgemeinschaft for support via a large number of research grants between 1990 and today.

References

1. U. Woggon: *Spectroscopy of Biexcitons and Trions in II-VI Quantum Dots*, this book
2. T.W. Berg, J.M. Hvam: *Semiconductor Quantum Dots for Optoelectronic Applications*, this book
3. C. Klingshirn: Adv. Materials for Optics and Electronics **3**, 103 (1994)
4. C.F. Klingshirn: *Semiconductor Optics*, 2nd corr. printing (Springer, Berlin Heidelberg New York 1997)
5. J.P. Doran, R.P. Stanley, J.F. Donegan, J. Hegarty, R. Fischer, E.O. Göbel, R.D. Feldmann, R.F. Austin: Physica B **185**, 566 (1993)
6. H. Kalt, J.H. Collet, Le Si Dang, J. Cibert, S.D. Baranowski, R. Saleh, M. Umlauff, K.P. Geyzers, M. Heuken, C. Klingshirn: Physica B **191**, 90 (1993)
7. T. Tsutsumi, J.Y. Jen, I. Souma, Y. Oka: Superlatt. Microstructures **16**, 41 (1994)
8. U. Neukirch, D. Weckendrup, W. Faschinger, P. Juza, H. Sitter: J. Crystal Growth **138**, 849 (1994)
9. U. Neukirch, D. Weckendrup, J. Gutowski, D. Hommel, G. Landwehr: J. Crystal Growth **138**, 861 (1994)
10. A. Schülzgen, E. Runge, F. Henneberger, R. Zimmermann: Mat. Sci. Forum **182–184**, 215 (1995)
11. E. Runge, A. Schülzgen, F. Hennebeger, R. Zimmermann: phys. stat. sol. (b) **188**, 565 (1995)
12. Y. Kawakami, A. Kaneta, K. Omae, A. Shikanai, K. Okamoto, G. Marutsuki, Y. Narukawa, T. Mukai, Sg. Fujita: phys. stat. sol. (b) **240**, 337 (2003)
13. U. Neukirch, G. Bley, J. Gutowski, D. Hommel: Phys. Rev. B **57**, 9208 (1998)
14. H. Zhao, S. Moehl, H. Kalt: phys. stat. sol. (b) **229**, 577 (2002)
15. H. Zhao, H. Kalt: *Hot Excitons in ZnSe Quantum Wells*, this book

16. S. Haacke, N.T. Pelekanos, H. Mariette, M. Zigone, A. Heberle, W.W. Rühle: Phys. Rev. B **47**, 16643 (1993)
17. S. Haacke, N.T. Pelekanos, H. Mariette, A. Heberle, W.W. Rühle, M. Zigone: J. Crystal Growth **138**, 831 (1994)
18. S. Ten, F. Henneberger, M. Rabe, M. Peyghambarian: Phys. Rev. B **53**, 12637 (1996)
19. W. Heiss, G. Prechtl, D. Stifter, H. Sitter, G. Springholz, T. Riemann, F. Bertram, D. Rudloff, J. Christen, G. Bley, U. Neukirch, J. Gutowski, J. Liu: Appl. Phys. Lett. **75**, 974 (1999)
20. Y. Yamada, T. Mishina, Y. Masumoto, Y. Kawakami, S. Yamaguchi, K. Ichino, Sz. Fujita, Sg. Fujita, T. Taguchi: Phys. Rev. B **51**, 2596 (1995)
21. J. Puls, H.-J. Wünsche, F. Henneberger: Chem. Phys. **210**, 235 (1996)
22. H. Grempel: Thesis, University of Bremen, Germany (1996)
23. H. Grempel, A. Diessel, W. Ebeling, J. Gutowski, K. Schüll, B. Jobst, D. Hommel, M.F. Pereira, K. Henneberger: phys. stat. sol. (b) **194**, 199 (1996)
24. Y. Yamada, T. Mishina, Y. Masumoto, Y. Kawakami, J. Suda, K. Ichino, Sz. Fujita, Sg. Fujita: Phys. Rev. B **52**, R2289 (1995)
25. F. Kreller, M. Lowisch, J. Puls, F. Henneberger: Phys. Rev. Lett. **75**, 2420 (1995)
26. F. Kreller, J. Puls, F. Henneberger: Appl. Phys. Lett. **69**, 2406 (1996)
27. V. Kozlov, P. Kelkar, A.V. Nurmikko, C.-C. Chu, D.C. Grillo, J. Han, C.G. Hua, R.L. Gunshor: Phys. Rev. B **53**, 10837 (1996)
28. V. Kozlov, P. Kelkar, A.V. Nurmikko, C.-C. Chu, D.C. Grillo, J. Han, C.G. Hua, R.L. Gunshor: Phys. Rev. B **54**, 13932 (1996)
29. O. Homburg, P. Michler, R. Heinecke, J. Gutowski, H. Wenisch, M. Behringer, D. Hommel: Phys. Rev. B **60**, 5743 (1999)
30. Mini-Conference on Trion Physics, Berlin (Germany) 2001, phys. stat. sol. (b) **227**, 297–417 (2001)
31. V. Ciulin, P. Kossacki, S. Haacke, J.-D. Ganière, B. Devaud, A. Esser, M. Kutrowski, T. Wojtowciz: Phys. Rev. B **62**, R16310 (2000)
32. V. Ciulin, P. Kossacki, S. Haacke, J.-D. Ganière, B. Devaud, A. Esser, M. Kutrowski, T. Wojtowciz: phys. stat. sol. (b) **227**, 307 (2001)
33. P. Kossacki, V. Ciulin, M. Kutrowski, J.-D. Ganière, T. Wojtowciz, B. Devaud: phys. stat. sol. (b) **229**, 659 (2002)
34. V. Ciulin, P. Kossacki, M. Kutrowski, J.-D. Ganière, T. Wojtowciz, B. Devaud: phys. stat. sol. (b) **229**, 627 (2002)
35. J. Puls, G.V. Mikhailov, F. Henneberger, D.R. Yakovlev, A. Waag, W. Faschinger: Phys. Rev. Lett. **89**, 287402 (2002)
36. J. Puls, G.V. Mikhailov, F. Henneberger, D.R. Yakovlev, A. Waag: phys. stat. sol. (b) **229**, 637 (2002)
37. H.P. Wagner, W. Langbein, J.M. Hvam, G. Bacher, T. Kümmell, A. Forchel: Phys. Rev. B **57**, 1797 (1998)
38. J. Shah: *Ultrafast Spectroscopy of Semiconductors and Semiconductor Nanostructures* (Springer, Berlin 1996)
39. D.S. Chemla, J. Shah: Nature **411**, 549 (2001)
40. C. Rulliere: *Femtosecond Laser Pulses* (Springer, Berlin 1998)
41. D. Fröhlich, A. Kulik, B. Uebbing, A. Mysyrowicz, V. Langer, H. Stolz, W. von der Osten: Phys. Rev. Lett. **67**, 2343 (1991)
42. P.M. Yu, M. Cardona: *Fundamentals of Semiconductors* (Springer, Berlin 1998)

43. U. Neukirch, K. Wundke, J. Gutowski, D. Hommel: phys. stat. sol. (b) **196**, 473 (1996)
44. K.-H. Pantke, P. Schillak, B.S. Razbirin, V.G. Lyssenko, J.M. Hvam: Phys. Rev. Lett. **70**, 327 (1993)
45. U. Neukirch, K. Wundke: Phys. Rev. B **55**, 15408 (1997)
46. T. Mishina, Y. Masumoto: Phys. Rev. Lett. **71**, 2785 (1993)
47. D.S. Kim, J. Shah, D.A.B. Miller, T.C. Damen, W. Schäfer, L. Pfeiffer: Phys. Rev. B **48**, 17902 (1993)
48. D.S. Kim, J. Shah, D.A.B. Miller, T.C. Damen, A. Vinattieri, W. Schäfer, L. Pfeiffer: Phys. Rev. B **50**, 18240 (1994)
49. T. Stroucken, A. Knorr, C. Anthony, A. Schulze, P. Thomas, S.W. Koch, M. Koch, S.T. Cundiff, J. Feldmann, E.O. Göbel: Phys. Rev. Lett. **74**, 2391 (1995)
50. G. Bartels, A. Stahl, V.M. Axt, B. Haase, U. Neukirch, J. Gutowski: Phys. Rev. Lett. **81**, 5880 (1998)
51. B. Haase, U. Neukirch, J. Gutowski, G. Bartels, A. Stahl, V.M. Axt, J. Nürnberger, W. Faschinger: *Proc. 24th Int. Conf. Phys. Semicond.*, ed by D. Gershoni (World Scientific, IVE3 1999)
52. K. Hazu, A. Shikanai, T. Sota, K. Suzuki, S. Adachi, S.F. Chichibu, T. Mukai: Phys. Rev. B **65**, 195202 (2003)
53. M. Koch, J. Feldmann, G. von Plessen, E.O. Göbel, P. Thomas, K. Köhler: Phys. Rev. B **69**, 3631 (1993)
54. H.G. Breunig: Thesis, University of Bremen, Germany (2003)
55. Y.Z. Hu, R. Binder, S.W. Koch, S.T. Cundiff, H. Wang, D.G. Steel: Phys. Rev. B **49**, 14382 (1994)
56. T.F. Albrecht, K. Bott, T. Meier, A. Schulze, M. Koch, S.T. Cundiff, J. Feldmann, W. Stolz, P. Thomas, S.W. Koch, E.O. Göbel: Phys. Rev. B **54**, 4436 (1996)
57. B. Haase, U. Neukirch, J. Gutowski, G. Bartels, A. Stahl, V.M. Axt, J. Nürnberger, W. Faschinger: Phys. Rev. B **59**, R7805 (1999)
58. W. Langbein, T. Meier, S.W. Koch, J.M. Hvam: J. Opt. Soc. Am. B **18**, 1318 (2001)
59. V.M. Axt, A. Stahl: Z. Phys. B **93**, 195 (1994)
60. V.M. Axt, S. Mukamel: Rev. Mod. Phys. **70**, 145 (1998)
61. S.R. Bolton, U. Neukirch, L.J. Sham, D.S. Chemla, V.M. Axt: Phys. Rev. Lett. **85**, 2002 (2000)
62. V.M. Axt, S.R. Bolton, U. Neukirch, L.J. Sham, D.S. Chemla: Phys. Rev. B **63**, 115303 (2001)
63. T.C. Weinacht, P.H. Bucksbaum: J. Opt. B: Quantum Semiclass. Opt. **4**, R35 (2002)
64. A.P. Heberle, J.J. Baumberg, K. Köhler: Phys. Rev. Lett. **75**, 2598 (1995)
65. J.J. Baumberg, A.P. Heberle, K. Köhler, K. Ploog: J. Opt. Soc. Am. B **13**, 1246 (1996)
66. J.J. Baumberg, A.P. Heberle, K. Köhler, A.V. Kavokin: phys. stat. sol. (b) **204**, 9 (1997)
67. J. Kunde, B. Baumann, S. Arlt, F. Morier-Genoud, U. Siegner, U. Keller: Appl. Phys. Lett. **77**, 924 (2000)
68. K. Komori, T. Sygaya, M. Watanabe, T. Hidaka: Jpn. J. Appl. Phys. **39**, 2347 (2000)

69. D.S. Yee, K.J. Yee, S.C. Hohng, D.S. Kim, T. Meier, S.W. Koch: Phys. Rev. Lett. **84**, 3474 (2000)
70. M.U. Wehner, J. Hetzler, M. Wegener: Phys. Rev. B **55**, 4031 (1997)
71. H.G. Breunig, T. Trüper, I. Rückmann, J. Gutowski, F. Jahnke: phys. stat. sol. (b) **229**, 621 (2002)
72. H.G. Breunig, T. Voss, I. Rückmann, J. Gutowski, V.M. Axt, T. Kuhn: J. Opt. Soc. Am. B **20**, 1769 (2003)
73. T. Voss, H.G. Breunig, I. Rückmann, J. Gutowski: Opt. Comm. **218**, 415 (2003)
74. H.G. Breunig, T. Voss, I. Rückmann, J. Gutowski: Phys. Rev. B **66**, 193302 (2002)

Quantum Kinetics and Femtosecond Spectroscopy – The Discovery of Slowness

Hartmut Haug

For time intervals which are shorter than characteristic periods of the medium, collisions can no longer be regarded as instantaneous. They have to be described instead by non-Markovian scattering integrals. In this early time regime the quantum mechanical coherence of the photo-excited carriers is not yet destroyed. In this regime quantum kinetics with its memory effects yields an excellent description of femtosecond four wave mixing (FWM) and differential transmission spectroscopy (DTS) experiments. The essential features of carrier-phonon scattering and the carrier-carrier scattering via a time-dependently screened Coulomb potential are discussed. Particularly the time-dependent build-up of screening by the optically excited carriers is itself a highly interesting quantum kinetic problem. The determination of the density-dependent dephasing time and the observation of the quantum kinetically predicted phonon-plasmon beats with coherent control techniques are discussed.

1 Quantum Kinetic Regime

In Sten Nadolny's novel[1] a young sailor is described with painfully slow reactions and decisions. Later when he became known as John Franklin, the explorer of the notorious North-West passage – he has discovered slowness as a virtue. As slow as he was for the human perception with a resolution of fractions of seconds, his actions would be instantaneous on a timescale of years. Similarly in physics, e.g., the electron scattering processes in solids seem to be instantaneous on time scales down to picoseconds, as it has been assumed as early as 1872 in the famous treaty of the molecular kinetics in rarefied gases by Ludwig Boltzmann [1]. Such instantaneous processes are also called Markov processes. However on a timescale which is shorter than the period of a lattice or plasma oscillation, the electron scattering events become "slow" processes, during which, e.g., one can accelerate an electron by an applied electric field. This process is called the intra-collisional field effect. The mentioned characteristic periods are in semiconductors typically

[1] "The discovery of slowness" is the title of a novel by Sten Nadolny. The original German title is "Die Entdeckung der Langsamkeit".

of the order of 100 femtoseconds. Thus the slowness of these processes becomes apparent for spectroscopic studies with a time resolution of only a few femtoseconds. In this regime totally new insights can be gained. The build-up of correlations, e.g., which traditionally is considered as being instantaneous, becomes temporally resolved and thus observable on a femtosecond timescale. Examples for the build-up of correlations are the formation of polarons connected with the polarization of the lattice in an ionic crystal by a carrier, or the rearrangement of carriers which give rise to the screening of their mutual interactions, and finally the renormalizations of the single-particle spectra which are connected with these correlations.

In this early time regime, the original coherence of the quantum mechanical waves of the excited carriers has not yet been destroyed by collisions. Thus in this regime, quantum coherence and the beginning of dephasing coexist and give rise to partially coherent phenomena such as quantum beats due to phonons and plasmons. At the same time, the quantum coherence changes the description of the collisions which now become scattering integrals. The memory kernels of these non-Markovian scattering integrals are determined by temporal quantum mechanical correlations contained in the retarded and advanced nonequilibrium Green functions. These functions – if self-consistently determined – allow to introduce temporal changes of the single-particle spectra into the kinetics which cannot be accounted for in the Boltzmann description. The memory depth is determined by the dephasing of the spectral Green functions, as the mutually related retarded and advanced Green functions are called also. Because of the dominance of quantum mechanical effects, the early-time kinetics is called quantum kinetics. The quantum kinetic scattering integrals describe correlation effects and the onset of dissipative effects such as the dephasing of the off-diagonal density matrix elements and the relaxation of the particle distributions described by the diagonal density matrix elements. An important prerequisite for the occurrence of dissipation is that the coherent system is coupled to an infinitely large "bath" with a continuous energy spectrum. With a discrete energy spectrum of the bath, the recurrency effect would occur. We will give an example in which one sees directly how the dephasing time gets shorter and shorter as the number of carriers in the system increases. This study can serve as an illustration for the fact that the dephasing of the off-diagonal density matrix elements (which contain the description of quantum coherence) gets extremely short in macroscopic systems. This fast dephasing or decoherence in macroscopic systems is the reason why the "Schrödinger cat" dilemma is normally not met in macroscopic objects.

With the development of titanium sapphire laser systems with pulse durations of only a few femtoseconds and frequencies which fit the fundamental gap of GaAs crystals, the quantum kinetic regime became accessible to experimental observation. Particularly, spectroscopy with two or more pulses with variable delay times yield the necessary femtosecond resolution for the

measurement of the temporal evolution of the off-diagonal and the diagonal density matrix elements. The off-diagonal elements which determine the optically induced interband polarization can be observed in a four-wave mixing (FWM) geometry. The kinetics of the carrier distributions in the various bands can best be observed by differential transmission spectroscopy (DTS) with a strong pump and a delayed weak probe pulse. The frequencies of the pump and probe pulses can be both of the optical range suited for the observation of carrier relaxation or can be different e.g. with an optical pump pulse and a far-infrared probe pulse. This later configuration allows to measure the build-up of correlations such as the carrier screening.
The relaxation and dephasing kinetics in polar semiconductors is dominated in the low-density regime by the scattering of the excited carriers, i.e. electrons (e) and holes (h), by longitudinal optical (LO) phonons and in the high-density regime by the carrier-carrier scattering. The energies of the exchanged quanta are for the model semiconductor GaAs for the LO-phonons $\omega_0 = 36$ meV, while the density-dependent plasmon energy equals the phonon energy at a density $n_e = n_h \simeq 10^{18}\mathrm{cm}^{-3}$. For time intervals shorter than the oscillation period of $T_0 = 2\pi/\omega_0 = 130$ fs the energy is no longer conserved in an individual collision, i.e. the Boltzmann kinetics with its Markovian structure and energy-conserving transition rates brakes down. Already an elementary derivation shows that in this early time regime the transition rate e.g. for the electron (e) distribution $n_k(t)$ in a spatially homogeneous system has the form

$$\begin{aligned}\left.\frac{\partial n_{\boldsymbol{k}}}{\partial t}\right|_{\mathrm{scatt}} &= -2\Re \sum_{\boldsymbol{q},\boldsymbol{k}'} \int_{-\infty}^{t} dt' \mathrm{e}^{-\mathrm{i}\Delta_{\boldsymbol{k},\boldsymbol{k}',\boldsymbol{q}}(t-t')} |W_q|^2 \\ &\times \Big(n_{\boldsymbol{k}}(t') n_{\boldsymbol{k}'}(t')(1-n_{\boldsymbol{k}+\boldsymbol{q}}(t'))(1-n_{\boldsymbol{k}'-\boldsymbol{q}}(t')) \\ &- (1-n_{\boldsymbol{k}}(t'))(1-n_{\boldsymbol{k}'}(t')) n_{\boldsymbol{k}+\boldsymbol{q}}(t') n_{\boldsymbol{k}'-\boldsymbol{q}}(t') \Big) , \end{aligned} \tag{1}$$

where W_q is the scattering potential with a momentum transfer $\boldsymbol{q}$. The energy difference before and after the collision is

$$\Delta_{\boldsymbol{k},\boldsymbol{k}',\boldsymbol{q}} = e_{\boldsymbol{k}} + e_{\boldsymbol{k}'} - e_{\boldsymbol{k}+\boldsymbol{q}} - e_{\boldsymbol{k}'-\boldsymbol{q}} \ . \tag{2}$$

The oscillatory factor is an expression of the coherence of the quantum mechanical waves of the four involved particle states. In this sense the non-Markovian structure of the simple quantum kinetic scattering integral is due to the quantum coherence of the excited carriers at small time intervals. Note that all population factors which contain the Pauli principle enter into the scattering integral at the earlier time t'. In order to understand the integral properly one has to take into account that the carriers are excited by a short pulse say at $t_0 = 0$. In the long-time limit $\omega_0 t \gg 1$ the population factors are slowly varying in comparison with the oscillatory factor and can be pulled out of the integral at the upper time t. The remaining integral gives approximately the energy conserving delta function of Fermi's golden rule. For times $\omega_0 t < 1$ this reduction to the Markovian kinetics is not possible.

2 Non-equilibrium Many-Body Theory and Quantum Kinetics

The description of optical excitations in semiconductors across the band gap needs the knowledge of the reduced density matrix

$$\rho_{\mu\nu,\boldsymbol{k}}(t) = \langle a^\dagger_{\nu,\boldsymbol{k}}(t) a_{\mu,\boldsymbol{k}}(t)\rangle \,, \tag{3}$$

where μ, ν are band indices. For the model of a two-band semiconductor they run over c, v, i.e. over the indices for the conduction and valence band.The diagonal elements describe the carrier distributions in both bands, the off-diagonal elements are the interband polarization components. Because we limit us to spatially homogeneous situations and neglect the photon momentum in comparison with the momenta of electrons, we can limit ourselves to diagonal matrix elements in k-space. The total optically induced polarization is

$$P(t) = \sum_k d_{\boldsymbol{k}} \rho_{cv,\boldsymbol{k}}(t) + \text{h.c.} \,, \tag{4}$$

where $d_{\boldsymbol{k}}$ is the optical matrix element.

Considering only resonant pulse excitation by a coherent classical laser $E(t) = E_0 \cos(\omega t)$ and the Coulomb mean-field effects, the density matrix (3) obeys the semiconductor Bloch equations, which can be written in the form

$$\begin{aligned}\left.\frac{\partial \rho_{\mu\nu,\boldsymbol{k}}}{\partial t}\right|_{\text{coh}} &= -\mathrm{i}(\varepsilon_{\mu,\boldsymbol{k}} - \varepsilon_{\nu,\boldsymbol{k}})\rho_{\sigma\nu,\boldsymbol{k}} \\ &- \mathrm{i}\sum_\sigma \left(\sigma^x_{\mu\sigma}(d_{\boldsymbol{k}}E(t) - \Sigma^{HF}_{\mu\sigma,\boldsymbol{k}})\rho_{\sigma\nu,\boldsymbol{k}} - \rho_{\mu\sigma,\boldsymbol{k}}(d_{\boldsymbol{k}}E(t) - \Sigma^{HF}_{\sigma\nu,\boldsymbol{k}})\sigma^x_{\sigma\nu}\right) \,,\end{aligned} \tag{5}$$

where

$$\sigma^x_{\mu\nu} = \begin{pmatrix} 0 & 1 \\ 1 & 0 \end{pmatrix}$$

is a Pauli matrix. The Hartree-Fock renormalized energies are $\varepsilon_{\mu,\boldsymbol{k}} = e_{\mu,\boldsymbol{k}} + \Sigma^{HF}_{\mu\mu,\boldsymbol{k}}$, with e.g. $\Sigma^{HF}_{cc,\boldsymbol{k}}(t) = -\sum V_q \rho_{cc,\boldsymbol{k}-\boldsymbol{q}}(t)$. The Rabi frequency $\omega_{R,\boldsymbol{k}}(t) = d_{\boldsymbol{k}}E(t)$ is renormalized by the interband matrix element of the Hartree-Fock self-energy $\Sigma^{HF}_{cv,\boldsymbol{k}}(t) = -\sum_q V_q \rho_{cv,\boldsymbol{k}-\boldsymbol{q}}(t)$, which is due to the attractive e-h interaction and gives rise to exciton resonances in linear spectroscopy.

Next the changes of the density matrix due to the scattering processes have to be evaluated

$$\frac{\partial \rho_{\mu\nu,\boldsymbol{k}}}{\partial t} = \left.\frac{\partial \rho_{\mu\nu,\boldsymbol{k}}}{\partial t}\right|_{\text{coh}} + \left.\frac{\partial \rho_{\mu\nu,\boldsymbol{k}}}{\partial t}\right|_{\text{scatt}} \,. \tag{6}$$

For the scattering term one has to derive a non-Markovian scattering integral with the structure of (1).

Two methods have been applied successfully to this task:

- **The hierarchy of density matrix equations-of-motion:**
  The coupling of the carriers to phonons results in terms proportional to $\langle a^\dagger ab \rangle$, the so-called phonon-assisted density matrices. Their equations-of-motion result in still higher moments, and so on. On truncates this system on a certain level by factorization of higher moments into products of lower density matrices. This approach which has been applied by Zimmermann [2], Kuhn [3], Axt and Stahl [4] results in coupled differential equations for which powerful integration algorithms exist. If one wants to treat carrier-carrier scattering, one has to include the four-point or two-particle density matrices. As first shown in plasma physics by Wyld and Fried [5], the equations for the four-point functions yield the Coulomb scattering integrals with time-dependent screening both for direct and exchange interactions. Only recently this extension has been applied to the study of the femtosecond carrier kinetics in semiconductors [6].
- **The non-equilibrium Keldysh-Green function theory:**
  The non-equilibrium Keldysh-Green functions (GF) are defined as

$$G_{\mu\nu,\boldsymbol{k}}(t_1,t_2) = -\mathrm{i}\langle T_c a_{\mu,\boldsymbol{k}}(t_1) a^\dagger_{\nu,\boldsymbol{k}}(t_2)\rangle \; . \tag{7}$$

The time-ordering takes place on a double contour which runs from $-\infty$ to $+\infty$ and back again to $-\infty$. Considering the four possibilities (both times on the upper (or lower) branch of the contour, or time 1 one the upper, time 2 on the lower branch or the reverse) one gets 4 different GF's. Two are independent. We choose as independent GF's the particle propagators $G^<$ and the retarded Green function G^r

$$G^<_{\mu\nu,\boldsymbol{k}}(t,t') = -\mathrm{i}\langle a^\dagger_{\nu,\boldsymbol{k}}(t') a_{\nu,\boldsymbol{k}}(t)\rangle , \tag{8a}$$

$$G^r_{\mu\nu,\boldsymbol{k}}(t,t') = -\mathrm{i}\Theta(t-t')\left\langle \left[a_{\mu,\boldsymbol{k}}(t), a^\dagger_{\nu,\boldsymbol{k}}(t')\right]_+ \right\rangle \; . \tag{8b}$$

One can construct from the Dyson equation of the time-order GF an equation of motion for the equal-time particle propagator: $G^<_{\mu\nu,\boldsymbol{k}}(t,t) = -\mathrm{i}\rho_{\mu\nu,\boldsymbol{k}}(t)$, which is up to a factor the density matrix. The scattering integral is

$$\begin{aligned}
&\left.\frac{\partial \rho_{\mu\nu,\boldsymbol{k}}}{\partial t}\right|_{\mathrm{scatt}} \\
&= \sum_\varrho \int_{-\infty}^{t} \mathrm{d}t' \Big[\Sigma^>_{\mu\varrho,\boldsymbol{k}}(t,t') G^<_{\varrho\nu,\boldsymbol{k}}(t',t) - \Sigma^<_{\mu\varrho,\boldsymbol{k}}(t,t') G^>_{\varrho\nu,\boldsymbol{k}}(t',t) \\
&\quad - G^>_{\mu\varrho,\boldsymbol{k}}(t,t') \Sigma^<_{\varrho\nu,\boldsymbol{k}}(t',t) + G^<_{\mu\varrho,\boldsymbol{k}}(t,t') \Sigma^>_{\varrho\nu,\boldsymbol{k}}(t',t) \Big] ,
\end{aligned} \tag{9}$$

where $\Sigma^<, \Sigma^>$ are the scattering self-energies. The non-Markovian nature of this scattering integral is obvious. One sees the scattering integral couples to the two-time propagators. In order to close the equation, one needs a connection between the two-time $G^<(t,t')$ and the one-time density matrix $\rho(t)$. For weak coupling, this relation is the generalized Kadanoff-Baym ansatz (GKBA) [7, 8]. For $t > t'$ the GKBA is

$$G^<_{\mu\nu,\boldsymbol{k}}(t,t') = \sum_\sigma G^r_{\mu\sigma,\boldsymbol{k}}(t,t')\rho_{\sigma\nu,\boldsymbol{k}}(t') . \tag{10}$$

In this weak-coupling approximation, $G^<(t,t')$ is given by the density matrix at the earlier time t', while the correlation between the times t and t' is given by the retarded GF $G^r(t,t')$. In the examples given below the retarded functions are calculated self-consistently in the mean-field approximation. By these means e.g. energy renormalizations are taken into account in the scattering integral. The dephasing of the retarded GF is approximated by the polaronic dephasing in equilibrium [10]. The coherent oscillations, e.g., of an LO-phonon are transported from the electron-phonon scattering self-energy into the interband polarization. In a four-wave mixing (FWM) configuration with two pulses, these oscillations can be found on the diffracted time-integrated FWM signal as a function of the delay time between these two pulses. The observation of these oscillations predicted by quantum kinetics has been the first convincing evidence for the relevance of the time-delayed non-Markovian quantum kinetics for femtosecond spectroscopy [9] (see Fig. 1). The oscillations can be understood as interferences between two interband polarization components coupled by the exchange of a phonon.
At least for the phonon scattering, it has been demonstrated that the numerical solution of the coupled equations for the two two-time GF's is possible [11]. Particularly for intermediate coupling, where the GKBA is no longer very good, this strategy has been used to describe DTS experiments in CdTe. In this treatment, the temporal build-up of the polaron is incorporated in the kinetics [12].

3 Time-Dependent Screening of the Phonon and Coulomb Interactions

In the following we will treat the scattering self-energies in the so-called GW-approximation

$$\Sigma^{\lessgtr}_{\mu\nu,\boldsymbol{k}}(t,t') = \mathrm{i}\sum_{\boldsymbol{q}} G^{\lessgtr}_{\mu\nu,\boldsymbol{k}-\boldsymbol{q}}(t,t')W^{\lessgtr}_q(t,t') , \tag{11}$$

where the effective interaction $W_q(t,t')$ is the screened phonon and Coulomb interaction potential [13]. Diagrammatically it is given by Fig. 2. Because the

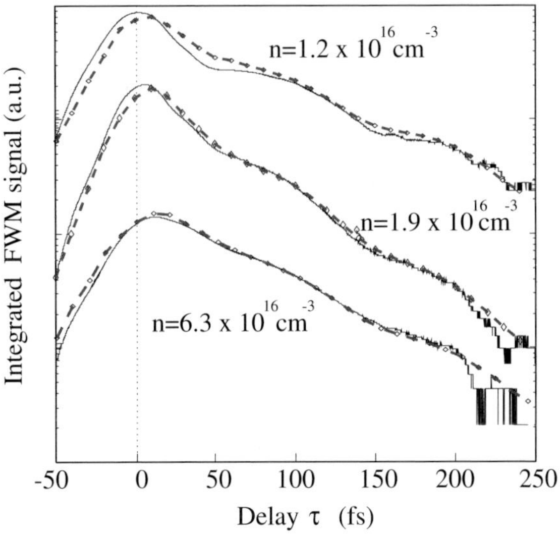

Fig. 1. Measured (solid lines) and calculated (dashed lines) time-integrated FWM signals for three excitation densities with LO-phonon scattering according to [9].

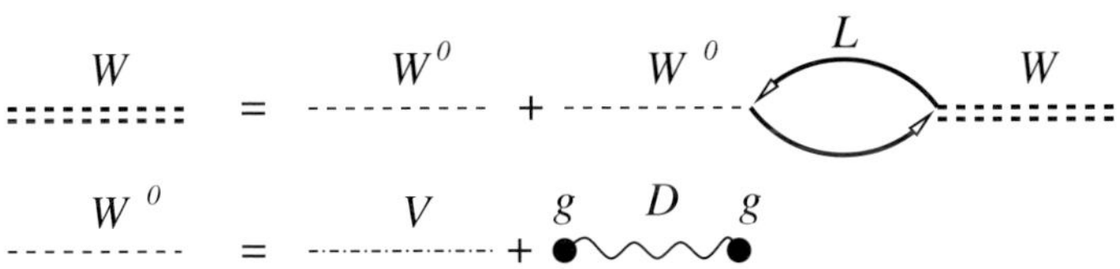

Fig. 2. Diagrams for screened interaction $W_q(t,t)$. $D_q(t,t')$ is the phonon propagator, V_q is the bare Coulomb potential.

carrier propagators enter in the intraband polarization bubble, the resulting potential can only be calculated self-consistently with the carrier kinetics. An example of the spectrum of the resulting interaction for various times t after a 11 fs excitation pulse is shown in Fig. 3 according to Ref. [13]. While the dispersion-less LO-phonon resonance existed also before the carriers have been excited, the plasmon resonance which at this density is slightly higher then the phonon resonance builds up in time. This scenario of the build-up of screening and of the plasmon pole is supported by recent optical-pump THz-probe experiments of Leitenstorfer et al. [14]. The density dependence of the two resonances yield the mixed phonon-plasmon modes. The question whether collective plasmon oscillations – similar to the LO-phonon quantum beats in Fig. 1 – can be observed in high-density FWM mixing experiments was one of the main motivations for the development of the Coulomb quantum kinetics. The delay in the build up of the plasmon resonance, their dispersion and density dependence made the task to observe these collective oscillations of the pulse excited e-h plasma considerably more difficult. Resonant FWM

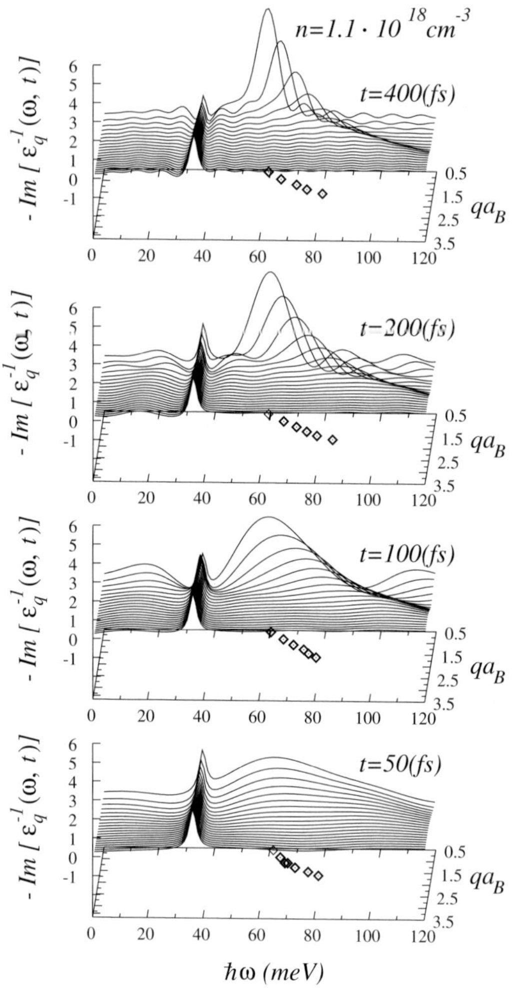

Fig. 3. Imaginary part $W_q(t,\omega) = \frac{V_q}{\epsilon_q(t,\omega)}$ for various times after the excitation pulse according to [13].

mixing with e-h-densities of up to $10^{18}\mathrm{cm}^{-3}$ showed (see Fig. 4) in agreement between theory and experiment photon-echo-like signals [15, 16]. Note that for small delay, the signals are to a very good approximation photon-echo like, as one expects it for an inhomogeneously broadened system to which the free electrons in the bands belong. Their energy dispersion provides the inhomogeneous broadening. For longer delay times and higher densities, plasma correlations build up. In the evolving plasma the electrons behave more and more collectively, which is seen clearly by the shifts of the FWM signals away from the echo line to earlier time. The signal evolves toward a simple induced absorption decay, as it is typical for a single resonance with homo-

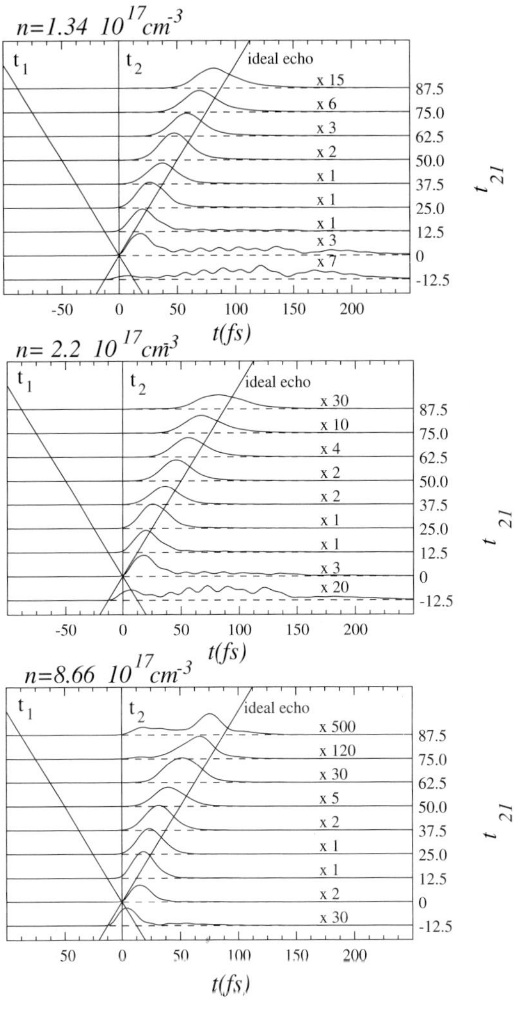

Fig. 4. Calculated time-resolved FWM signals for GaAs at 300 K for various delay times t_{21} and three excitation densities. The left and center line mark the centers of the two pulses, the right line is the ideal echo line. According to [15].

geneous linewidth.
The time-integrated FWM yielded a density-dependent dephasing . The calculated and measured dephasing times (see Fig. 5) could be fitted with

$$\frac{1}{\tau} = \gamma_0 + an^{\frac{1}{3}} \; . \tag{12}$$

The same power-law dependence has already been found in a pioneering paper of Shank et al. [17]. Surprisingly we found that this law holds also for quasi-two-dimensional quantum wells. If one argues that the mean distance

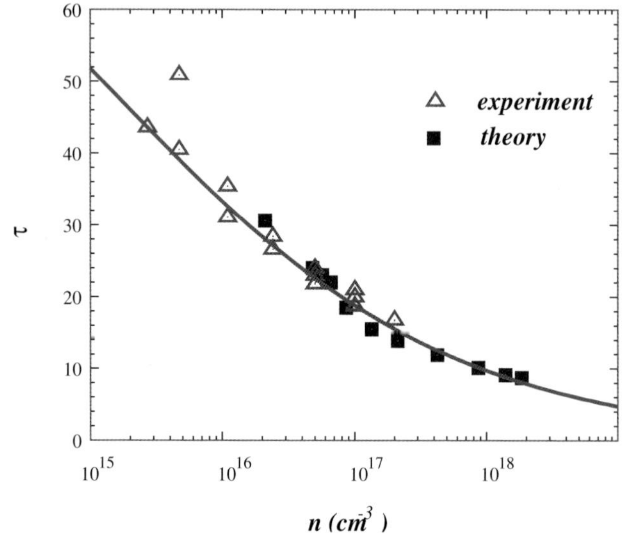

Fig. 5. Calculated and measured dephasing times for GaAs as a function of the excitation density. According to Hügel et al. [16].

between particles $d \propto n^{-1/3}$ in 3D and $d \propto n^{-1/2}$ in 2D determines the dephasing, one would expect a square root dependence of the dephasing time in 2D, which earlier – but less accurate – experiments [18] suggested.

Based on our calculations for FWM with coherent control, M. Wegener et al. used with 11 fs pulses this coherent control technique to look for the predicted phonon-plasmon-mixed mode oscillations. The results of these rather involved experiments and not less involved numerical calculations are shown in Fig. 6 according to [19]. Figure 6 shows that the obtained oscillations belong to the upper branch of the mixed phonon-plasmon modes.

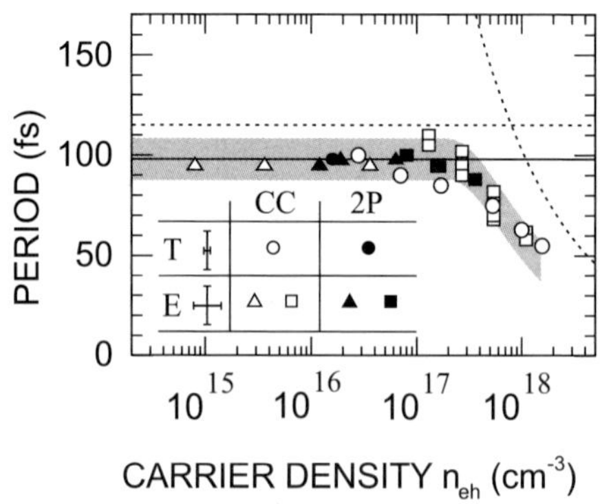

Fig. 6. Observed and calculated LO-phonon-plasmon oscillations in the resonant 11 fs FWM with coherent control. According to [19].

The observation and analysis of the coherent collective plasma oscillations as a function of the delay time superimposed on the rapidly dephased interband polarization caused by the same plasma was certainly a highlight in the joint experimental and theoretical effort to study quantum kinetics.
All these results clearly show the wealth of physics which becomes accessible if one opens the regime of "slowness" by femtosecond spectroscopy and by quantum kinetics.

Acknowledgment

I dedicate this contribution to the 60th birthday of my friend and colleague, Claus Klingshirn. I gratefully remember the fruitful scientific cooperations and the pleasant mineral collection tours during the years of C.K. in Frankfurt. This time also marked the early beginning of yet another cooperation between experiment and theory, namely between Martin Wegener – at that time a student of C.K. – and me. Many results of our joint work are described in this contribution.

References

1. L.W. Boltzmann: Ber. Wien Akad. **66**, 275 (1872)
2. R. Zimmermann: Phys. Status Solidi (b) **59**, 317 (1990)
3. T. Kuhn: *Ladungsträgerdynamik in Halbleitersystemen fern vom Gleichgewicht: Elektronisches Rauschen und kohärente Prozesse* (Shaker, Aachen 1994)
4. V.M. Axt, A. Stahl: Z. Phys. B **93**, 195 and 205 (1994)
5. R.W. Wyld, B.D. Fried: Annals of Physics **23**, 374 (1963)
6. T. Woltering, V.M. Axt, T. Kuhn: Phys. Rev. B **67**, 115311 (2003)
7. P. Lipavsky, V. Spicka, B. Velicky: Phys. Rev. B **34**, 6933 (1986)
8. H. Haug, A.P. Jauho: *Quantum Kinetics in Transport and Optics of Semiconductors* (Springer, Berlin 1996)
9. L. Banyai, D.B. Tran Thoai, E. Reitsamer, H. Haug, D. Steinbach, M.U. Wehner, M. Wegener, T. Marschner, W. Stolz: Phys. Rev. Lett. **75**, 2188 (1995)
10. H. Haug, L. Banyai: Solid State Commun. **100**, 303 (1996)
11. P. Gartner, L. Bányai, H. Haug: Phys. Rev. B **60**, 14234 (1999)
12. M. Betz, G. Göger, A. Laubereau, P. Gartner, L. Bányai, H. Haug, K. Ortner, C.R. Becker, A. Leitenstorfer: Phys. Rev. Lett. **86**, 4684 (2001)
13. Q.T. Vu, H. Haug: Phys. Rev. B **62**, 7179 (2000)
14. R. Huber et al.: Nature **414**, 286 (2001)
15. H. Haug: Phys. Stat. Sol. (b) **221**, 179 (2000)
16. W.A. Hügel, M.F. Heinrich, M. Wegener, Q.T. Vu, L. Bányai, H. Haug: Phys. Rev. Lett. **83**, 3313 (1999)
17. B.C. Becker, M.T. Portella, R.W. Schoenlein, J.F. Cunningham, C.V. Shank: Phys. Rev. Lett. **61**, 1647 (1988)
18. Y. Bigot, H.L. Fragito, C.H. Brito Cruz, R.L. Fork, J.F. Cunningham, C.V. Shank: Phys. Rev. Lett. **67**, 636 (1991)
19. Q.T. Vu, H. Haug, W.A. Hügel, S. Chatterjee, M. Wegener: Phys. Rev. Lett. **89**, 3508 (2000)

Extreme Nonlinear Optics in Semiconductors

Martin Wegener

In *traditional nonlinear optics*, the absolute changes of the optical properties are tiny if one follows them versus time on a timescale of a cycle of light (e.g., 2.9 fs for the GaAs band gap). This simple fact is the basis of many concepts and approximations of traditional nonlinear optics. Today, about 40 years after the invention of the laser, the shortest optical pulses generated are about 1.5 cycles of light in duration. This comes close to the ultimate limit of a single optical cycle. In our own experiments described below, we employ two-cycle (5 fs) pulses. Moreover, laser pulses with peak intensities around 10^{13} W/cm^2 are available directly from mode-locked laser oscillators. Under these conditions, one has substantial or even extreme changes of the optical properties on the timescale of a cycle of light. In general, one can say that whenever an energy associated to the light intensity becomes comparable to or even larger than a characteristic energy of the material or system under investigation, the laws of traditional nonlinear optics fail and something new is expected to happen. We want to call this regime *extreme nonlinear optics* or carrier-wave nonlinear optics. For the special case of resonant interband semiconductor optics, it is entered if the Rabi energy becomes comparable to the carrier photon energy.

In Sect. 1, we give an introduction into extreme nonlinear optics on the basis of two-level systems, which can easily be followed by experimentalists. Intraband effects are briefly addressed within the effective mass approximation for pedagogical purposes in Sect. 2. In Sect. 3, we review some of our own recent corresponding experimental as well as corresponding modelling work on GaAs and ZnO from Refs. [1–6] (also see references cited therein). Finally, we summarize in Sect. 4.

1 Extreme Nonlinear Optics of Two-Level Systems

Semiconductor *interband* nonlinear optics has many interesting and pronounced differences with respect to that of two-level systems. Nevertheless, we chose to focus our discussion on the extreme nonlinear optics of two-level systems here, because this is not "well-known" at all and because this allows to obtain an intuitive understanding. In this simplistic approach, the

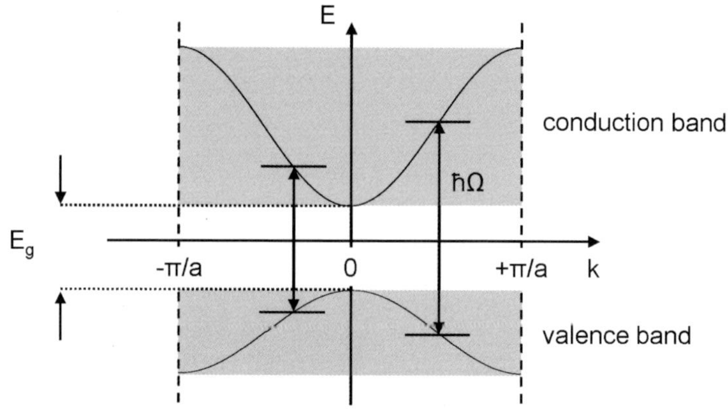

Fig. 1. Scheme of the valence and the conduction band of a direct gap semiconductor in the first Brillouin zone, i.e. for wave numbers, k, in the interval $[-\pi/a, +\pi/a]$, a is the lattice constant. At each k, the optical interband transition resembles that of a two-level system with transition energy $\hbar\Omega = E_{\mathrm{c}}(k) - E_{\mathrm{v}}(k)$. E_{g} is the band gap energy. Near the center of the first Brillouin zone ($k = 0$), the bands are nearly parabolic and the effective mass approximation can be employed.

two-level system ground state mimics the valence band and its excited state corresponds to the conduction band of the semiconductor (see Fig. 1).

1.1 Definition of the Problem

Within the dipole approximation, but without employing the rotating wave approximation and without transverse or longitudinal damping, the Bloch equations of a two-level system with transition frequency Ω for the Bloch vector $(u, v, w)^{\mathrm{T}}$ can be written in matrix form as

$$\begin{pmatrix} \dot{u} \\ \dot{v} \\ \dot{w} \end{pmatrix} = \begin{pmatrix} 0 & +\Omega & 0 \\ -\Omega & 0 & -2\,\Omega_{\mathrm{R}}(t) \\ 0 & +2\,\Omega_{\mathrm{R}}(t) & 0 \end{pmatrix} \begin{pmatrix} u \\ v \\ w \end{pmatrix} . \tag{1}$$

The dots denote the derivative with respect to time t. Here we have introduced the (instantaneous) Rabi frequency $\Omega_{\mathrm{R}}(t)$ with

$$\hbar\Omega_{\mathrm{R}}(t) = dE(t) \tag{2}$$

with the dipole matrix element d and the laser electric field

$$E(t) = \tilde{E}(t)\cos(\omega_0 t + \phi) . \tag{3}$$

$\tilde{E}(t)$ is the electric field envelope, ω_0 the carrier frequency of light and ϕ the so-called carrier-envelope phase. Note that the Rabi frequency itself oscillates with the carrier frequency of light and periodically changes sign. We shall call the peak of the Rabi frequency Ω_{R} (rather than $\Omega_{\mathrm{R}}(t)$) with $\hbar\Omega_{\mathrm{R}} = d\tilde{E}_0$, where $\tilde{E}_0$ is the peak of the electric field envelope. As usual, $w(t)$ is the inversion of the two-level system. The (real) macroscopic optical polarization $P(t)$ entering into the Maxwell equations is given by the dipole density of two-level systems $d\,N_{2\mathrm{LS}}/V$ times the (real) component $u(t)$ of the Bloch vector, i.e.,

$$P(t) = d\,\frac{N_{2\mathrm{LS}}}{V}\,u(t)\,. \tag{4}$$

The light intensity I_{rad} radiated by the two-level system is proportional to the square modulus of the second temporal derivative of the polarization, hence proportional to $|\omega^2\,u(\omega)|^2$ in the Fourier domain, where ω is the spectrometer frequency. It is natural to relate all frequencies to the carrier frequency of light ω_0, in which case the dependence of the radiated intensity on the three dimensionless parameters

- **transition frequency** Ω/ω_0
- **Rabi frequency** $\Omega_{\mathrm{R}}/\omega_0$
- **spectrometer frequency** ω/ω_0

has to be studied. In all calculations, we start from the ground state of the two-level system, i.e., from Bloch vector $(0,0,-1)^{\mathrm{T}}$.

Before we outline these calculations in the coming subsection, it is interesting to relate the ratio $\Omega_{\mathrm{R}}/\omega_0$ to the famous Keldysh parameter γ_{K}, which is frequently employed in extreme nonlinear optics of atoms. The Keldysh parameter is given by the ratio of the carrier frequency of light ω_0 and the so-called tunnelling frequency Ω_{tun} (which must not be confused with the tunnelling rate). The latter is simply the inverse of the time t_{tun} an electrons spends *within* the barrier in a semiclassical treatment of tunnelling. The appeal of the Keldysh parameter is that one can employ the rules of electrostatic tunnelling for $\gamma_{\mathrm{K}} \ll 1$ (high intensities). It can be expressed as

$$\gamma_{\mathrm{K}} = \frac{\omega_0}{\Omega_{\mathrm{tun}}} = \frac{\omega_0\,\sqrt{2m_{\mathrm{e}}E_{\mathrm{b}}}}{\mathrm{e}\,E_0}\,, \tag{5}$$

where E_{b} is the electron binding energy (ionization potential). Interpreting an optical transition from the valence to the conduction band as tunnelling of a crystal electron through the forbidden band gap, i.e. replacing $E_{\mathrm{b}} \to E_{\mathrm{g}}$, and using the usual expression for the dipole matrix element d from $k \cdot p$ perturbation theory, i.e.

$$d = \sqrt{\frac{\hbar^2\mathrm{e}^2}{2E_{\mathrm{g}}}\left(\frac{1}{m_{\mathrm{e}}} - \frac{1}{m_0}\right)}\,, \tag{6}$$

with the elementary charge $\mathrm{e} = +1.6 \times 10^{-19}$ As, one can easily show that the relation

$$\frac{\Omega_R}{\omega_0} = \frac{1}{\gamma_K}\sqrt{1-\frac{m_e}{m_0}} \approx \frac{1}{\gamma_K} \tag{7}$$

holds. Corrections to the simple expression on the RHS are on the order of the conduction band effective mass m_e divided by the free electron mass m_0. For the model semiconductor GaAs with $m_e/m_0 = 0.07$, the deviation is less than 5%. For atoms (where $m_e = m_0$ and $E_b \approx 10\,\mathrm{eV}$), $\gamma_K = 1$ according to Eq. (5) typically corresponds to peak laser pulse intensities of $I = 10^{14}\,\mathrm{W/cm^2}$ at optical frequencies ω_0. We will see below that in semiconductors the condition $\gamma_K = 1 = \Omega_R/\omega_0$ is reached at "only" several $10^{12}\,\mathrm{W/cm^2}$.

1.2 Overview Based on Exact Numerical Calculations

The simplest and cleanest situation is given when the envelope Rabi frequency $\tilde{\Omega}_R(t)$ with $\hbar\tilde{\Omega}_R(t) = d\tilde{E}(t)$ is either constant ($\hbar\tilde{\Omega}_R = d\tilde{E}_0 = \hbar\Omega_R$) or zero. For the special case of resonant excitation ($\Omega/\omega_0 = 1$), Fig. 2 gives a first impression of the resulting physics in the time domain.

To get an overview about what can be expected in measurable optical spectra, we can either fix Ω/ω_0 and depict the radiated intensity versus Ω_R/ω_0 and ω/ω_0 (Fig. 3) or, alternatively, fix Ω_R/ω_0 and plot the signals versus Ω/ω_0 and ω/ω_0 (Fig. 4). In these calculations we employ box-shaped optical pulses which are $N = 30$ optical cycles in duration (for, e.g. $\hbar\omega_0 = 1.5\,\mathrm{eV}$, this would roughly correspond to a 90 fs long pulse).

For $\Omega_R/\omega_0 \ll 1$ on the vertical axis of Fig. 3(a), where $\Omega/\omega_0 = 1$ (resonant excitation), conventional Rabi flopping [7] occurs and the well-known Mollow triplet can be seen at $\omega/\omega_0 = 1$ on the horizontal axis. At larger Ω_R/ω_0 approaching unity, carrier-wave Rabi flopping takes place and additional carrier-wave Mollow triplets [1] appear around odd integers ω/ω_0. Beyond $\Omega_R/\omega_0 = 1$, the Mollow sidebands are "repelled" by the central peaks of the adjacent Mollow triplets. They oscillate around even integer values of ω/ω_0 and finally converge towards these values in the limit $\Omega_R/\omega_0 \gg 1$. On the way, they periodically cross at even integers ω/ω_0 with a period versus Ω_R/ω_0 of $\pi/2$ for $\Omega_R/\omega_0 \gg 1$ (whereas the first crossing occurs at $\Omega_R/\omega_0 \approx 1$ [2]). For off-resonant excitation, e.g. $\Omega/\omega_0 = 5$ in Fig. 3(b), the behavior is different for $\Omega_R/\omega_0 < 1$ and $\Omega_R/\omega_0 \approx 1$, but becomes similar to Fig. 3(a) for $\Omega_R/\omega_0 \gg 1$.

The other way to look at the parameter space is to fix the Rabi frequency Ω_R/ω_0. For large Ω/ω_0 but not too large peak Rabi frequencies Ω_R/ω_0 in Fig. 4, well separated high harmonics are observed, as expected from traditional nonlinear optics. On the diagonal, where $\omega = \Omega$, very large resonant enhancement effects are observed. This is also true for the adjacent harmonics at spectrometer frequencies $\omega = \Omega \pm 2M\omega_0$ with integer M, which altogether leads to a band of enhancement around the diagonal in Fig. 4. Especially note that large contributions can occur at the spectral positions of even harmonics. We will explain the underlying physics below. These contributions are especially pronounced for even integer values of Ω/ω_0.

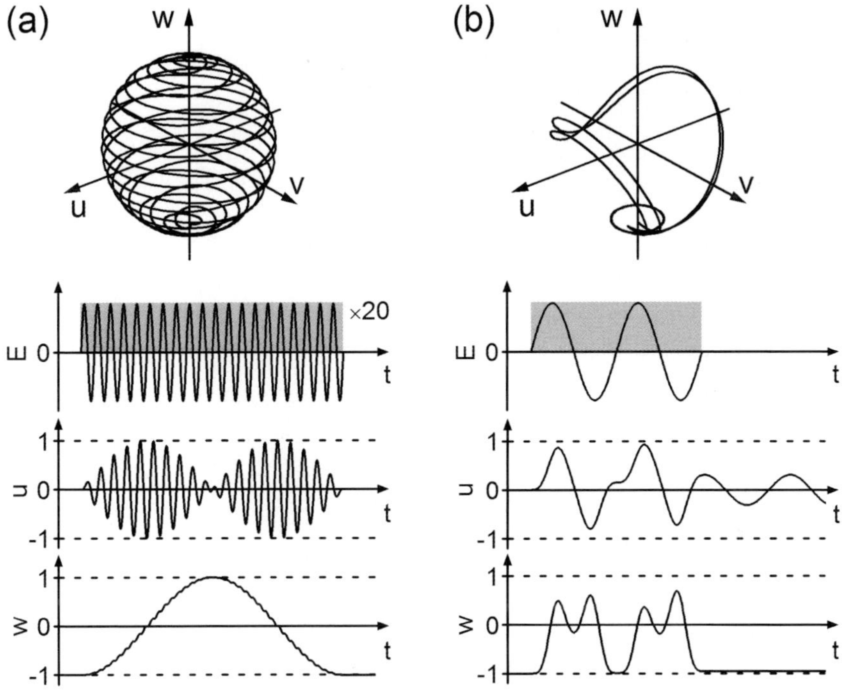

Fig. 2. Illustration of (a) conventional Rabi flopping and (b) carrier-wave Rabi flopping. In (a), the exciting box-shaped electric field pulse $E(t)$ contains $N = 20$ cycles of light and is weak, i.e., the peak Rabi frequency is given by $\Omega_R/\omega_0 = 1/20$. As a result, the inversion w (illustrated by the insets) nearly reproduces the classical result of Rabi according to $w(t) = -\cos(\tilde{\Omega}_R t)$. A closer inspection does, however, reveal a small superimposed, staircase-like structure which is due to deviations from the rotating wave approximation used by Rabi. These minor structures become large and prominent qualitative changes in (b), where we depict the behavior for a $N = 2$ cycle long, intense pulse with $\Omega_R/\omega_0 = 1$ (shown on a stretched time scale). Here, both the inversion $w(t)$ as well as the optical polarization $P(t) \propto u(t)$ exhibit significant dynamics on the timescale of one cycle of light - in contrast to conventional Rabi flopping (a). For, e.g., a photon energy of $\hbar\omega_0 = 1.5\,\text{eV}$, a cycle of light has a period of 2.8 fs and the pulse envelope (grey area) in (b) corresponds to a duration of 5.6 fs.

In practice, values Ω_R/ω_0 (much) larger than unity are not accessible for typical semiconductor sample damage thresholds. On the other hand, we will see below that the condition $\Omega_R/\omega_0 \approx 1$ is indeed approached in our experiments. Using about 800 nm wavelength light corresponding to $\hbar\omega_0 = 1.5\,\text{eV}$, the GaAs room temperature band gap translates into a transition energy of about $\Omega/\omega_0 = 1$, that of ZnO to about $\Omega/\omega_0 = 2$. Thus, these two situations are further illustrated in Figs. 5 and 6, respectively. Parameters have been chosen to correspond to our experiments described below. The

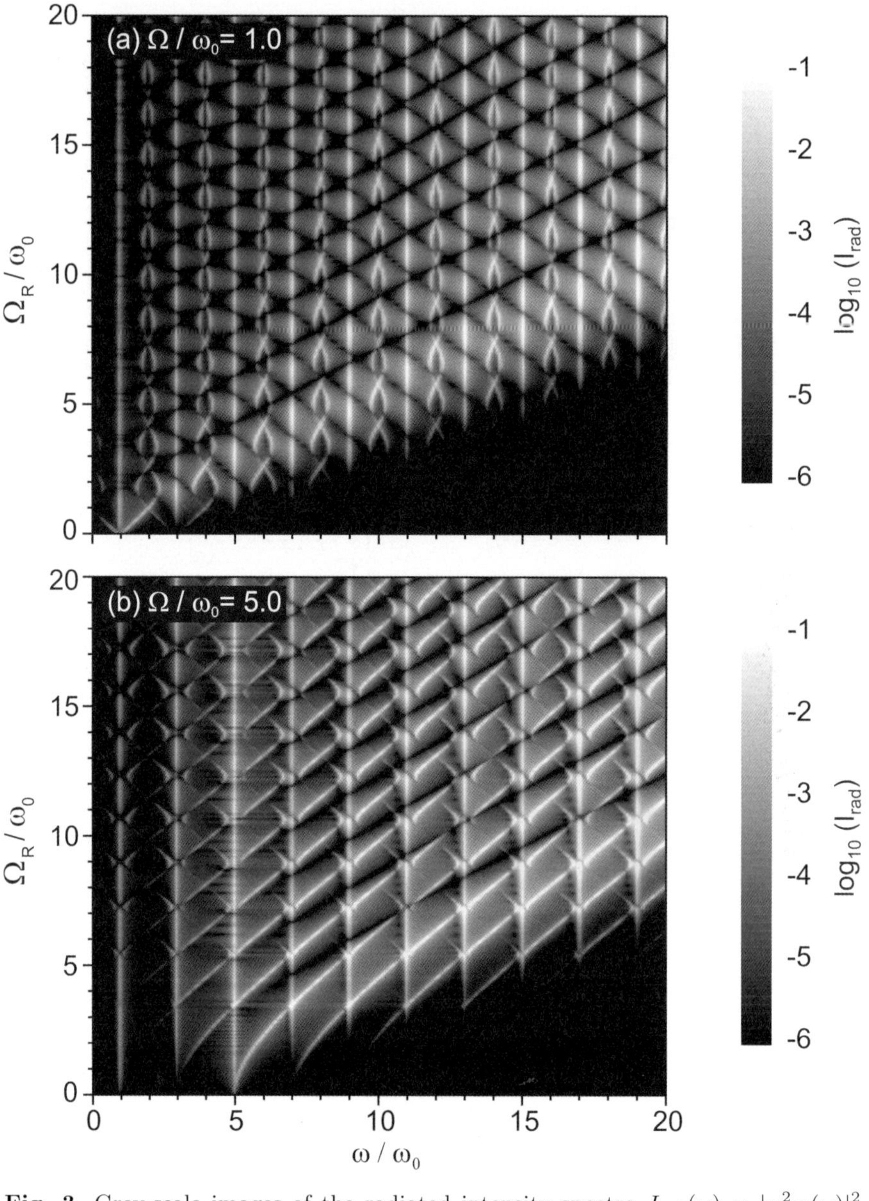

Fig. 3. Grey-scale images of the radiated intensity spectra $I_{\mathrm{rad}}(\omega) \propto |\omega^2\, u(\omega)|^2$ (normalized and on a logarithmic scale) from exact numerical solutions of the two-level system Bloch equations (1). The peak Rabi frequency Ω_{R} of the exciting $N = 30$ cycles long box-shaped optical pulses is plotted along the vertical axis. The transition frequency Ω is parameter. (a) $\Omega/\omega_0 = 1$, (b) $\Omega/\omega_0 = 5$. ω_0 is the carrier frequency of the laser pulses. Taken from Ref. [5].

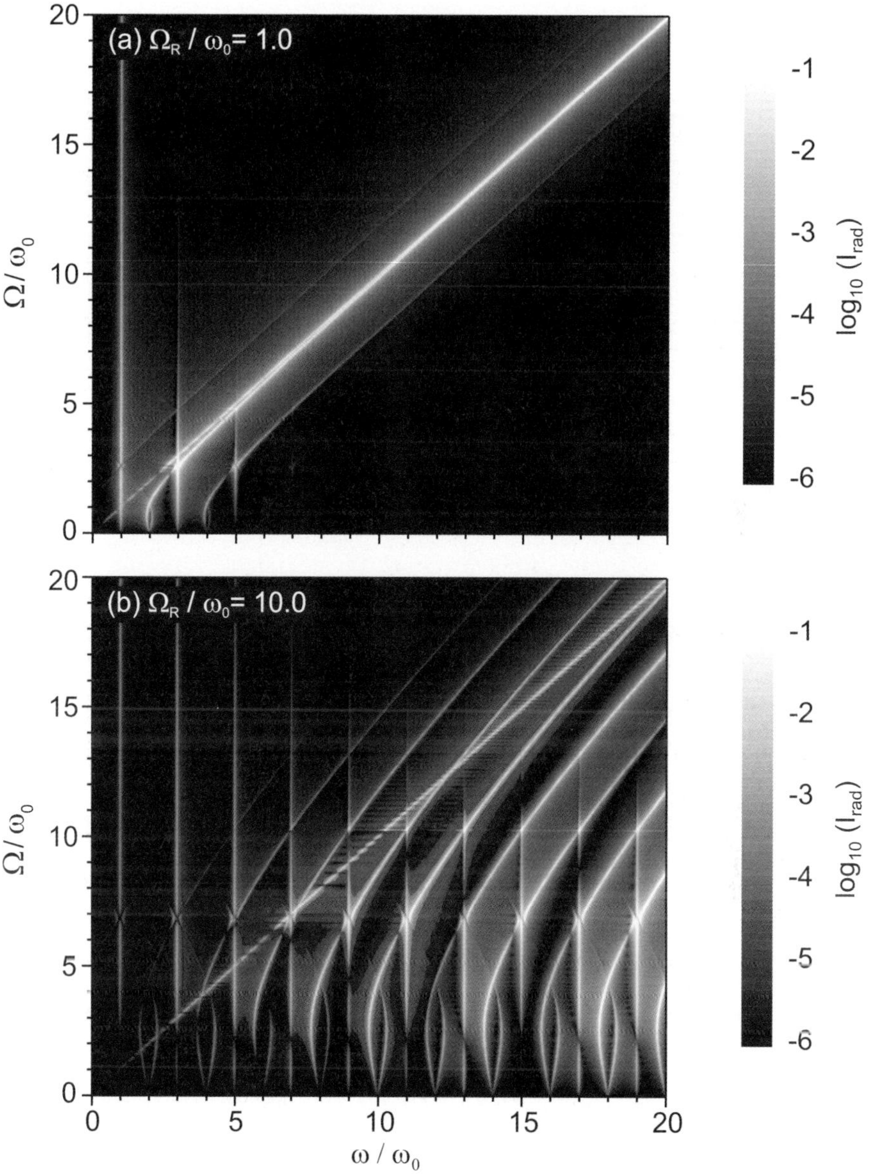

Fig. 4. As Fig. 3, but versus transition frequency Ω for two fixed values of the peak Rabi frequency Ω_R. (a) $\Omega_R/\omega_0 = 1$, (b) $\Omega_R/\omega_0 = 10$. Taken from Ref. [5].

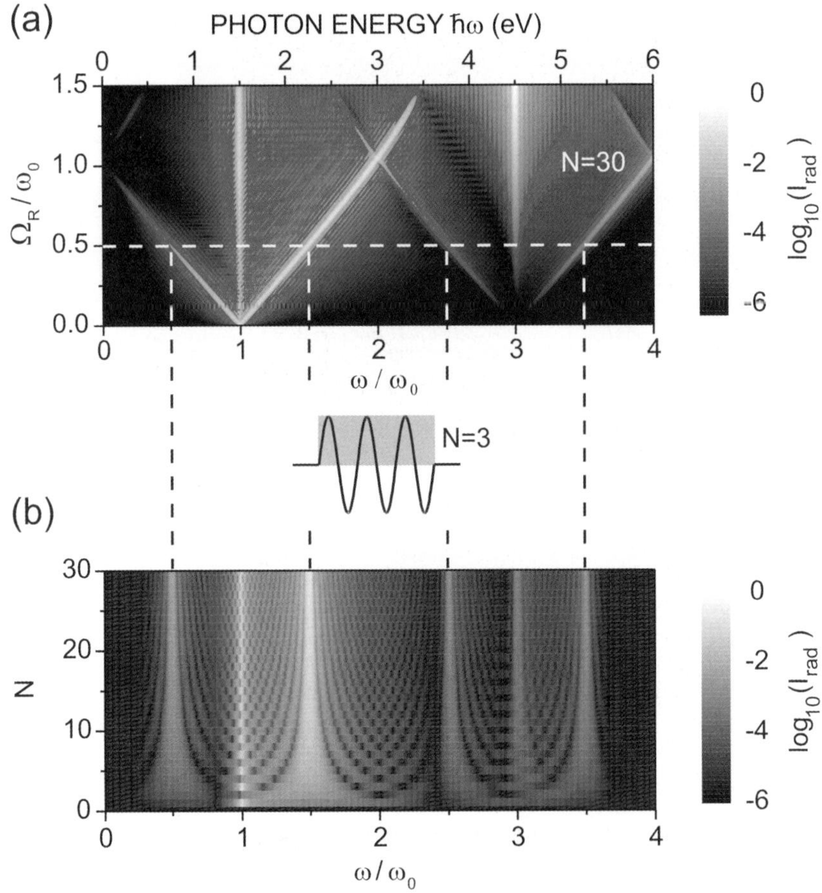

Fig. 5. Grey-scale plots of the radiated light intensity $I_{\rm rad}$ (i.e., normalized square modulus of the Fourier transform of the second temporal derivative of the optical polarization $P(t) \propto u(t)$ from the Bloch equations) versus spectrometer frequency ω. (a) Position of the peaks of the fundamental and the third harmonic Mollow triplet versus peak Rabi frequency $\Omega_{\rm R}$ in units of the laser carrier frequency ω_0 for $\Omega/\omega_0 = 1$ with $\hbar\omega_0 = 1.5\,{\rm eV}$ and for a $N = 30$ cycle long box-shaped optical pulse (see center inset for $N = 3$). (a) is nothing but a magnification of the lower LHS corner of Fig. 3. (b) Dependence on the integer number of cycles $N = 1, 2, ...30$ in the pulse for fixed $\Omega_{\rm R}/\omega_0 = 0.5$. Note the occurrence of additional side maxima for few-cycle pulses.

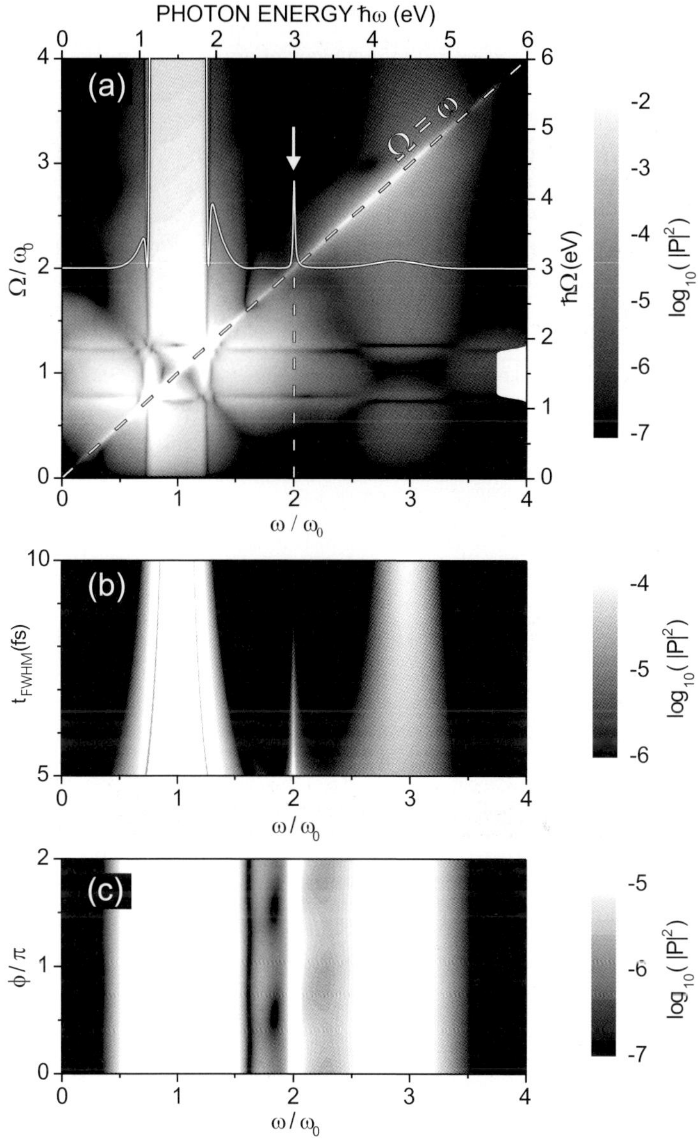

Fig. 6. (a) Grey-scale image of $|P|^2$ (normalized) versus spectrometer frequency ω and versus transition frequency Ω, both in units of ω_0 with $\hbar\omega_0 = 1.5\,\text{eV}$. The peak Rabi frequency Ω_R of the exciting $\text{sinc}^2(t)$ pulses with duration $t_\text{FWHM} = 5\,\text{fs}$ and $\phi = 0$ is $\Omega_\text{R}/\omega_0 = 0.76$. The white curve is a cut through the data at $\Omega/\omega_0 = 2$ (linear scale). The laser pulse spectrum is shown as the grey area on the RHS. (b) $|P(\omega)|^2$ versus pulse duration t_FWHM for fixed $\Omega/\omega_0 = 2$ and $\phi = 0$. (c) $|P(\omega)|^2$ versus ϕ for fixed $\Omega/\omega_0 = 2$ and $t_\text{FWHM} = 5\,\text{fs}$. Taken from Ref. [4].

main aspect of Fig. 6 is that a peak at the spectral position of the second harmonic of the laser carrier frequency can occur in the optical spectra (see arrow). Within traditional nonlinear optics, such frequency doubling is not expected from the two-level system because of its inversion symmetry. In extreme nonlinear optics, this peak arises from a resonant enhancement of the low-energy spectral tail of third-harmonic generation (THG) at the two-level system transition frequency. We have named this unusual phenomenon *THG in disguise of SHG* [4]. It requires both extremely short pulses and high intensities. As this contribution stems from THG with carrier frequency $3\omega_0$, its phase is 3ϕ. Thus, interference with the fundamental with phase ϕ leads to a difference phase 2ϕ, hence to a beat period of π (see Fig. 6c). In contrast to this, interference of true SHG (with carrier frequency $2\omega_0$ and phase 2ϕ) and the fundamental would lead to 2π beat period. These distinctly different beat periods or beat frequencies, respectively, will allow for an unambiguous experimental distinction between "true" second-harmonic generation and *THG in disguise of SHG* in our experiments on ZnO.

In certain limits, the "static-field approximation" or the "square-wave approximation" allow for simple analytical calculations. Interested readers are referred to Ref. [5].

2 Intraband Effects

In addition to the *interband* effects discussed in the previous section, the laser electric field can also lead to *intraband* effects (see Fig. 1). Classically speaking, intraband excitations correspond to acceleration of electrons within the bands (Drude model without damping). It is well known that a classical electron in a laser field with constant envelope $\tilde{E}(t) = \tilde{E}_0$ acquires a cycle-averaged kinetic energy, the ponderomotive energy $\langle E_{\mathrm{kin}} \rangle$ (or quiver or wiggle energy), which is given by

$$\langle E_{\mathrm{kin}} \rangle = \frac{1}{4} \frac{\mathrm{e}^2 \tilde{E}_0^2}{m_\mathrm{e} \omega_0^2} \,. \tag{8}$$

For holes, the electron effective mass m_e has to be replaced by the hole effective mass m_h, i.e. $m_\mathrm{e} \to m_\mathrm{h}$. The corresponding quantum mechanical problem has been solved by D. M. Volkov in 1935 using the so-called radiation gauge, in which the laser electric field is encoded solely into the vector potential. A "readable" derivation is given in Ref. [8]. For a laser electric field with constant envelope, the time-dependent Schrödinger equation in one dimension becomes

$$\mathrm{i}\hbar \frac{\partial}{\partial t} \psi(x,t) = \frac{1}{2m_\mathrm{e}} \left(-\mathrm{i}\hbar \frac{\partial}{\partial x} - \frac{\mathrm{e}\tilde{E}_0}{\omega_0} \sin(\omega_0 t + \phi) \right)^2 \psi(x,t) \,. \tag{9}$$

The solution is given by the Volkov states

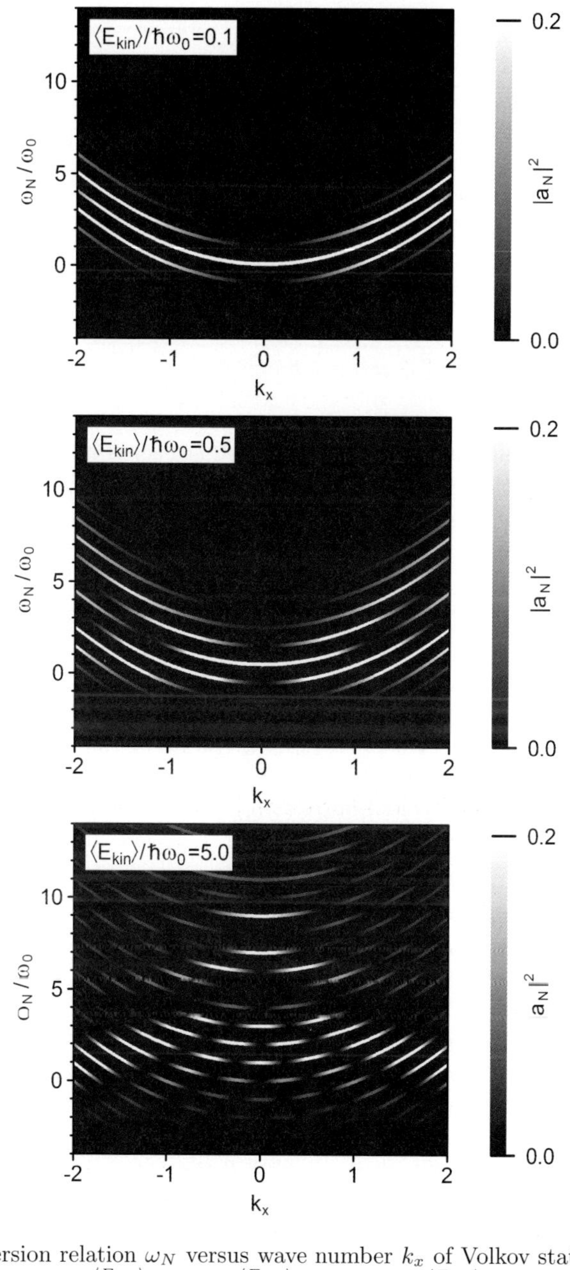

Fig. 7. Dispersion relation ω_N versus wave number k_x of Volkov states for a ponderomotive energy of $\frac{\langle E_{\text{kin}}\rangle}{\hbar\omega_0} = 0.1$, $\frac{\langle E_{\text{kin}}\rangle}{\hbar\omega_0} = 0.5$ and $\frac{\langle E_{\text{kin}}\rangle}{\hbar\omega_0} = 5.0$ (as indicated) according to Eqs. (10) and (12). The grey scale corresponds to the square modulus of the amplitude, i.e. to $|a_N|^2$, on a linear scale. A normalized wave vector with $k_x^2 = 4$ corresponds to an actual electron kinetic energy of $\frac{\hbar^2 k_x^2}{2m_e} = 4\,\hbar\omega_0$.

$$\psi(x,t) = \mathrm{e}^{\mathrm{i}k_x x} \sum_{N=-\infty}^{+\infty} a_N \, \mathrm{e}^{-\mathrm{i}(\omega_N t + N\phi)} \,. \tag{10}$$

The frequencies ω_N are given by

$$\hbar\omega_N = \frac{\hbar^2 k_x^2}{2m_\mathrm{e}} + \langle E_\mathrm{kin}\rangle + N\,\hbar\omega_0 \,, \tag{11}$$

the amplitudes are

$$a_N = \mathrm{e}^{+\mathrm{i}\frac{\pi}{2}N} \sum_{M=-\infty}^{+\infty} J_M\left(\frac{\langle E_\mathrm{kin}\rangle}{2\,\hbar\omega_0}\right) J_{N-2M}\left(-\frac{k_x \tilde{E}_0 \mathrm{e}}{m_\mathrm{e}\omega_0^2}\right) . \tag{12}$$

J_N is the N-th order Bessel function of the first kind.

The resulting modified "band structure" is visualized in Fig. 7. Qualitatively, the modulation of the electron with the laser carrier frequency ω_0 leads to sidebands at energies $\pm N\,\hbar\omega_0$ with integer N – no surprise. The larger the amplitude of the modulation, the larger the amplitude of the sidebands. If the perturbative regime with $\frac{\langle E_\mathrm{kin}\rangle}{\hbar\omega_0} \ll 1$ is left, this strength is taken away from the fundamental ($N=0$) conduction band and "holes" in the parabola occur.

Whether interband or intraband effects dominate the physics depends on whether either the ratio $\Omega_\mathrm{R}/\omega_0$ or the ratio $\frac{\langle E_\mathrm{kin}\rangle}{\hbar\omega_0}$ approaches unity first with increasing laser intensity. Obviously, the ponderomotive energy is proportional to the square of the laser electric field, whereas the Rabi energy is directly proportional to it. At infrared frequencies ω_0, the ratio $\frac{\langle E_\mathrm{kin}\rangle}{\hbar\omega_0}$ usually dominates because it scales as $1/\omega_0^3$. On the other hand, for the following examples from the optical regime, interband effects dominate, but intraband effects may well provide relevant corrections. The combined effect of the two has not been investigated theoretically yet to the best of our knowledge.

3 Two Experimental Examples

3.1 GaAs

Using a Ti:sapphire laser with an approximate carrier photon energy of $\hbar\omega_0 = 1.5\,\mathrm{eV}$, the band edge of GaAs can be excited resonantly. The 5 fs laser pulses are derived directly from the mode-locked laser oscillator and are tightly focused to a spot of about one micron radius by means of high-numerical-aperture, reflective microscope lenses. In most of our experiments, we excite the samples with a pair of 5 fs pulses with time delay τ derived from a balanced Michelson interferometer, actively stabilized via the Pancharatnam skrew [9]. In the examples discussed here, the GaAs samples consist of GaAs film of thickness l, directly epitaxially grown on a sapphire substrate.

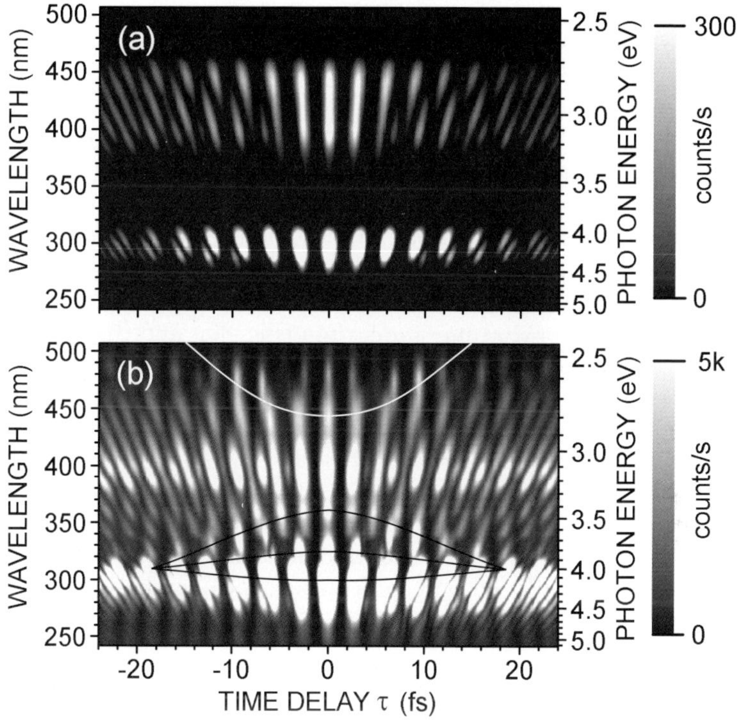

Fig. 8. Experiment: Spectra of light (linear scale) emitted into the forward direction by a $l = 100$ nm thin GaAs film on sapphire substrate resonantly excited by a pair of 5 fs pulses with time delay τ. The CEO phase ϕ of the laser pulses is not stabilized. (a) Excitation intensity $I = 0.24 \times 10^{12}\,\mathrm{W/cm^2}$, (b) $I = 2.8 \times 10^{12}\,\mathrm{W/cm^2}$ (both arms at $\tau = 0$). The contribution centered around $\lambda = 425$ nm wavelength is due to surface SHG. The single peak in (a) centered around $\lambda = 300$ nm wavelength (the third harmonic of the GaAs band gap) evolves into three peaks in (b), which are attributed to the carrier-wave Mollow triplet. The corresponding three black lines are a guide to the eye. The white curve at the top of (b) (another guide to the eye) indicates the position of the high-energy peak of the fundamental Mollow triplet. For (b) we estimate that the peak Rabi energy *within* the GaAs film (and accounting for reflection losses at the air-GaAs interface) is given by $\Omega_R/\omega_0 = 0.76$. Taken from Ref. [6].

Corresponding rather realistic calculations have been discussed in [3]. Similar experiments on a high-quality GaAs/AlGaAs double heterostructure have been published in Ref. [1].

An example is shown in Fig. 8. (a) corresponds to low excitation intensity, (b) to high excitation. The intensity in (b) at $\tau = 0$ translates into a peak Rabi energy of $\Omega_R/\omega_0 = 0.76$. At high excitation (Fig. 8(b)), the emitted light intensity around the third harmonic of the GaAs band gap splits and overlaps with the second-harmonic generation (SHG) signal. From the dependence on

l (not shown) we conclude that the SHG has a large surface contribution (or is even completely generated at the two GaAs surfaces), while the third harmonic is consistent with a bulk effect. At $\tau = 0$, the spectrum exhibits three peaks around the third harmonic which evolve with time delay τ because the peak Rabi frequency changes with τ. The solid lines are guides to the eye and indicate that the splitting decreases with increasing $|\tau|$. These three peaks are interpreted as the carrier-wave Mollow triplet. Note also that a contribution from the fundamental moves into the picture from the top. Following the above theory, this is expected to be the high-energy peak of the fundamental Mollow triplet. The data of the $l = 50\,\text{nm}$ sample (not shown) are compatible with the $l = 100\,\text{nm}$ sample data (Fig. 8(a) and (b)), however - as already discussed above - the second-harmonic contribution is more prominent with respect to the third harmonic in the $l = 50\,\text{nm}$ case as compared to the $l = 100\,\text{nm}$ case due to a larger surface contribution. This significantly reduces the visibility of the low-energy peak of the third-harmonic Mollow triplet.

At the spectral positions of interference of the fundamental high-energy Mollow sideband with the SHG and of the third-harmonic low-energy Mollow sideband with the SHG, we observe a dependence on the carrier-envelope offset phase ϕ [6] (not shown here).

3.2 ZnO

Using the same laser pulses and conditions as in the section on GaAs, the band gap of ZnO is about twice as large as the carrier photon energy $\hbar\omega_0$, which allows to study off-resonant excitation. While we have discussed both, bulk ZnO single crystals as well as ZnO epitaxial films [3], here we only mention our experiments on $l = 350\,\text{nm}$ thin ZnO layers epitaxially grown on sapphire substrate [4].

Fig. 9(a) shows measured spectra of light emitted into the forward direction versus time delay τ, exhibiting a rich structure. (b) shows corresponding radio-frequency (RF) power spectra S_{RF}. The peak at the laser oscillator repetition frequency f_{r} is expected and trivial. As the carrier-envelope offset phase ϕ changes from pulse to pulse with the carrier-envelope offset frequency f_ϕ in an unstabilized laser oscillator, the temporal variation of ϕ translates into peaks in the RF spectrum labelled by f_ϕ. The peak at f_ϕ corresponds to a 2π periodicity of the signal versus ϕ and is expected for interference of the fundamental with phase ϕ and the second harmonic with phase 2ϕ, leading to a difference phase ϕ. The peak at frequency $(f_{\text{r}} - f_\phi)$ is simply a mixing product. Surprisingly, however, also a peak at $2f_\phi$ is visible, corresponding to a π periodicity of the signal versus ϕ. This is exactly what we have discussed above for *SHG in disguise of THG* in the framework of two-level systems. Indeed, the value of $\Omega_{\text{R}}/\omega_0 = 0.76$ chosen in Fig. 6 corresponds to the electric field of $\tilde{E}_0 = 6 \times 10^9\,\text{V/m}$ in Fig. 9. One might argue, however, that the semiconductor is not really a single two-level system but rather has bands with

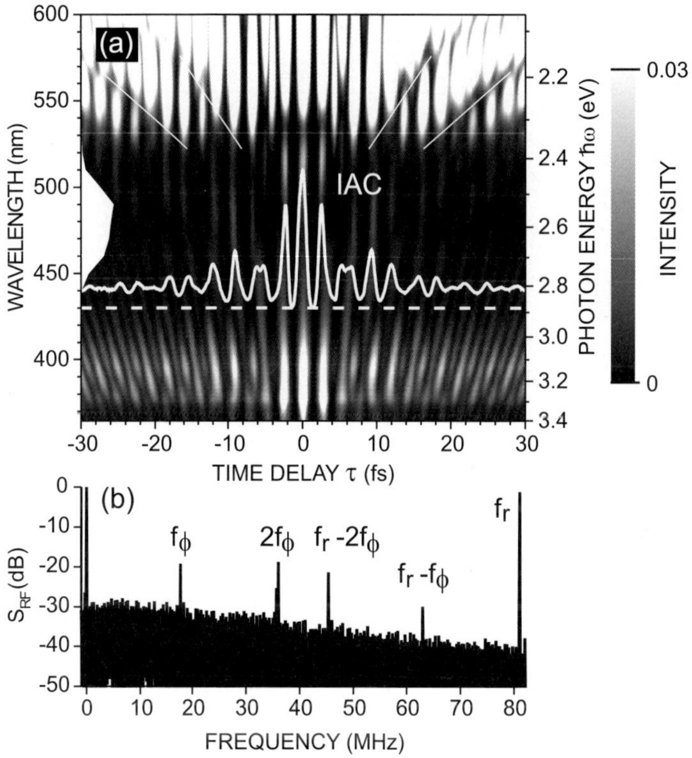

Fig. 9. Experiment on the 350 nm ZnO film, $\tilde{E}_0 = 6 \times 10^9$ V/m at $\tau = 0$. (a) Grey-scale image of the light intensity (normalized) emitted into the forward direction versus spectrometer photon energy $\hbar\omega$ and time delay τ. ϕ is not stabilized. The white thin lines are a guide to the eye. The white curve labelled IAC is the interferometric autocorrelation of the laser pulses, obtained from an independent measurement using a β-barium borate SHG crystal. (b) Radio-frequency power spectrum S_{RF} of the intensity in the spectral interval indicated by the grey area on the LHS in (a), $\tau = 0$. Note the good agreement with the theory calculated under the same conditions (Fig. 10). Taken from Ref. [4].

a complicated band structure. Also, propagation effects might obscure the results. Detailed model calculations [4] depicted in Fig. 10 do show, however, that the experimental results can almost be reproduced quantitatively with the absolute value of the incident peak electric field $\tilde{E}_0$ put into the theory, which accounts for these aspects.

4 Conclusions and Outlook

The regime of *extreme nonlinear optics* is currently intensely discussed in solids, atoms and even for the Dirac vacuum. For atoms excited by intensi-

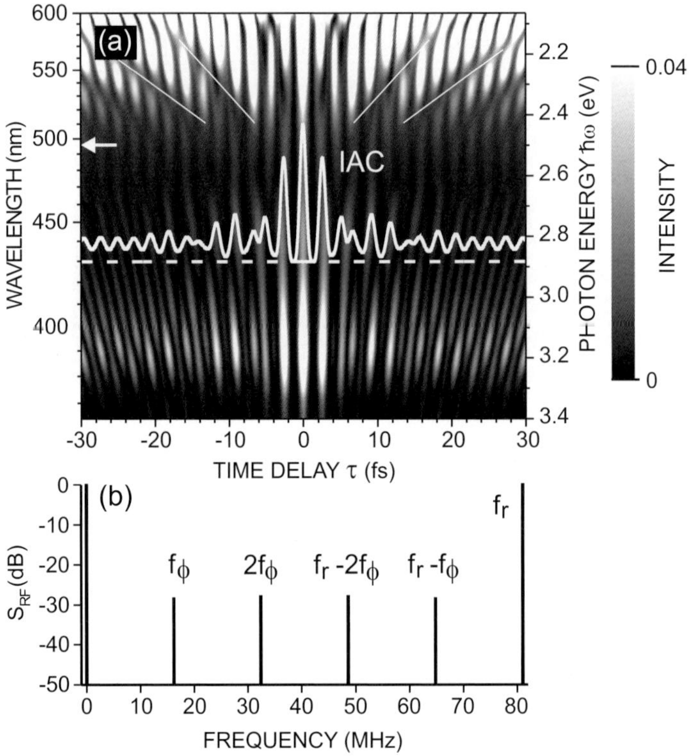

Fig. 10. Theory, $\tilde{E}_0 = 6 \times 10^9$ V/m at $\tau = 0$. (a) Grey-scale image of the light intensity (normalized) emitted into the forward direction versus spectrometer photon energy $\hbar\omega$ and time delay τ. The data are averaged over $\phi = 0...2\pi$. The white thin lines are a guide to the eye. The white curve labelled IAC is the interferometric autocorrelation of the laser pulses. (b) Radio-frequency power spectrum S_{RF} of the intensity at the spectral position indicated by the arrow on the LHS in (a), $\tau = 0$. Taken from Ref. [4].

ties $I = 10^{14} - 10^{16}\,\mathrm{W/cm^2}$, high-harmonic generation deep in extreme UV as well as attosecond pulses result, the electron dynamics becomes relativistic at around $I = 10^{18}\,\mathrm{W/cm^2}$, leading to nonlinear Thomson scattering. The nonlinear optics of the Dirac vacuum starts to get into reach with the highest intensities available on earth of about $I = 10^{22}\,\mathrm{W/cm^2}$. For the case of interband optical transitions in semiconductors, the regime of extreme nonlinear optics is entered if the Rabi energy becomes comparable to the carrier photon energy ($I = 10^{12} - 10^{13}\,\mathrm{W/cm^2}$). *Carrier-wave Rabi flopping* and *third-harmonic generation in disguise of second-harmonic generation* are two examples which we have discussed in this brief review. For both examples, one obtains a dependence of the nonlinear signals on the so called carrier-envelope offset phase, i.e., on the phase between the rapid carrier-wave oscillation and the electric field envelope of a pulse. In *traditional nonlinear optics*, one expects strictly no (!) dependence on this carrier-envelope phase at all. Such dependence might open the door to measuring the actual value of this phase for laser pulses (and not only its change with time) – which has not been possible by any means so far. Notably, it is usually argued in quantum mechanics that the corresponding phase for an electron wave packet cannot be measured. In order to make semiconductors attractive for such applications, we still need an improved theoretical understanding of the underlying microscopic physics under these unusual and extreme conditions, where the laser electric field corresponds to potential drops over one GaAs lattice constant on the order of electron Volts and where the carrier density changes by values as large as 10^{20} /cm^3 over a timescale of just a few femtoseconds.

Acknowledgments

The scientific results described in this article have been obtained by Oliver D. Mücke and Thorsten Tritschler. We thank Uwe Morgner (Electrical Engineering, Universität Karlsruhe (TH), Germany and presently Max Planck Institut für Kernphysik, Heidelberg, Germany) and Franz X. Kärtner (Electrical Engineering, Universität Karlsruhe (TH), Germany and later Massachusetts Institute of Technology, Boston, U.S.A.) for intense support concerning the building of the five femtosecond laser system in the initial phase of the experiments and for stimulating discussions, especially on the role of the carrier-envelope offset phase.

We thank Wolfgang Stolz (Universität Marburg, Germany) as well as Galina Khitrova and Hyatt Gibbs (Optical Sciences Center, Tucson, U.S.A.) for several high quality GaAs samples specifically made for our experiments, Claus Klingshirn (Universität Karlsruhe (TH), Germany) for providing the ZnO and CdS single crystal platelets and Heinz Kalt (Universität Karlsruhe (TH), Germany) for the thin, epitaxially grown ZnO films.

The theoretical part of this work has largely benefited from discussions with the group of Hartmut Haug (Universität Frankfurt, Germany).

This work is financially supported by the DFG via the Leibniz Award 2000 (project DFG-We 1497/9) and by project DFG-We 1497/11-1. The latter is a cooperation of the group of M.W. and the group of Franz X. Kärtner (Electrical Engineering, Universität Karlsruhe (TH), Germany).

References

1. O.D. Mücke, Th. Tritschler, M. Wegener et al: Phys. Rev. Lett. **87**, 057401 (2001)
2. O.D. Mücke, Th. Tritschler, M. Wegener et al: Phys. Rev. Lett. **89**, 127401 (2002)

3. O.D. Mücke, T. Tritschler, M. Wegener et al: Opt. Lett. **27**, 2127 (2002)
4. T. Tritschler, O.D. Mücke, M. Wegener et al: Phys. Rev. Lett. **90**, 217404 (2003)
5. T. Tritschler, O.D. Mücke, M. Wegener: Phys. Rev. A **68**, 033404 (2003)
6. T. Tritschler, O.D. Mücke, M. Wegener et al: unpublished
7. W. Schäfer, M. Wegener: *Semiconductor Optics and Transport Phenomena*, Advanced Texts in Physics (Springer, Berlin Heidelberg New York 2002)
8. H.R. Reiss: Phys. Rev. A **22**, 1786 (1980)
9. M.U. Wehner, M.H. Ulm, M. Wegener: Opt. Lett. **22**, 1455 (1997)

Nonlinear Semiconductor Microcavities

Hyatt M. Gibbs

1 Perturbative Regime: Optical Bistability

A lossless Fabry-Pérot interferometer in its simplest form consists of two parallel mirrors, and it converts changes in the roundtrip phase into a modulation of the transmitted intensity. The general formula for a Fabry-Pérot depends also on the intensity reflectivity coefficient of each of the mirrors as well as the absorption within the spacer (Sect. 3.1.6 of [1]). If one knows the dependencies of the refractive index n and absorption coefficient α upon the intracavity intensity (output intensity over output mirror transmission), one can solve for the input intensity versus the output intensity. When that function is camelbacked, then the output is multi-valued over a certain range of inputs. If the upper and lower branches are stable the system is said to exhibit optical bistability (Chap. 23 of [1]). The first optical bistability was seen in lasers, where a hysteresis in lasing occurred as a function of cavity frequency (see Sect. 1.3 in [2]). If the spacer medium is not inverted then the self-action effect is referred to as passive optical bistability, the first demonstration in sodium vapor being motivated largely by Sam McCall's quest for an optical transistor [2, 3]. Perhaps the first nonlinear semiconductor etalons were ordinary narrowband interference filters [4]; even commercial ones often exhibit low-power optical bistability based on the thermal increase of both n and the physical length L of the spacer. Electronic nonlinearities such as the carrier density dependent changes in band edge susceptibility of GaAs or InSb are much faster ($\approx$ 1 ns), but the holding power is much higher. During this same period a systematic investigation of the nonlinear optical mechanisms in II-VI compounds, especially CdS, was also conducted (Chap. 20 of [1]).

After the first few observations of bistability and NOR-gate action, much of the research on semiconductor nonlinear etalons was driven by the hope that the "massive parallelism" of optics and low-energy optical logic elements would enable optical computing. This also motivated or at least leveraged funding for much of the research on semiconductor optical nonlinearities. Attempts were made to increase the finesse of the nonlinear etalons and to operate them as optical gates requiring no holding power [5], but the power/speed tradeoff exacted its toll. This was understood by some already two decades ago. "The devices which work continuous with lasers and which have very low switching powers in the milliwatt area until now at least tend to be rather slow with switching times in the 10 to 100 nanosecond region or even longer. The physical mechanisms which promise switchings in picoseconds generally need rather high intensities. For example, if you go to, say, biexcitons in copper chloride, Hanamura has shown that you can, from a theoretical point of view, expect picosecond switching times; but you need then megawatts per square centimeter, and with this you can not apply a continuously working beam on the sample – you can apply it only in short pulses [6]."

Perhaps research on nonlinear semiconductor etalons for switching and logic did contribute to the development of the vertical-cavity surface-emitting laser (VCSEL). Not much notice was taken when the first optically bistable etalon (GaAs) lased [7], because it had to be pumped very hard due to the very low (90%) reflectivity R of its mirrors. The growth of a complete high-finesse microcavity structure in one growth operation evolved out of the nonlinear etalon development [8]. The new high-Q logic elements that followed were immediately shown to be low-threshold lasers when optically pumped [9]. This evolutionary improvement of all-optical nonlinear etalons was one path to the VCSEL, complementing the approach of Iga in Japan [10].

As the hard realities made hope for optical computing vaporize, we and many other researchers not developing VCSEL's returned to the semiconductor physics of quantum wells and microcavities. More and more experiments were performed at low temperatures, with the applications requirement of room temperature operation no longer relevant.

2 Nonperturbative Regime: Normal-Mode Coupling

Weisbuch, Arakawa et al. discovered normal-mode coupling (NMC) between GaAs quantum wells and the single mode of a microcavity with AlGaAs/AlAs Bragg mirrors (R = 98%) designed for low temperature use [11]. The double-peaked transmission characteristic of NMC is predicted by the Fabry-Pérot formula when the quantum well linear absorption and refractive index are inserted. Alternatively, the complete multi-layer structure can be computed by multiplying transfer matrices. NMC breathed new life into the nonlinear semiconductor etalon, partially because of the confusion possible with two peaks instead of one and because of misconception about the nature of the coupling. Most atomic physicists distinguish between nonpertubative normal-mode coupling, where the splitting arises from the susceptibility of many atoms, and genuine strong coupling where the splitting is due to a single atom. Many semiconductor physicists do not distinguish the two, suggesting they do not really appreciate the difference.

2.1 Nonresonant Excitation: Carrier Dependent Nonlinear NMC

High Densities: Loss of Oscillator Strength Collapses NMC

The first nonlinear studies were performed in NMC samples with broad exciton transitions and therefore small NMC splitting-to-linewidth ratios [12, 13]. Because of the large inhomogeneous broadening, the NMC linewidths were insensitive to carrier-dependent broadening and the transmission of the etalon was already low. Consequently increased excitation density caused the NMC splitting to decrease, due to the reduction of the oscillator strength of the exciton resonance.

Low Densities: Exciton Broadening Kills NMC Transmission

Later another nonlinear response was found at lower densities in microcavities containing narrow linewidth InGaAs quantum wells and exhibiting very large splitting-to-linewidth ratios. The transmission of the two NMC peaks went down markedly with very little change in oscillator strength [14]. This was an early example where the local nature of the nonlinear susceptibility at the two NMC peaks was shown to determine the transmission and linewidth of each NMC peak. Here the integrated absorption (proportional to the oscillator strength) only determines the NMC splitting. Each of the NMC peaks probes the local nonlinear susceptibility. The sensitivity of the transmission at the energies of the NMC peaks highlights the susceptibility at these two detuned energies. Although the nonlinear behavior of a nonresonantly excited microcavity is basically understood by computing the microcavity transmission with each nonlinear susceptibility, the microcavity may put the spot light on different aspects. This means the microcavity is very sensitive to the broadening of the exciton resonance at densities below those needed for appreciable loss of oscillator strength.

In the pure dephasing limit, where only diagonal dephasing due to carrier-carrier scattering out of a state of given in-plane momentum k is considered, the broadening is strongly overestimated and an artificial redshift of the exciton resonance occurs at high densities. Off-diagonal dephasing that accounts for scattering from all other states k' back into state k compensates diagonal dephasing to a large extent and eliminates the artificial shift. The full dephasing calculation is in good agreement with the density dependence of the observed nonlinear NMC transmission [14, 15].

Linear NMC Linewidths: No Evidence for "Motional Narrowing"

A curious question was raised about NMC linewidths [16]. Are light-coupling effects able to modify the influence of structural disorder on the excitonic quasiparticles within their quantum wells? If one computes the effective mass of a polariton from the coupled system E(k) curves, one finds very light masses and hence, if such masses are assumed to be the physical mass of the polariton, large deBroglie wavelengths. This suggests that cavity polaritons should result in averaging over a much larger diameter (1000 nm instead of an exciton Bohr radius) in a microcavity than in a single quantum well. Evidence cited for this "motional narrowing" was the narrower linewidth of the lower NMC branch [16]. In fact, the measured linear susceptibility in transfer matrix computations reproduces the linewidth of both branches very well [17]. This again emphasizes that the NMC linewidths are determined by n and α in the vicinity of the two NMC peaks. The fact that α of a disordered quantum well drops off faster on the low energy side makes the lower branch linewidth narrower [17, 18]. In principle, disorder and propagation should be treated simultaneously [19], but the nature of the disorder in quantum wells that we studied does not "entangle" the two. Consequently, the measured linewidths are explained well by first measuring the disorder-averaged susceptibility and then using it in the propagation computation.

Neither do we find any evidence for the claim [20, 21] of disorder related "motional narrowing" of quantum wells outside of a microcavity. Light coupling between quantum wells does result in linewidth narrowing (and broadening) effects, but this occurs even for homogeneously broadened (disorderless) lines [22].

Threshold-Like Behavior, but No Boser

When an NMC microcavity is pumped nonresonantly with ps pulses and the pump photon energy exceeds the exciton resonance, one sees a threshold-like rapid increase of the upper-branch photoluminescence (PL) with increased carrier density which overtakes and greatly exceeds the lower-branch PL. This dramatic behavior was first interpreted as Boser action: rapid formation of cavity polaritons which undergo final-state stimulation to the upper branch transition that then emits coherent light from its macroscopically coherent (BEC) state [23]. Subsequent experiments revealed that the crossover was occurring at densities that collapse NMC in the nonperturbative to weak coupling transition [15, 24, 25]. It was also shown that the results could be understood completely by an electron-hole plasma (no excitonic population) emitting within a nonlinear etalon. I.e, the emission of the plasma was inhibited and enhanced in the usual manner of controlled spontaneous emission by the photonic bandstructure of the etalon. That such a plasma could emit at all at the energetic positions of the NMC peaks, i.e. well below the unexcited bandgap in the vicinity of the bare exciton resonance, was implicit. A later publication [26] emphasized that this means that even at low densities PL is emitted at the ls exciton resonance by a pure plasma, a surprising prediction that stirred considerable controversy that we investigate next.

Excitonic Light Does Not Have to Come from Excitons

At low temperatures and with above bandgap excitation, high quality quantum wells reradiate most of the input power at the 1s exciton resonance. Photoluminescence at the continuum edge is down three or four orders of magnitude from the peak 1s PL. The latter occurs close to the peak of the exciton resonance in absorption, with a Stokes shift to lower energy that is smaller the less the disorder. Consequently, it has been assumed for many years that PL emitted at the 1s resonance (here referred to as excitonic PL) is emitted by a population of excitons. It was never questioned that excitonic PL might have a different origin. In fact, its buildup and decay were interpreted to map out the dynamical behavior of the exciton population. However, when a quantized field was used to properly treat PL in a many-body plasma [25, 26], PL was found to be emitted by a pure Coulomb correlated plasma where no population of excitons was included in the theory. Previously most of us used a two-particle argument against such a possibility: since the electron and hole were in the plasma, their energy must exceed the bandgap energy; therefore, energy conservation forbade emission at the 1s resonance. A many-body description, however, can satisfy energy conservation easily by transferring the excess energy of the recombining pair to the remainder of the plasma which is heated by the process. The fact that excitonic PL can be emitted by a plasma automatically raises questions about the interpretation of many experiments on exciton formation and possible condensation. Consequently, we set out to perform definitive experiments to determine the exciton fraction in our InGaAs/GaAs quantum wells following nonresonant excitation.

The basic approach is the following [27]. Theoretical spectra show that for a given density and temperature, the spectrum emitted by an exciton population has a larger ratio of 1s to continuum-edge PL than does a correlated plasma. This can be understood because 1s PL requires overlap of the wavefunctions of the recombining pair with the exciton wavefunctions. The idea was to compare a measured PL spectrum at a known density and temperature with that of the plasma theory and then to add a population of excitons if needed to make them agree. Therefore, experimentally we needed to measure the excitation density and carrier temperature. The experiment was performed using picosecond pulses cut from the 100-fs output pulses from a modelocked Ti:Sa laser. The excitation energy was chosen to be above the continuum edge and below other resonances, particularly the light hole. The PL spectrum was measured with a streak camera at the output plane of a spectrometer. The decay of the PL was so slow that it was not finished before the next pulse arrived 12.5 ns later. Consequently a pulse picker and a 2-MHz detection rate were used. In order to determine the carrier temperature, each PL spectrum was taken several meV into the continuum. Long data accumulation times were required because of the weak continuum emission. The carrier temperature for a given time following the excitation pulse was then determined from the slope of the continuum PL. The excitation density was extracted from nonlinear absorption. A double-pass delay line of length > 1 m was used to delay the 100-fs probe pulse in order to measure its transmission over the same times as the PL measurements. From the measured absorption at the pump wavelength, the measured pump power, repetition rate, etc. the initial carrier density could be determined and associated with a given change in absorption spectrum. For later times in the decay, the excitation density was determined from the measured change in absorption at that time, using the lookup table of density versus peak height of the nonlinear absorption. Because we wanted to compare our data with a theory assuming quasi-equilibrium, we concentrated the analysis on data taken 1 ns after the excitation pulse. The largest decrease in peak 1s absorption at 1 ns was 25%, so that over the whole range of densities considered the exciton resonance was very pronounced and its linewidth much less than its binding energy.

Because we wanted to compare many spectra for many different carrier densities, we decided to introduce a parameter β to summarize each spectrum. It is defined at the measured peak 1s PL divided by that expected from thermal equilibrium, found by extrapolating the continuum emission back to the 1s energy. The results are shown in Fig. 1, where β values are compared with the plasma predictions.

We were able to reach several conclusions. Clearly the measured β values are approximated very well by the pure plasma theory for all densities at the higher lattice temperature. And the high density end of the low temperature data also agrees. For high temperatures, clearly 1s PL is monitoring the plasma, not excitons. But for the low temperature, lower densities the measured β's are clearly much larger. In order to increase the predicted 1s PL to agree with that measured, excitons had to be added. Fig. 2 shows the fraction of excitons needed: a very small fraction of bright excitons is able to outshine the plasma at the 1s resonance. Even at low temperature and densities low enough that the 1s PL is mostly emitted by excitons, the exciton fraction may be tiny; this means that the system dynamics may still be dominated by the plasma.

Although we were able to learn quite a bit from this study, it does not give the *total* exciton fraction, that is bright plus dark excitons. Of course, only bright excitons contribute directly to the 1s PL. One must know something about the exciton distribution as a function of exciton in-plane center-of-mass momentum q, to extract information on the total exciton fraction. The uncertainty comes from the lack of information on the depth of the hole around $q \approx 0$ created by the competition between radiative decay of bright excitons with their formation from the plasma or by scattering from dark excitons. If one assumes that there is no hole, then the peak bright exciton fraction in Fig. 2 corresponds to a total exciton fraction of 4%. A 50% hole would yield an 8% fraction, etc. The total exciton fraction can be determined in principle by direct absorption of THz radiation by the exciton 1s to 2p

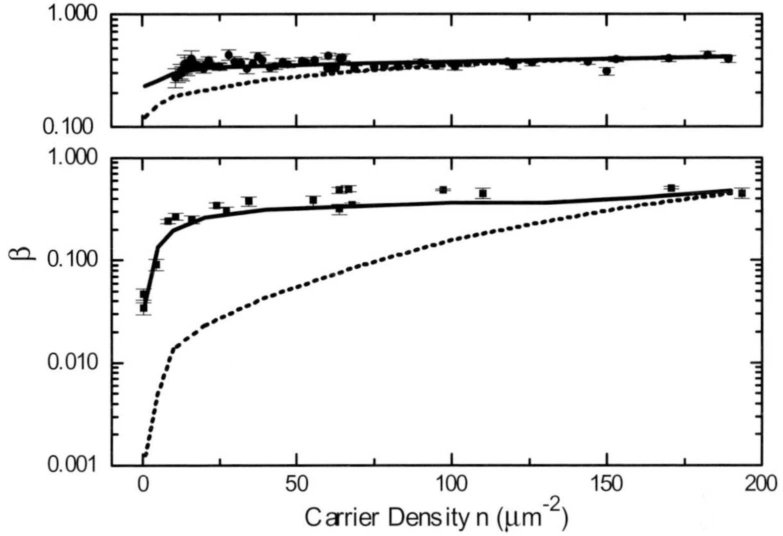

Fig. 1. β versus carrier density. Full squares are experimental values taken at 1 ns after nonresonant excitation at a lattice temperature of 4 K; the densities are also at 1 ns. The extracted carrier temperatures at 1 ns range from 14 K at the lowest density to 21 K at the highest. The theoretical values (dotted line) are calculated using the PL formula for a pure correlated plasma with a carrier temperature of 16±2 K. The solid line shows a theoretical fit including a bright exciton contribution. Top: same for 50 K lattice temperature and extracted carrier temperature of 50±3 K; the theoretical curves are computed for a carrier temperature of 48 K.

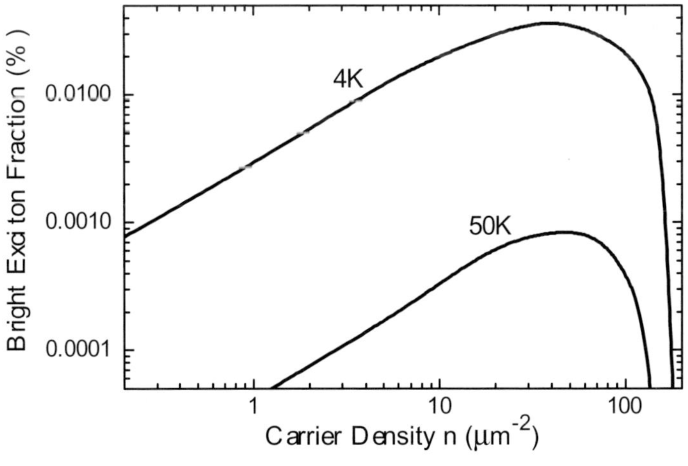

Fig. 2. Optically active $q \approx 0$ exciton contributions as a function of the carrier density.

transition as pointed out by [28]. There have been some such measurements [29, 30] that are so far inconclusive, because of the difficulty of obtaining good statistics and the lack of corresponding density measurements.

This subsection has been a digression from microcavities to determine the exciton/plasma ratio in quantum wells for various nonresonant excitation densities and carrier temperatures. Clearly, it is relevant to the large number of claims to see coherent emission from semiconductor microcavities at densities well below the threshold density for conventional lasing. There would seem to be no problem with those reports that involve a driven parametric process that is phase matched by excitation at the "magic" angle, since they can be understood using coherent polarizations without the concept of condensation of cavity polaritons; see [31, 32] and references therein. Evidence for stimulated polariton-polariton scattering has been reported in [33–36]. Based on our exciton/plasma ratio work, the claims of incoherent polariton accumulation at the bottleneck region of the lower polariton branch are puzzling [33, 37–39]. Especially interesting and needing further study is the report of condensation of lower cavity polaritons due to stimulated polariton-polariton scattering following lower branch pumping close to the exciton resonance at an angle deliberately much larger than the magic angle [40]. BEC of excitons or of cavity polaritons will require more convincing tests.

Our exciton/plasma ratio conclusions were drawn from experiments on ordinary type-I quantum wells (InGaAs/GaAs) as usually used in microcavity condensation studies. The low-density PL decay time was typically 1 ns. Some condensation claims have been made in type-II double quantum wells in an electric field where the PL decay time can be lengthened to many ns by separating the electron and hole, giving much more time for exciton formation and condensation [41, 42]. Even in those cases, the study here should serve as a warning that excitonic PL is not a good monitor of exciton density; the PL may be coming from the plasma or it may be coming from excitons embedded in a much higher density of plasma. Further experiments such as performed here or THz absorption could help clarify the exciton density and evaluate the condensation evidence.

2.2 Resonant Excitation: Polarization-Dependent Nonlinear NMC

Third Peak in NMC Transmission

At the end of 1995 we observed a very curious albeit disturbing third transmission peak that emerged between the two NMC peaks as we increased the energy of the single beam. We were destined to struggle with this effect for nearly five years before explaining it. The semiclassical approach so successful with the nonresonant case failed to predict a third peak. Both optical bistability and the nonresonant nonlinear NMC transmission were understood very well by first measuring the nonlinear behavior of the medium (nonlinear α

and Kramers-Kronig for nonlinear n) and then using that in the propagation solution (Fabry-Pérot formula or transfer matrix code). That procedure yielded no hint of a third peak. Furthermore, the effect was not small, as can be seen in Fig. 3(a). At the lowest intensities, one sees no third peak. But as the exciton broadening brings down the transmission of the two NMC peaks as described in subsection 2.1, the third peak emerges until it becomes dominant - while the splitting has just begun to decrease. To try to unravel this strange effect, we compiled a list of facts about it, mostly during the postdoctoral stint of Martin Hübner. It only occurs for pulsed resonant excitation; even resonant excitation that is so broad that it generates lots of carriers may diminish or destroy the third peak [43]. If the cavity resonance is detuned from the exciton resonance, the third peak in transmission follows the cavity resonance. When a narrowband pump is used, the third peak in the probe transmission follows the energetic position of the pump. The third peak is spectrally somewhat narrower than the two NMC peaks, and spectrally resolved upconversion of transmission reveals that the third peak emission lasts longer than that from the two NMC peaks. The third peak persists to 30 K, ruling out any dreams of condensation. And the third peak is not unique to one sample as first thought, but is seen in samples with different numbers of quantum wells and various mirror reflectivities and cavity lengths.

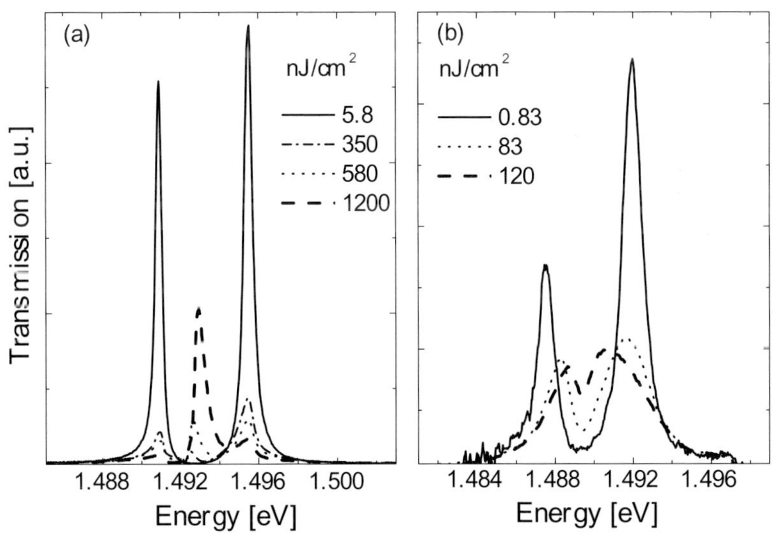

Fig. 3. Single-beam transmission for various pulse intensities $[nJ/cm^2]$ (a) for the planar microcavity, and (b) for the oxide-aperture 3D microcavity. Inset: energetic position of the third peak (squares), the upper (triangles up) and lower (triangles down) normal-mode peak, the bare exciton (dotted) and cavity peak (straight) as a function of the exciton-cavity mode detuning.

Our Marburg theorists, especially Professors Stephan Koch and Mackillo Kira, realized that the three dimensional (3D) aspects of a planar microcavity could be responsible [15]. The coupling to the classical coherent fields in the normal direction occurs in the nonperturbative regime as described above. But coupling of the quantum well excitations into guided modes occurs in the perturbative regime. Then the transmission and emissions in the normal direction could be affected by coupling back from the guided modes. At first they considered disorder, which could couple to the guided modes [44]. But when we grew an NMC microcavity with quantum wells having broader linewidths due to increased disorder introduced by the higher In concentration, the third peak was suppressed. Then they applied their fully quantum treatment to the problem. Quantizing the field does more than introducing field-interband correlations like $\langle B_q^\dagger e_{k-q} h_k \rangle$ that account for spontaneous emission by creating a photon as an electron and hole are destroyed [45]. It also introduces field-intraband correlations of the form $\langle B_q^\dagger e_{k-q}^\dagger e_k \rangle$ that entangle a photon with an electron. Not only did the theory that included these correlations reproduce all of the list of facts about the third peak, it also made predictions about the outcome of two experiments we had not thought of trying. Since the field-intraband correlations are dominated by guided modes because of their very large density of states, the third peak was predicted to vanish in a laterally confined microcavity. Fortunately, we had available just such a 3D microcavity for studies of the saturation photon number as described below. The oxide aperture reduced the coupling into the guided modes by a factor of two; that was enough to eliminate the third peak. NMC collapsed into a single peak with no third peak along the way as shown in Fig. 3(b). The second prediction was to reduce the probe intensity and see the third peak *increase* relative to the two NMC peaks. Indeed that is what we saw [45]. It clearly shows that the third peak is not opening up increased transmission of the probe, but it is a redirection of pump energy into the probe direction.

But how does the field/intraband-coherence term arise? Its dynamical equation has terms proportional to $(f_{k+q}^e + f_{k+q}^h)P_k$; i.e. a finite carrier density and a coherent polarization are required to drive it. That is consistent with seeing it only when the excitation is resonant. The third peak is an example of an experiment in which one uses a classical coherent beam to perform an experiment one expects to be understood semiclassically, later to discover it cannot be understood without the field-carrier correlations that arise only if the field is quantized.

Oscillations of Intraband Coherences

Even though the third peak was seen well before, its understanding evolved from the experiment we describe now. Yun-Shik Lee and Ted Norris were performing coherent control experiments on one of our NMC microcavities. When they used two pump pulses followed by a probe pulse, they discovered coherent effects that lasted several picoseconds - longer than the interband polarization decay time at that excitation level [46]. This suggested that intraband coherences (such as $\langle e^{\dagger}_{k-q} e_k \rangle$ with $q \neq 0$) , known to live somewhat longer, might be contributing. The pump (probe) pulse was resonant with the lower (upper) mode, with pulse durations of 190 fs. The pump pulse was split into two excitation pulses with a variable delay using a Michelson interferometer having a piezoelectric translator in one arm to allow continuous control of the relative phase between the two pulses. Oscillations with a 2-ps period were seen in the differential reflectivity, corresponding to the energy difference between the excitonic resonance and the upper normal mode. As revealed above, our Marburg theorists found terms that involved correlations between the field and intraband coherences originating from the quantized nature of the field. When they analyzed the Michigan experiments including such terms, they found remarkable agreement in all of the observed features. If the quantum correlations were switched off in the theory, then the long-lived 2-ps oscillations disappeared, indicating that they are caused by the intraband coherences. Coherent control of the microcavity normal modes using the phase-locked pump pulses enabled considerable enhancement of the intraband oscillations due to quantum interference, with the detailed dependence upon phase difference reproduced well by the theory. The primary role of the microcavity was to enhance the change in reflectivity due to the quantum correlations; the oscillations would be roughly 2 orders of magnitude smaller in a bare quantum well.

3 Toward the Strong Coupling Regime

The long-term goal is to achieve genuine strong coupling between a single quantum dot (SQD) and a small-volume 3D semiconductor nanocavity. In this regime, absorption of a single photon is enough to change the absorption spectrum for a second probe photon. Such a device could exhibit quantum entanglement or provide a single photon on demand, two operations needed for quantum information processing and quantum cryptography. Whether or not such a device is ever useful for quantum information, it is the ultimate evolution of LED's and VCSEL's toward quantum devices. Nonperturbative normal-mode coupling between one or more quantum wells and the single mode of a planar microcavity exhibits many interesting nonlinear optical effects, some of which are summarized above. However such an etalon is far from the strong coupling regime. In fact, for a typical beam diameter of 50 μm, we measured that a large number ($\approx$200,000) of photons must be absorbed to appreciably affect the coupling [15]. The following sections describe two steps toward genuine strong coupling, the first a quantum well in a 2-μm-diameter oxide-aperture cyclindrical microcavity, and the second a single quantum dot in a photonic crystal nanocavity.

3.1 NMC in a 3D Aluminum-Oxide-Aperture Microcavity

Recent progress in microfabrication involving etching processes makes it possible to engineer three-dimensional microcavities from MBE-grown planar Fabry-Pérot microcavities [47–49]. Here the optical mode is confined laterally by implementing a thin dielectric (native oxide of aluminum) aperture layer on top of the cavity spacer [49]. The sample under investigation consists of a 16 period GaAs/AlAs bottom mirror, a 1-λ GaAs spacer, a 1 period mirror layer pair ($\lambda/4$ layer of AlAs and $Al_{0.8}Ga_{0.2}As$/GaAs and a $\lambda/4$ layer of GaAs) in which the oxide aperture is etched, and a 5 period ZnSe/MgF_2 dielectric mirror. The aperture diameters range from 1 to 7 μm. A high-quality 8.5 nm InGaAs single quantum well is located in the anti-node of the spacer. The effect of 3D confinement on the bare cavity mode is revealed in transmission spectra [50]. Several transverse modes can be observed for each aperture size. As expected, the mode spacing increases with decreasing aperture diameter. To confirm the quantized nature of the photon mode, we performed additional measurements with various angles of incidence with respect to the normal. The spectral position of each transverse mode is preserved, but the amplitudes of the high-order modes increase with increasing angle. The cavity quality factors $Q \equiv \lambda/\delta\lambda$ for all aperture diameters are high enough to provide clear normal-mode coupling even though Q degrades as the aperture size decreases. The high quality of the cavity together with the narrow exciton absorption linewidth of 0.6 meV FWHM results in well resolved NMC and a splitting-to-linewidth ratio of 4.9 at minimum splitting. A typical NMC anticrossing behavior is seen using different temperatures to tune the exciton resonance through the cavity mode; see Fig. 4.

The observation of such well-defined NMC allows controlled nonlinear measurements. We investigated microcavities in which the optical mode is confined by oxide apertures with diameters between 2 and 7 μm. Pump-probe experiments on these microcavities revealed that the number of absorbed photons required to saturate the normal-mode-coupling peaks scales with the transverse area of the cavity mode and drops to 300 photons for an aperture diameter of 2 μm [50, 51]; see Fig. 5. The straight line yields a saturation number of 90 photons/μm^2; this extrapolates for a single photon to an area of 1/90 μm^2 or a diameter of 119 nm. Note that this is approaching the diameter of a larger quantum dot. And it is not much smaller than the mode diameter of a photonic crystal nanocavity, suggesting that even a single quantum well in such a nanocavity may exhibit statistical effects such as flipping back and forth between bistable states.

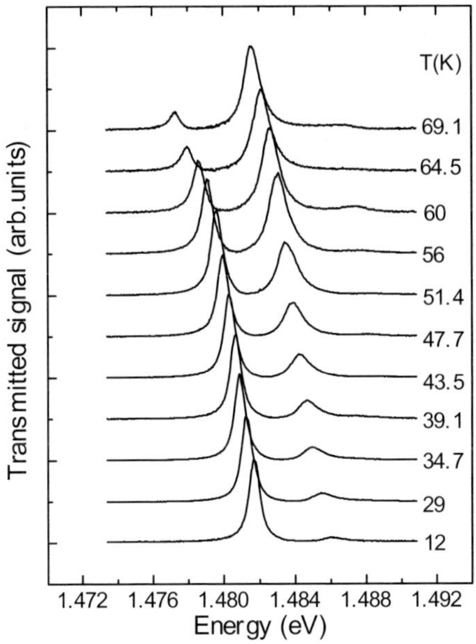

Fig. 4. NMC transmission spectra of a 2-μm oxide–aperture 3D microcavity for various temperatures.

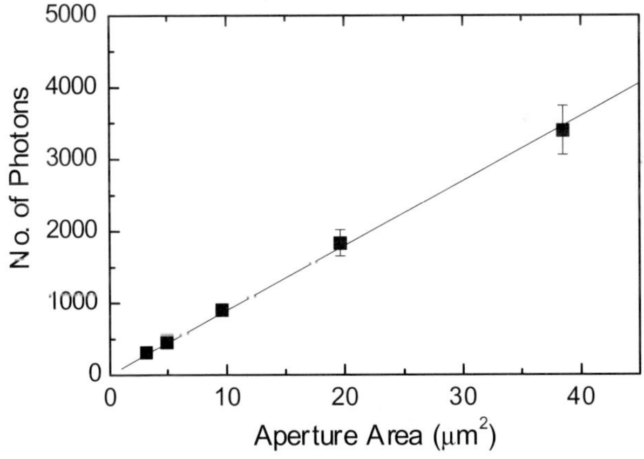

Fig. 5. Number of photons needed for the saturation of the normal-mode peaks as a function of the aperture area. Straight line: 90 photons/μm^2.

3.2 Onset of Intermediate Coupling: Single Quantum Dot in a Photonic Crystal Nanocavity

A quantum dot in a three-dimensional nanocavity is the obvious semiconductor system for demonstrating genuine strong coupling. Due to the fermionic nature of quantum dot transitions, it is the true analog of the single-atom/cavity system. Genuine strong coupling in semiconductors has not been demonstrated yet. Strong coupling is characterized by the vacuum Rabi splitting $\hbar g = 2 \mid \langle \boldsymbol{d} \cdot \boldsymbol{E} \rangle \mid$, where $\boldsymbol{d}$ is the dipole moment of the quantum dot and $\boldsymbol{E}$ is the vacuum electric field at the dot location, which must exceed the sum of the decay rates of the cavity and dot polarization. To increase the likelihood of achieving strong coupling, we chose a large InAs self-assembled quantum dot because of it large dipole moment and a photonic crystal nanocavity [52] because of its near-minimum mode volume. In order to motivate our experiment, consider the Jaynes-Cummings Hamiltonian for an atom-cavity system and the eigenvalues that come from the solution of the corresponding master equation [53]:

$$2\alpha_{\pm} = -[(\gamma_{\text{dot}} + \gamma_{\text{cav}})/2 + i\delta] \pm \sqrt{[(\gamma_{\text{dot}} - \gamma_{\text{cav}})/2 + i\delta]^2 - 4\hbar^2 g^2}. \qquad (1)$$

γ_{dot} and γ_{cav} are the FWHM energy linewidths of the uncoupled dot transition and nanocavity mode; $\delta = \hbar(\omega_{\text{cav}} - \omega_{\text{dot}})$. The real parts of the eigenvalues $\alpha_{\pm}$ give the linewidths of the emission from the two coupled oscillators, and the imaginary parts are the eigenenergies. The idea of the present experiment is shown in Fig. 6: in the intermediate-coupling regime for Q increasing towards Q_{split} the eigenenergies of the quantum dot and nanocavity are degenerate and the linewidths approach each other; for $Q > Q_{\text{split}}$, the eigenenergies split and the linewidths are equal. Analysis of our dot/nanocavity system shows that our Q results in intermediate coupling.

Here we treat only the "bad cavity" case for which $\gamma_{\text{cav}} \gg \gamma_{\text{dot}}$, experimentally verified for our dot/nanocavity. In the weak-coupling limit, $4\hbar g \ll (\gamma_{\text{cav}} - \gamma_{\text{dot}})$, the PL spectrum reduces to a single Lorentzian line when the dot transition and cavity mode are in resonance. The emission spectrum is broadened by the sum of γ_{dot} and γ_{se}, where $\gamma_{\text{se}} = 4\hbar^2 g^2/\gamma_{\text{cav}}$. This relation is used to define the Purcell enhancement factor F_P by $\gamma_{\text{se}} = F_P \gamma_0$ where $\gamma_0/\hbar$ is the radiative decay rate of the uncoupled quantum dot transition in the medium [54].

The PL measurements were performed in a liquid helium cryostat (Cryovac) with temperature control and internal x-y nanopositioner. The samples were optically pumped by the 770 nm output of a Ti:Sapphire cw laser. The pump beam was focused by a 0.5 NA reflecting microscope objective onto the sample resulting in a spot size of 1 µm. The emission from the sample was collected by the same microscope objective, then analyzed with a spectrometer, and detected by a CCD array. Because temperature shifts the quantum dot resonance faster than the cavity mode, a quantum dot transition can be temperature scanned through the cavity resonance. Fig. 7 shows the extracted

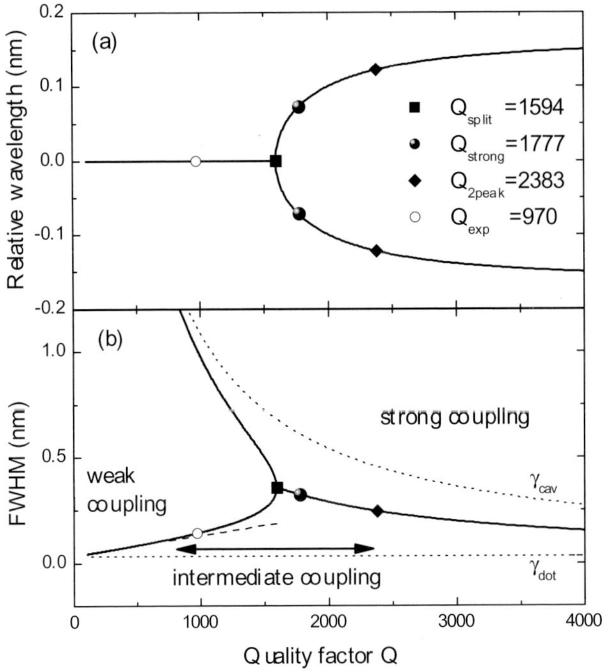

Fig. 6. Energies (a) and FWHM linewidths (b) calculated for zero detuning and, from the fit to the data in Fig. 7, $\gamma_{\text{dot}} = 0.035$ nm and $\hbar g = 0.161$ nm. The dashed line in (b) shows the dot linewidth broadened by the enhanced spontaneous emission rate in the weak-coupling regime.

FWHM linewidth of the quantum dot (black squares) as a function of the wavelength detuning $\delta\lambda = \lambda_{\text{dot}} - \lambda_{\text{cav}}$ [55]. The extracted maximum linewidth is $\gamma' = 0.148$ meV $= 0.141$ nm at resonance. Is this enhanced emission in the weak-coupling regime, or are there clear signatures of intermediate coupling marking the crossover to strong coupling?

We computed spontaneous emission spectra for various detunings, convolved each with the experimental instrument function, and compared the extracted linewidths with the experimental data; see Fig. 7 dotted and solid curves. The fit, done with $\gamma_{\text{cav}} = 1.18$ meV $= 1.12$ nm as measured for the uncoupled cavity ($Q = 970$), yields $\gamma_{\text{dot}} = 0.037$ meV $= 0.035$ nm and $\hbar g =$ 0.170 meV $= 0.161$ nm. These fit values give for the enhanced spontaneous emission linewidth $\gamma_{\text{se}} = 4\hbar^2 g^2/\gamma_{\text{cav}} = 0.098$ meV $= 0.093$ nm, whereas the coupled linewidth is $\gamma' = 0.148$ meV $= 0.141$ nm at resonance. Therefore, γ' $= 0.141$ nm $> (\gamma_{\text{dot}} + \gamma_{\text{se}}) = 0.128$ nm indicating the onset of intermediate coupling; i.e., the linewidth has begun to increase with Q more rapidly than the linear Purcell enhancement in the weak coupling regime [55].

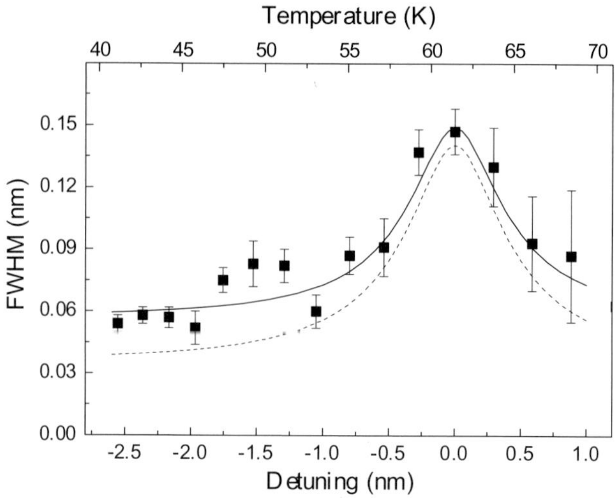

Fig. 7. FWHM linewidth of the quantum dot (black squares) as a function of the dot-cavity detuning $\delta\lambda$ compared to calculated linewidths without (dotted line) and with (solid line) convolution with the spectrometer function.

The analysis to this point has involved only three parameters: γ_{cav}, the FWHM measured uncoupled cavity photoluminescence linewidth, and the two fit parameters for the measured dot linewidth in Fig. 7, the uncoupled dot linewidth γ_{dot} and the coupling strength $\hbar g$. We now show that the fit value of $\hbar g$ is not unexpected. Above it was found that the weak-coupling quantum dot linewidth in the bad-cavity limit was given by $\gamma_{\mathrm{se}} = 4\hbar^2 g^2/\gamma_{\mathrm{cav}} = F_P \gamma_0$. The coupling energy can be written as $\hbar g = d|\,\boldsymbol{E}\,| = d\sqrt{\hbar\omega_{\mathrm{cav}}/2\varepsilon_0 n^2 V_{\mathrm{pc}}}$, where d is related to γ_0 by $\gamma_0/\hbar = 1/\tau = (4/3)(n/4\pi\varepsilon_0)(d^2\omega^3/\hbar c^3)$ and MKS units are used throughout [54]. It follows that $F_P = (3/4\pi^2)(\lambda_{\mathrm{cav}}/n)^3 Q/V_{\mathrm{pc}}$. Using a 3D finite difference time domain code, the simulated mode volume was found to be $V_{\mathrm{pc}} = 0.4(\lambda_{\mathrm{cav}}/n)^3$. This yields a maximum value of F_P of 184. But the actual F_P is proportional to $|\,\boldsymbol{E}\,|^2$ at the quantum dot, computed to be no more than 0.25 that at the peak which occurs in the middle of the center hole where no quantum dot can be. This reduces F_P to 46. Then one can use the fact that $\gamma_{\mathrm{se}} = F_P\gamma_0 = F_P\hbar/\tau = 0.098$ meV to extract the lifetime $\tau \leq 309$ ps and then the dipole moment $d = 63$ Debye. The latter is a reasonable value for a quantum dot transition, even somewhat large for an excited state transition as is the case here [56].

In conclusion, a ground state transition of a quantum dot in the field maximum of a $Q = 3000$ photonic crystal slab nanocavity should exhibit clear strong coupling. That should be achievable, but so far we have failed to resolve ground state transitions in the present quantum dot samples or to find single quantum dot transitions close enough to couple to photonic crystal modes with Q's approaching 3000.

4 Conclusions

Perhaps the most important conclusion from these studies is trivial: in order to understand the nonlinear optical properties of a semiconductor microcavity, one must understand the linear and nonlinear optical properties of the quantum well(s) or dot(s) that it contains. The optical part of the problem, i.e. propagation through the structure, is well understood. It is the nonlinear behavior of the quantum wells that is so difficult to compute because of the hierarchy problem introduced by the Coulomb interaction. For nonresonant excitation, the carrier dependent susceptibility can be used in a transfer matrix computation of the propagation, to determine the reflectivity, transmission, and true absorption at each value of the carrier density. This approach, using the measured nonlinear susceptibility, has been shown to work very well both in the weak coupling regime of optical bistability and in the nonperturbative regime of normal-mode coupling. Once enough terms in the theory are retained to explain the measured nonlinear susceptibility, then the measured and computed nonlinear properties of the microcavity also agree. This conclusion is for densities well below lasing threshold, so that no coherence builds up at the cavity peak(s) and PL into the cavity mode(s) can be neglected.

The case of resonant excitation is, of course, more difficult. Inclusion of a quantized field results in the appearance not only of interband field-carrier correlations giving rise to incoherent PL but also to new intraband field-carrier correlations that grow when a carrier density and polarization are present simultaneously. One might think that it would be difficult to see the effect of such correlations. Instead, very large effects, namely the third NMC peak and long-lived oscillations in coherent control experiments, were seen before the intraband correlations needed to explain them were known to arise naturally in a quantized-field treatment. They couple the dynamics in the normal direction, where the quantum-well/cavity coupling is nonperturbative, and in the transverse directions (guided modes), where the coupling is perturbative. This is one case where the use of a microcavity is advantageous: its frequency selectivity sometimes enhances the correlation effects by as much as one hundred over similar effects in quantum wells. Quantized-field effects in semiconductor microcavities are likely to lead to many new results in the coming years.

Another area of rapid development likely to continue is the interpretation of emission from quantum wells and microcavities. It has been long assumed that the buildup and decay of the 1s excitonic PL monitors directly the population of excitons, but recent theoretical findings concluded that it can also come from a correlated plasma. Comparison of our PL spectra with theoretical spectra for the corresponding densities has shown that indeed sometimes the 1s PL is dominated by the plasma and sometimes by excitons. This calls for a reinterpretation of many experiments including those employing a microcavity. The search for condensation of excitons and the evaluation of claims thereto are being aided by these theoretical advances and by techniques such as direct absorption by the 1s–2p exciton transition.

The nanotechnological limit of a quantum well 3D microcavity is a single quantum dot in a nanocavity of close-to-minimum volume. This is a very challenging regime for the experimentalist and a rather boring one for the theorist, the strong coupling regime of a single oscillator having been studied many years ago in atomic physics. Progress is rapid in the fabrication of photonic crystal nanocavities with ever higher Q's and antinodes accessible to quantum dots and in the growth of larger quantum dots with larger dipole moments and narrower linewidths. Once strong coupling is achieved, one can study departures from two-level-atom behavior and pursue possible technological applications.

Acknowledgments

By the time Claus Klingshirn and I attended the International Conference on Excited States and Multiresonant Nonlinear Optical Processes in Solids, March 18-20, 1981, in Aussois, France, he was already firmly established as a pre-eminent experimental semiconductor physicist. His article with Hartmut Haug on the optical properties of highly excited direct gap semiconductors is an impressively scholarly work that further showed that professors can collaborate effectively even when they keep the door locked between their sections of the building [57]. At that time I was chasing nonlinear semiconductor etalons for their practical applications to controlling light by light, doing most of my fundamental research in atomic systems. Perhaps Claus will find interesting this account of some of the fundamental semiconductor science that we have discovered studying such etalons since we met.

Support from NSF (AMOP and EPDT), AFOSR, JSOP (AFOSR and ARO), COEDIP, and a Humboldt Research Award is appreciated.

References

1. C.F. Klingshirn: *Semiconductor Optics*, 2nd corr. printing (Springer, Berlin Heidelberg New York 1997)
2. H.M. Gibbs: *Optical Bistability: Controlling Light with Light* (Academic, Orlando 1985)
3. H.M. Gibbs, S.L. McCall, T.N.C. Venkatesan: Phys. Rev. Lett. **36**, 1135 (1976)
4. F.V. Karpushko, A.S. Kireev, I.A. Morozov, G.V. Sinitsyn, N.V. Strizhenok: J. Appl. Spectrosc. USSR **26**, 204 (1977)
5. J.L. Jewell, Y.H. Lee, M. Warren, H.M. Gibbs, N. Peyghambarian, A.C. Gossard, W. Wiegmann: Appl. Phys. Lett. **46**, 918 (1985)
6. Answer by Prof. Dr. Klingshirn after his talk "Optically Non-linear and Bistable Behaviour of Direct Gap Semiconductors", AGARD meeting on Digital Optical Circuit Technology, Schliersee, Germany, 11–12 September, 1984

7. A. Passner, H.M. Gibbs, A.C. Gossard, S.L. McCall, T.N.C. Venkatesan: IEEE J. Quantum Electron. **QE-16**, 1362 (1979)
8. J.L. Jewell, Y.H. Lee, S.L. McCall, J.P. Harbison, L.T. Florez: Appl. Phys. Lett. **53**, 640 (1988)
9. J.L. Jewell, S.L. McCall, Y.H. Lee, A. Scherer, A.C. Gossard, J.H. English: Appl. Phys. Lett. **54**, 1400 (1988)
10. H. Soda, K. Iga, C. Kitahara, Y. Suematsu: Japan J. Appl. Phys. **18**, 2329 (1979)
11. C. Weisbuch, M. Nishioka, A. Ishikawa, Y. Arakawa: Phys. Rev. Lett. **69**, 3314 (1992)
12. T.B. Norris, J.-K. Rhee, D.S. Citrin, M. Nishioka, Y. Arakawa: Nuovo Cimento **17D**, 1295 (1995)
13. R. Houdré, J.L. Gibernon, P. Pellandini, R.P. Stanley, U. Oesterle, C. Weisbuch, J. O'Gorman, B. Roycroft, M. Ilegems: Phys. Rev. B **52**, 7810 (1995)
14. F. Jahnke, M. Kira, S.W. Koch, G. Khitrova, E.K. Lindmark, T.R. Nelson, Jr., D.V. Wick, J.D. Berger, O. Lyngnes, H.M. Gibbs, K. Tai: Phys. Rev. Lett. **77**, 5257 (1996)
15. G. Khitrova, H.M. Gibbs, F. Jahnke, M. Kira, S.W. Koch: Rev. Mod. Phys. **71**, 1591 (1999)
16. D.M. Whittaker, P. Kinsler, T.A. Fisher, M.S. Skolnick, A. Armitage, A.M. Afshar, M.D. Sturge, J.S. Roberts: Phys. Rev. Lett. **77**, 4792 (1996)
17. C. Ell, J. Prineas, T.R. Nelson, Jr., H.M. Gibbs, G. Khitrova, S.W. Koch, R. Houdré: Phys. Rev. Lett. **80**, 4795 (1998)
18. A.V. Kavokin: Phys. Rev. B **57**, 3757 (1998)
19. D.M. Whittaker: Phys. Rev. Lett. **80**, 4791 (1998)
20. J.J. Baumberg, A.P. Heberle, A.V. Kavokin, M.R. Vladimirova, K. Köhler: Phys. Rev. Lett. **80**, 3567 (1998)
21. A.V. Kavokin, J.J. Baumberg: Phys. Rev. B **57**, R12697 (1998)
22. J.P. Prineas, C. Ell, E.S. Lee, G. Khitrova, H.M. Gibbs, S.W. Koch: Phys. Rev. B **61**, 13863 (2000)
23. S. Pau, H. Cao, J. Jacobson, G. Björk, Y. Yamamoto, A. Imamoḡlu: Phys. Rev. A **54**, R1789 (1996)
24. H. Cao, S. Pau, J.M. Jacobson, G. Björk, Y. Yamamoto, A. Imamoḡlu: Phys. Rev. A **55**, 4632 (1997)
25. M. Kira, F. Jahnke, S.W. Koch, J.D. Berger, D.V. Wick, T.R. Nelson Jr., G. Khitrova, H.M. Gibbs: Phys. Rev. Lett. **79**, 5170 (1997)
26. M. Kira, F. Jahnke, S.W. Koch: Phys. Rev. Lett. **81**, 3263 (1998)
27. S. Chatterjee, C. Ell, S. Mosor, G. Khitrova, H.M. Gibbs, W. Hoyer, M. Kira, S.W. Koch, J.P. Prineas, H. Stolz: Phys. Rev. Lett. **92**, 067402 (2004)
28. M. Kira, W. Hoyer, T. Stroucken, S.W. Koch: Phys Rev. Lett. **87**, 176401 (2001)
29. R.A. Kaindl, M.A. Carnahan, D. Hägele, R. Lövenich, D.S. Chemla: Nature **423**, 734 (2003)
30. R. Chari, I. Galbraith, J. Phillips, S. Pellegrini, C.J. Dent, A.K. Kar, G.S. Buller, C.R. Pidgeon, D.G. Clarke, B.N. Murdin, J. Allam, A.F.G. van der Meer, G. Strasser: *Conference on Quantum Electronics and Laser Science* (Optical Society of America, Washington, 2003) postdeadline
31. P.G. Savvidis, J.J. Baumberg, R.M. Stevenson, M.S. Skolnick, D.M. Whittaker, J.S. Roberts: Phys. Rev. B **62**, R13278 (2000)

32. S. Savasta, O. Di Stefano, R. Girlanda: Phys. Rev. Lett. **90**, 096403 (2003)
33. P. Snellart, J. Bloch: Phys. Rev. Lett. **82**, 1233 (1999)
34. R.M. Stevenson, V.N. Astratov, M.S. Skolnick, D.M. Whittaker, M. Emam-Ismail, A.I. Tartakovskii, P.G. Savvidis, J.J. Baumberg, J.S. Roberts: Phys. Rev. Lett. **85**, 3680 (2000)
35. F. Boeuf, R. André, R. Romestain, Le Si Dang, E. Péronne, J.F. Lampin, D. Hulin, A. Alexandrou: Phys. Rev. B **62**, R2279 (2000)
36. R. Huang, F. Tassone, Y. Yamamoto: Phys. Rev. B **61**, R7854 (2000)
37. Le Si Dang, D. Heger, R. André, F. Bœuf, R. Romestain: Phys. Rev. Lett. **81**, 3920 (1998)
38. G. Dasbach, T. Baars, M. Bayer, A. Larionov, A. Forchel: Phys. Rev. B **62**, 13076 (2000)
39. R. Huang, Y. Yamamoto, R. André, J. Bleuse, M. Muller, H. Ulmer-Tuffigo: Phys. Rev. B **65**, 165314 (2002)
40. H. Deng, G. Weihs, C. Santori, J. Bloch, Y. Yamamoto: Science **298**, 199 (2002)
41. L.V. Butov, A.C. Gossard, D.S. Chemla: Nature **418**, 751 (2002)
42. L.V. Butov: phys. stat. sol. (b) **238**, 381 (2003)
43. O. Lyngnes, J.D. Berger, J.P. Prineas, S. Park, G. Khitrova, H.M. Gibbs, F. Jahnke, M. Kira, S.W. Koch: Solid State Commun. **104**, 297 (1997)
44. B. Grote, C. Ell, S.W. Koch, H.M. Gibbs, G. Khitrova, J.P. Prineas, J. Shah: Phys. Rev. B **64**, 45330 (2001)
45. C. Ell, P. Brick, M. Hübner, E.S. Lee, O. Lyngnes, J.P. Prineas, G. Khitrova, H.M. Gibbs, M. Kira, F. Jahnke, S.W. Koch, D.G. Deppe, D.L. Huffaker: Phys. Rev. Lett. **85**, 5392 (2000)
46. Y.-S. Lee, T.B. Norris, M. Kira, F. Jahnke, S.W. Koch, G. Khitrova, H.M. Gibbs: Phys. Rev. Lett. **83**, 5338 (1999)
47. J.M. Gérard, B. Sermage, B. Gayral, B. Legrand, E. Costard, V. Thierry-Mieg: Phys. Rev. Lett. **81**, 1110 (1998)
48. T. Gutbrod, M. Bayer, A. Forchel, J.P. Reithmaier, T.L. Reinecke, S. Rudin, P.A. Knipp: Phys. Rev. B **57**, 9950 (1998)
49. L.A. Graham, D.L. Huffaker, D.G. Deppe: Appl. Phys. Lett. **74**, 2408 (1999)
50. E.S. Lee, S. Park, P. Brick, C. Ell, Ch. Spiegelberg, H.M. Gibbs, G. Khitrova, D.G. Deppe, D.L. Huffaker: In: *Directions in Quantum Optics*, ed by H.J. Carmichael, R.J. Glauber, M.O. Scully (Springer Verlag, New York, 2001) Lecture Notes in Physics **561**, p 229
51. E.S. Lee, C. Ell, P. Brick, Ch. Spiegelberg, H.M. Gibbs, G. Khitrova, D.G. Deppe, D.L. Huffaker: J. Appl. Phys. **89**, 807-809 (2001)
52. T. Yoshie, J. Vuĉković, A. Scherer: Appl. Phys. Lett. **79**, 4289 (2001)
53. H.J. Carmichael, R.J. Brecha, M.G. Raizen, J. Kimble, P.R. Rice: Phys. Rev. A **40**, 5516 (1989)
54. A. Thränhardt, C. Ell, G. Khitrova, H.M. Gibbs: Phys. Rev. B **65**, 035327 (2002)
55. T. Yoshie, A. Scherer, G. Rupper, C. Ell, S. Chatterjee, J. Hendrickson, J. Xu, G. Khitrova, H.M. Gibbs, O.B. Shchekin, D.G. Deppe: unpublished
56. A. Thränhardt, C. Ell, G. Khitrova, H.M. Gibbs: Eur. Phys. J. B **27**, 571 (2002)
57. C. Klingshirn, H. Haug: Physics Reports **70**, 315 (1981)

All-Optical Control of Charge and Spin in GaAs: Densities and Currents

Martin J. Stevens, Ravi D.R. Bhat, Ali Najmaie, Henry M. van Driel, John E. Sipe, and Arthur L. Smirl

1 Introduction

Charge and spin are two of the fundamental properties and defining characteristics of electrons and holes in semiconductors. The manipulation of charge – the density and the current – has formed the basis of the ongoing solid-state electronic revolution of the last fifty years. Recently, it has been realized that a deeper understanding and more complete control of spin could lead to novel data processing and storage schemes. This interest has sparked the rapid growth of the field of semiconductor "spintronics", where electron spin, as opposed to charge, is used to carry information [1–4].

One challenge of spintronics is finding a way to inject and control spin densities and spin currents without perturbing the charge densities[1] or charge currents. In this chapter, we will show that quantum interference between the transition amplitude for one photon absorption of a second harmonic pulse and the transition amplitude for two photon absorption of the corresponding fundamental pulse can be used to separately or collectively control each of these parameters. Specifically, we will demonstrate that the phase of the incident light can be used to control four quantities: charge density, charge current, spin density, and spin current. More importantly, we will show how the polarization and the crystal orientation can be used to effectively decouple the spin and charge degrees of freedom, allowing independent control of each of these four quantities.

1.1 Coherent Control of Charge Density and Spin Density

The coherent optical control of electron populations in semiconductors is an important tool that can help us develop a greater understanding of the fundamental interactions – including the phase relationship – between optical fields and electrons. For example, Ahn et al. have used a two-color technique to study the relative phase of real and virtual transition amplitudes in a semiconductor [5]. The relative phases of two or more optical fields have been used to coherently control electron populations in semiconductors in a

[1] By charge density, we mean the total density of charge carriers (i.e., free electron-hole pairs), regardless of their spin.

variety of ways. Many of these schemes used one photon absorption of multiple phase-related optical pulses having a single frequency (i.e., a single color) [6–8]. Some of these one-color coherent control schemes have also used the polarization of the incident pulses to control carrier spin [6, 7]. Additionally, Gupta et al. reported using the optical Stark effect to optically manipulate the direction of electron spin, although this technique did not rely on the relative optical phase of the pulses [9].

Two-color Quantum Interference Control (QUIC) [10] techniques also have been used to control charge density in semiconductors [11, 12]. By using phase-controlled optical pulses with frequencies ω and 2ω propagating along the [111] direction in bulk GaAs, Fraser et al. observed coherent control of the charge density [11, 12]. The source of this control is quantum interference between the transition amplitudes associated with one photon absorption of 2ω and two photon absorption of ω connecting the same initial valence band and final conduction band states. However, control of the charge density was demonstrated for only one combination of 2ω and ω polarizations, and no control of the spin density was reported [11, 12].

Recently, we have demonstrated that, by using the appropriate polarizations and crystal orientations, the two-color technique of Fraser et al. [11, 12] can be extended to achieve coherent control of the *spin density* without disturbing the charge density, of the charge density without disturbing the spin density, or of both spin and charge densities simultaneously in (111)-oriented GaAs [13]. Section 3 of the present work reviews theoretical predictions for separate control of carrier charge density and spin density, and Sect. 5 details our experimental results.

1.2 Optically Injected Charge Currents and Spin Currents

The generation of spin-polarized currents in semiconductors is one of the most widely pursued topics in spintronics. A variety of methods have been demonstrated, including incoherent electrical [14–16] and Zener tunnelling [17, 18] injection from magnetic semiconductors, incoherent electrical [19, 20] and tunnelling [21] injection from ferromagnetic metals, and ballistic injection from scanning tunnelling microscopes [22, 23]. Optical injection presents an alternative method for creating spin-polarized carriers [24], and optics has played a key role in the injection and study of spin-polarized carriers in semiconductors [1–4]. A single color, circularly polarized optical excitation beam can produce a spin-polarized carrier population in a direct-gap semiconductor such as GaAs [24]. A population that is optically injected in this way is distributed *symmetrically* in k-space; therefore, even though individual carriers can have large momenta when the material is excited well above the band gap, there is no net charge current without an external field or some underlying material asymmetry. Nevertheless, optically injected carriers have been pulled by an external electric field to create a spin-polarized charge current [25, 26], and such a current has even been transferred through a barrier between two different semiconductor materials [27].

In addition to dragging optically injected carriers in an external electric field, currents have also been *directly* optically injected into semiconductors *without* applying an external bias. This has been accomplished using two-color QUIC techniques similar to those used to control charge density and spin density. In initial demonstrations, Haché et al. showed that ω and 2ω fields propagating along the [001] direction in unbiased bulk GaAs produce a ballistic charge current if ω and 2ω have parallel linear polarizations [10, 28, 29]. This current arises because ω and 2ω fields act in concert through QUIC to produce a charge distribution that is asymmetric in k-space – without changing the total charge density. By adjusting the relative phase of the ω and 2ω pulses, Haché et al. demonstrated coherent control of this charge current, but did not report a spin current [10, 28, 29]. Also, Dupont et al. used a two-color QUIC scheme to inject charge currents via intersubband transitions in a doped quantum well structure [30].

In contrast, Laman et al. [31] used one-color excitation to generate an optically injected charge current, but this technique relied on the noncentrosymmetric wurtzite structure of CdSe. Additionally, Ganichev et al. have demonstrated one-color optical injection of spin-polarized charge currents using the circular photogalvanic effect in off-axis-grown and off-axis-excited quantum wells [32, 33] and the spin-galvanic effect in quantum wells with an applied magnetic field [33, 34].

Recently, Bhat and Sipe [35] predicted that it should also be possible to separate the charge and spin degrees of freedom for currents by combining QUIC with the optical selection rules for carrier generation (i.e., with the polarization of the light and the sample symmetry). By doing so, they predict that one could create charge currents in the absence of spin currents, spin currents in the absence of charge currents, or even spin-polarized charge currents. In contrast to the unpolarized charge current that had been demonstrated earlier [28], Bhat and Sipe proposed that, by using ω and 2ω excitation fields having the same circular polarization, it should be possible to optically inject and control a *spin-polarized* ballistic charge current. The direction of this current is expected to depend on the relative phase of the ω and 2ω pulses, but its magnitude should be independent of this phase difference [35].

Bhat and Sipe also predicted that, by using ω and 2ω pulses with orthogonal linear polarizations, it should be possible to inject spin currents *without* injecting a charge current, which they referred to as a pure spin current [35]. In this case, spin-up and spin-down currents are expected to have equal magnitudes but travel in opposite directions, yielding no net charge current. Furthermore, the direction and magnitude of this pure spin current should be controlled by the polarizations and relative phases of the ω and 2ω pulses [35]. Since there is no accompanying charge current, this pure spin current is quite different from all the spin-polarized charge currents we have discussed up to this point [25–27, 32–35].

Each of the two spin currents predicted by Bhat and Sipe [35] has recently been observed [36–38]. In this chapter, we will review our experimental demonstrations and separations of all three currents in GaAs: an unpolarized charge current [36, 39], a spin-polarized charge current [36], and a pure spin current [37]. To detect these currents, we use two techniques. The first relies on a pair of electrodes to monitor net charge currents, and is discussed in Sect. 7. The second is a spatially resolved optical detection technique that can track the motion of charge and spin, and is discussed in Sect. 8. It should be noted that Hübner et al. have provided independent confirmation of the existence of pure spin currents by spatially resolving the photoluminescence from such currents in ZnSe [38].

1.3 Organization

This chapter presents a review of some of our recent work in two-color control of the density and motion of charge and spin in semiconductors. This review illustrates how we have successfully decoupled the charge and spin degrees of freedom and are able to independently control both density and motion of both charge and spin. In Sect. 2, we review the theoretical foundations of QUIC and present some simple calculations that give insight into the underlying physics. The remainder of the chapter covers two topics: (i) coherent control of charge density and spin density in Sects. 3-5, and (ii) coherent control of charge currents and spin currents in Sects. 6-8. Section 9 summarizes the work.

2 Theoretical Foundations of Quantum Interference

Here, we review in very general terms how the one- and two-photon absorption of two *phase-coherent* optical fields can be used to establish interference between different quantum mechanical pathways connecting the same initial and final states in a semiconductor. We will qualitatively discuss how this quantum interference can be applied to the coherent control of carrier charge densities and carrier spin densities, as well as ballistic charge currents and spin currents. Detailed analysis of the specific crystal orientations and excitation polarizations required for each type of control are found in Sects. 3 and 6.

Throughout this chapter, we consider a two-color optical field of the form

$$\mathbf{E}(\boldsymbol{r},t) = \mathbf{E}_{\omega} e^{-i(\omega t - \boldsymbol{K}_{\omega}\cdot\boldsymbol{r})} + \mathbf{E}_{2\omega} e^{-i(2\omega t - \boldsymbol{K}_{2\omega}\cdot\boldsymbol{r})} + \text{c.c.}, \tag{1}$$

where $\mathbf{E}_{\omega} = |E_{\omega}| e^{i\phi_{\omega}} \hat{\mathbf{e}}_{\omega}$ and $\mathbf{E}_{2\omega} = |E_{2\omega}| e^{i\phi_{2\omega}} \hat{\mathbf{e}}_{2\omega}$. The unit vectors $\hat{\mathbf{e}}_{\omega}$ and $\hat{\mathbf{e}}_{2\omega}$ denote the polarization states, $|E_{\omega}|$ and $|E_{2\omega}|$ are the slowly varying field

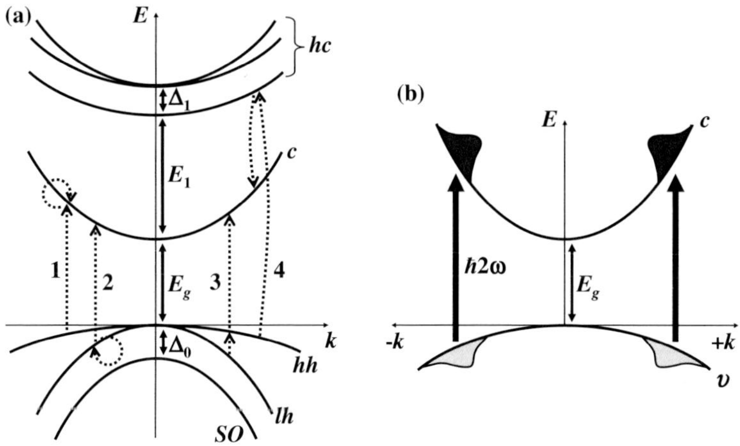

Fig. 1. (a) Schematic 14-band model of GaAs bandstructure, showing four types of two photon absorption pathway (1-4). (b) One-color excitation of GaAs with the 2ω field: dark (light) gray shading indicates electron (hole) distributions.

amplitudes, and ϕ_ω and $\phi_{2\omega}$ are the slowly varying phases. In this chapter, we take the propagation directions, $\boldsymbol{K}_\omega$ and $\boldsymbol{K}_{2\omega}$, to both point along $+\hat{\mathbf{z}}$. This field excites GaAs, whose bandstructure can be approximated by the parabolic band model sketched qualitatively in Fig. 1(a). For the discussion here, the exact values of the energy gaps are not important; to give an idea of scale, in bulk GaAs at 0 K they can be approximated as: $E_g \approx 1.52$ eV, $E_1 \approx 2.97$ eV, $\Delta_0 \approx 0.34$ eV, and $\Delta_1 \approx 0.17$ eV [40]. The excitation fields satisfy the relation $\hbar\omega < E_g < \hbar 2\omega < (E_g + \Delta_0)$ so that the ω and 2ω pulses acting independently generate carriers through two- and one-photon absorption, respectively.

Most of the calculations in this section follow an approach to the 8-band Kane model [41] that assumes parabolic, isotropic bands [24]. This approach is perturbative, with approximations taken to lowest order in $\boldsymbol{k}$ [35, 42, 43]. The bands considered are the lowest four bands (each with two degenerate spin subbands) in Fig. 1(a), including the lowest conduction band (c) and three types of valence bands: heavy hole (hh), light hole (lh), and split-off (SO). The 8-band model possesses inversion symmetry, even though GaAs does not. As a result, it predicts no coherent control of charge density or spin density. To properly capture the lack of inversion symmetry in GaAs, one can employ a more complete 14-band model, which includes the next three higher conduction (hc) bands, each of which has two spin subbands [40, 42, 44, 45].

Most of the early theory on QUIC processes conceptually separated the optical injection and control of densities and currents from the relaxation and transport of these quantities. In this section, we focus on the optical injection microscopically using Fermi's golden rule and macroscopically using symmetry considerations. We note that relaxation and transport have been approached with an effective circuit model [29], hydrodynamic equations [46, 47], a Boltzmann treatment [48], a non-equilibrium Green function formalism [49], and the semiconductor Bloch equations [38].

2.1 Fermi's Golden Rule Approach

Figure 1(b) is a simplified picture of the band structure model from Fig. 1(a), showing only one pair of valence bands (v) and the lowest pair of conduction bands (c). To determine the distribution of charge and spin resulting from optical excitation, one can calculate the transition rate of electrons from an initial state in the valence band, $\psi_{v\boldsymbol{k}}$, to a final state in the conduction band, $\psi_{c\boldsymbol{k}}$. Here a contracted notation is used, with the subscript v (c) representing the band and spin indices of the initial valence band (final conduction band) state, and $\boldsymbol{k}$ is the crystal wave vector. The initial state can be in either the lh or hh band, and is energetically separated from the final state in the lowest conduction band by $\hbar\omega_{cv}(\boldsymbol{k})$.

To find the electron contribution to the time derivative of the expectation value of an arbitrary single particle operator $\hat{\theta}$, one can perform a Fermi's golden rule calculation, giving [10, 35, 46, 50]

$$\frac{d}{dt}\left\langle\hat{\theta}\right\rangle = \frac{2\pi}{L^3}\sum_{c,c',v,\boldsymbol{k}} a_{c'v}(\boldsymbol{k})^* a_{cv}(\boldsymbol{k})\left\langle\hat{\theta}(\boldsymbol{k})\right\rangle \delta\left(2\omega - \omega_{cv}(\boldsymbol{k})\right), \tag{2}$$

where

$$\left\langle\hat{\theta}(\boldsymbol{k})\right\rangle = \left\langle\psi_{c\boldsymbol{k}}\right|\hat{\theta}\left|\psi_{c'\boldsymbol{k}}\right\rangle, \tag{3}$$

$a_{cv}(\boldsymbol{k})$ is the transition amplitude, and L^3 is the normalization volume. Here the indices c and c' denote two degenerate states in the conduction band that can have different spin indices. The sum over c, c', and v in Eq. (2) indicates the sum over all states and all spin subbands in each of the bands. One can also write an expression for the *hole* contribution that is similar to Eq. (2) [35, 50].

For two-color excitation, the total transition amplitude in Eq. (2) is the sum of transition amplitudes for one- and two-photon absorption,

$$a_{cv}(\boldsymbol{k}) = a_{cv}^{2\omega}(\boldsymbol{k}) + a_{cv}^{\omega}(\boldsymbol{k}). \tag{4}$$

The one-photon transition amplitude is

$$a_{cv}^{2\omega}(\boldsymbol{k}) \propto \left\langle\psi_{c\boldsymbol{k}}\right| H^{(1)}\left|\psi_{v\boldsymbol{k}}\right\rangle. \tag{5}$$

In the usual semiclassical treatment under the dipole approximation, the first order perturbation Hamiltonian in the Coulomb gauge ($\boldsymbol{\nabla}\cdot\boldsymbol{A} = 0$) is [51]:

$$H^{(1)} = -\frac{e}{m}\boldsymbol{A}_{2\omega}\cdot\boldsymbol{p}. \tag{6}$$

In the Coulomb gauge and in the absence of charge sources, the vector potential $\boldsymbol{A}$ is parallel to the electric field vector $\boldsymbol{E}$, allowing us to write the transition amplitude for one-photon absorption as

$$a^{2\omega}_{cv}(\boldsymbol{k}) \propto \langle \psi_{c\boldsymbol{k}} | \, \boldsymbol{E}_{2\omega} \cdot \boldsymbol{p} \, | \psi_{v\boldsymbol{k}} \rangle = |E_{2\omega}| \, e^{i\phi_{2\omega}} \, [\hat{\mathbf{e}}_{2\omega} \cdot \boldsymbol{p}_{cv}(\boldsymbol{k})] \, , \tag{7}$$

where the momentum matrix element between the valence and conduction bands is

$$\boldsymbol{p}_{cv}(\boldsymbol{k}) = \langle \psi_{c\boldsymbol{k}} | \, \boldsymbol{p} \, | \psi_{v\boldsymbol{k}} \rangle \, . \tag{8}$$

Similarly, the two-photon transition amplitude is [51]

$$a^{\omega}_{cv}(\boldsymbol{k}) \propto |E_{\omega}|^2 \, e^{i2\phi_{\omega}} \sum_m \frac{[\hat{\mathbf{e}}_{\omega} \cdot \boldsymbol{p}_{cm}(\boldsymbol{k})] \, [\hat{\mathbf{e}}_{\omega} \cdot \boldsymbol{p}_{mv}(\boldsymbol{k})]}{\hbar\omega - \hbar\omega_{mv}(\boldsymbol{k})} , \tag{9}$$

where m denotes an intermediate state $\psi_{m\boldsymbol{k}}$ that is energetically separated from the initial state by $\hbar\omega_{mv}(\boldsymbol{k})$. There are many possible intermediate states that must be summed over, several of which are indicated in Fig. 1(a). The transition amplitudes in Eqs. (7) and (9) are proportional to the complex fields, including their phases, and contain dot products of the field polarization vectors and the momentum matrix elements. Consequently, $a^{2\omega}_{cv}(\boldsymbol{k})$ and $a^{\omega}_{cv}(\boldsymbol{k})$ depend on field polarizations and crystal symmetry.

In a crystal with inversion symmetry, or within the isotropic 8-band model, the bands can be chosen so that $\boldsymbol{p}_{nm}(-\boldsymbol{k}) = -\boldsymbol{p}_{nm}(\boldsymbol{k})$, for arbitrary band labels n and m [52]. This implies from Eq. (7) that the one photon amplitude is an odd function in $\boldsymbol{k}$,

$$a^{2\omega}_{cv}(-\boldsymbol{k}) = -a^{2\omega}_{cv}(\boldsymbol{k}) \, . \tag{10}$$

Additionally, since $\omega_{nm}(-\boldsymbol{k}) = \omega_{nm}(\boldsymbol{k})$ and $\boldsymbol{p}_{nm}(-\boldsymbol{k}) = -\boldsymbol{p}_{nm}(\boldsymbol{k})$, one can see from Eq. (9) that the two photon amplitude is an even function in $\boldsymbol{k}$,

$$a^{\omega}_{cv}(\ \boldsymbol{k}) = a^{\omega}_{cv}(\boldsymbol{k}) \, . \tag{11}$$

Charge Density

Many physical quantities of interest can be written in terms of these transition amplitudes using Eq. (2). For example, substituting the expectation value of the carrier charge density operator

$$\langle \hat{n}(\boldsymbol{k}) \rangle = \langle \psi_{c\boldsymbol{k}} | \, \psi_{c'\boldsymbol{k}} \rangle = \delta_{cc'} , \tag{12}$$

into Eq. (2) leads to the charge density injection rate [50]

$$\dot{n} \propto \sum_{c,v,\boldsymbol{k}} | \, a_{cv}(\boldsymbol{k}) \, |^2 , \tag{13}$$

where $\delta\left(2\omega - \omega_{cv}\left(\boldsymbol{k}\right)\right)$ is implied in the remainder of Sect. 2. We use the term "charge density" to designate the total density of charge carriers, regardless of carrier spin. Equations (4) and (13) yield

$$\dot{n} \propto \sum_{c,v,\boldsymbol{k}} \left\{ \left| a_{cv}^{2\omega}\left(\boldsymbol{k}\right) \right|^2 + \left| a_{cv}^{\omega}\left(\boldsymbol{k}\right) \right|^2 + a_{cv}^{2\omega}\left(\boldsymbol{k}\right)^* a_{cv}^{\omega}\left(\boldsymbol{k}\right) + a_{cv}^{2\omega}\left(\boldsymbol{k}\right) a_{cv}^{\omega}\left(\boldsymbol{k}\right)^* \right\}. \tag{14}$$

The first and second terms in Eq. (14) are the injection rates for one- and two-photon absorption acting alone, $\dot{n}_{2\omega}$ and $\dot{n}_{\omega}$, respectively. The total charge density injection rate can thus be written as the sum of three terms,

$$\dot{n} = \dot{n}_{2\omega} + \dot{n}_{\omega} + \dot{n}_{I}, \tag{15}$$

where the last two terms in Eq. (14) represent the QUIC contribution to charge density:

$$\dot{n}_{I} \propto \sum_{c,v,\boldsymbol{k}} a_{cv}^{2\omega}\left(\boldsymbol{k}\right)^* a_{cv}^{\omega}\left(\boldsymbol{k}\right) + \text{c.c.} \tag{16}$$

In terms of the fields and momentum matrix elements in Eqs. (7) and (9), the transition amplitude cross term in Eq. (16) is

$$a_{cv}^{2\omega}\left(\boldsymbol{k}\right)^* a_{cv}^{\omega}\left(\boldsymbol{k}\right) \propto \left|E_{2\omega}\right| \left|E_{\omega}\right|^2 e^{i\Delta\phi} \times \left[\hat{\mathbf{e}}_{2\omega} \cdot \boldsymbol{p}_{cv}\left(\boldsymbol{k}\right)\right]^* \sum_{m} \frac{\left[\hat{\mathbf{e}}_{\omega} \cdot \boldsymbol{p}_{cm}\left(\boldsymbol{k}\right)\right]\left[\hat{\mathbf{e}}_{\omega} \cdot \boldsymbol{p}_{mv}\left(\boldsymbol{k}\right)\right]}{\hbar\omega - \hbar\omega_{cm}\left(\boldsymbol{k}\right)} \tag{17}$$

where $\Delta\phi \equiv 2\phi_{\omega} - \phi_{2\omega}$. The explicit dependence on the phase difference between ω and 2ω allows for coherent control. The key to understanding QUIC is recognizing that, because the interaction of light with matter is a coherent process, the total transition amplitude when both ω and 2ω fields are present is found by taking the *sum* of the individual transition amplitudes for ω and 2ω in Eq. (4) *before* multiplying by the complex conjugate, resulting in the cross term in Eq. (17). Through the dot products of the field vectors and momentum matrix elements, $\dot{n}_{I}$ also depends on the polarizations of ω and 2ω and the crystal symmetry and orientation.

In the presence of inversion symmetry, Eqs. (10) and (11) predict that $\left|a_{cv}^{2\omega}\left(\boldsymbol{k}\right)\right|^2$ and $\left|a_{cv}^{\omega}\left(\boldsymbol{k}\right)\right|^2$ are both even functions of $\boldsymbol{k}$. For the QUIC cross term in Eq. (17), by contrast, $a_{cv}^{2\omega}\left(\boldsymbol{k}\right)^* a_{cv}^{\omega}\left(\boldsymbol{k}\right)$ is an *odd* function of $\boldsymbol{k}$. The sum over all $\boldsymbol{k}$ of an odd function is zero; thus, in the presence of inversion symmetry, $\dot{n}_{I} = 0$ and the charge density cannot be controlled.

The lack of inversion symmetry in GaAs can be taken into account with transitions that involve the higher conduction bands as intermediate states (pathway 4 in Fig. 1(a)) [53]. The momentum matrix elements between the valence (v), conduction (c), and higher conduction (h) bands can be found in Tables I and III of [42]. To lowest order in $\boldsymbol{k}$, the elements $\boldsymbol{p}_{cv}\left(\boldsymbol{k}\right)$, $\boldsymbol{p}_{ch}\left(\boldsymbol{k}\right)$, and $\boldsymbol{p}_{hv}\left(\boldsymbol{k}\right)$ are independent of $k = |\boldsymbol{k}|$, whereas $\boldsymbol{p}_{cc}\left(\boldsymbol{k}\right)$, $\boldsymbol{p}_{vv}\left(\boldsymbol{k}\right)$, and $\boldsymbol{p}_{vv'}\left(\boldsymbol{k}\right)$

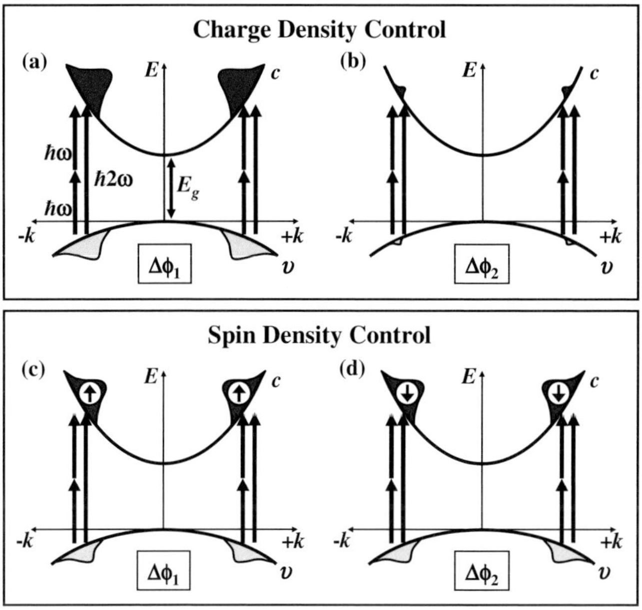

Fig. 2. Schematic of two-color excitation and QUIC of (a, b) charge density only and (c, d) spin density only. Each type of control is shown for two values of $\Delta\phi$.

are proportional to k. Consequently, $a_{cv}^{2\omega}(\boldsymbol{k})^* a_{cv}^{\omega}(\boldsymbol{k})$ is proportional to k for two-photon pathways 1-3, but independent of k for pathway 4. Thus, for a pathway 4 transition, $a_{cv}^{2\omega}(\boldsymbol{k})^* a_{cv}^{\omega}(\boldsymbol{k})$ can be approximated by its value at the Γpoint ($\boldsymbol{k}=0$), which is proportional to

$$\frac{|E_{2\omega}|\,|E_\omega|^2\,e^{i\Delta\phi}}{\hbar\omega-\hbar\omega_{hv}}\left[\hat{\mathbf{e}}_{2\omega}^*\cdot\boldsymbol{p}_{vc}(\Gamma)\right]\left[\hat{\mathbf{e}}_\omega\cdot\boldsymbol{p}_{ch}(\Gamma)\right]\left[\hat{\mathbf{e}}_\omega\cdot\boldsymbol{p}_{hv}(\Gamma)\right]. \qquad (18)$$

It is independent of $\boldsymbol{k}$, and therefore even in $\boldsymbol{k}$. For the appropriate choice of ω and 2ω polarizations, none of the three terms in square brackets is zero, and there is no cancellation in the sum over c, v, and h. In this case, charge density *can* be coherently controlled; we often refer to this as charge control. Figure 2(a)-(b) shows the total excited state population for this case, where $a_{cv}^{2\omega}(\boldsymbol{k})^* a_{cv}^{\omega}(\boldsymbol{k})$ is even in $\boldsymbol{k}$ and depends on $\Delta\phi$. For a specific phase difference $\Delta\phi_1$, $\dot{n}_I$ is positive, causing the total carrier density n to *increase* in both directions in k-space, as shown in Fig. 2(a). For another specific phase difference $\Delta\phi_2$, $\dot{n}_I$ is negative, and carrier density *decreases* in both directions in k-space, as shown in Fig. 2(b).

Pathways 1 and 2 of Fig. 1(a) are two-band transitions, where the intermediate state is the same as either the final state (1) or the initial state (2). Since these are the most likely pathways in GaAs when exciting above the band edge [54], yet they do not contribute to charge control, $\dot{n}_I$ is small compared to $\dot{n}_{2\omega}+\dot{n}_\omega$.

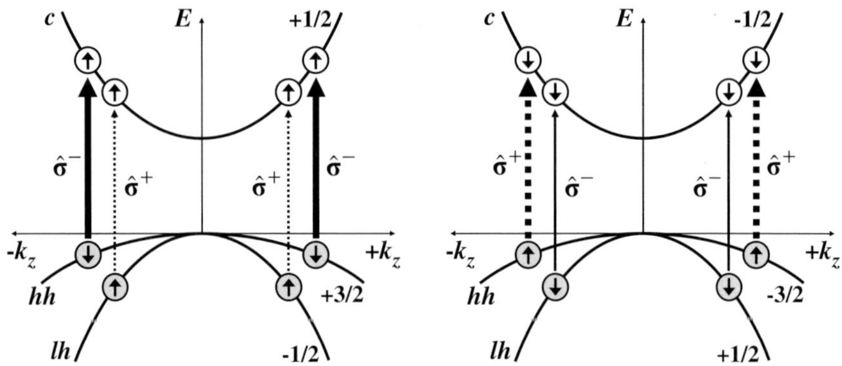

Fig. 3. Allowed one photon absorption transitions for $\hat{\sigma}^-$ ($\hat{\sigma}^+$) excitation in GaAs shown as solid (dashed) arrows. Bands are labelled by name (lh, hh, c) and projection of angular momentum along the z-axis, m_j ($\pm 1/2$, $\pm 3/2$). The arrows inside the white (gray) circles represent the spin of the optically injected electrons (holes). The thickness of each transition arrow indicates the relative strength of that transition.

Spin Density

Likewise, one can find the total net spin of the injected carriers – which we refer to as the spin density – using Fermi's golden rule [50]:

$$\dot{\boldsymbol{S}} \propto \sum_{c,c',v,\boldsymbol{k}} \langle \boldsymbol{S}(\boldsymbol{k}) \rangle_{cc'} \left\{ a_{c'v}^{2\omega}(\boldsymbol{k})^* a_{cv}^{2\omega}(\boldsymbol{k}) + a_{c'v}^{\omega}(\boldsymbol{k})^* a_{cv}^{\omega}(\boldsymbol{k}) + \left[a_{c'v}^{2\omega}(\boldsymbol{k})^* a_{cv}^{\omega}(\boldsymbol{k}) + \text{c.c.} \right] \right\} \tag{19}$$

where $\langle \boldsymbol{S}(\boldsymbol{k}) \rangle_{cc'}$ is the spin matrix element between c and c' at $\boldsymbol{k}$, following Eq. (3). The first two terms in Eq. (19) are the spin injection rates for one- and two-photon absorption acting alone, $\dot{\boldsymbol{S}}_{2\omega}$ and $\dot{\boldsymbol{S}}_{\omega}$. The total spin density injection rate can thus be written as the sum of three terms,

$$\dot{\boldsymbol{S}} = \dot{\boldsymbol{S}}_{2\omega} + \dot{\boldsymbol{S}}_{\omega} + \dot{\boldsymbol{S}}_{I} . \tag{20}$$

For a left-hand ($\hat{\sigma}^- = (\hat{\mathbf{x}} - i\hat{\mathbf{y}})/\sqrt{2}$) circularly polarized 2ω (ω) field acting alone, $\dot{\boldsymbol{S}}_{2\omega}$ ($\dot{\boldsymbol{S}}_{\omega}$) is along $+\hat{\mathbf{z}}$, which we take to be "spin-up". For right-hand ($\hat{\sigma}^+ = (\hat{\mathbf{x}} - i\hat{\mathbf{y}})/\sqrt{2}$) circular excitation, by contrast, $\dot{\boldsymbol{S}}_{2\omega}$ and $\dot{\boldsymbol{S}}_{\omega}$ point along $-\hat{\mathbf{z}}$ ("spin-down"). The results for 2ω follow the usual circularly polarized optical selection rules [24], which are illustrated in Fig. 3. The results for ω can be found from a Kane model calculation [35, 55–57]. For a circularly polarized 2ω or ω, the spin injected through one-color excitation is distributed symmetrically in k-space.

The last two terms in Eq. (19) represent the interference, or spin density control, terms:

$$\dot{\boldsymbol{S}}_I \propto \sum_{c,c',v,\boldsymbol{k}} \langle \boldsymbol{S}(\boldsymbol{k}) \rangle_{cc'} \left[a_{c'v}^{2\omega}(\boldsymbol{k})^* a_{cv}^{\omega}(\boldsymbol{k}) + \text{c.c.} \right]. \tag{21}$$

As with charge density control, there is no spin density control in the presence of inversion symmetry. In the 14-band model, however, terms similar to Eq. (18) appear that can be non-zero for certain polarizations. Due to the presence of the spin matrix element in Eq. (21) and the sum over c', there are cases where there is a cancellation in the charge control ($\dot{n}_I = 0$) but not the spin control ($\dot{S}_I^i \neq 0$) and vice versa. Figure 2(c)-(d) illustrates such a spin density control signal when $\dot{n}_I = 0$. For a specific phase difference $\Delta\phi_1$, QUIC results in a net carrier spin pointing up for both directions in k-space, as Fig. 2(c) shows. For another specific phase difference $\Delta\phi_2$, QUIC results in a net carrier spin pointing down, as in Fig. 2(d).

Charge Current

The electron charge current density injection rate is obtained by using Eq. (2) with the current operator $\boldsymbol{J} = q\hat{\mathbf{v}}$, where we take the velocity operator as $\hat{\mathbf{v}} = \boldsymbol{p}/m$ and q and m are the electron charge and effective mass:[2]

$$\dot{\boldsymbol{J}} \propto q \sum_{c,v,\boldsymbol{k}} \boldsymbol{p}_{cc}(\boldsymbol{k}) \left\{ \left| a_{cv}^{2\omega}(\boldsymbol{k}) \right|^2 + \left| a_{cv}^{\omega}(\boldsymbol{k}) \right|^2 + \left[a_{cv}^{2\omega}(\boldsymbol{k})^* a_{cv}^{\omega}(\boldsymbol{k}) + \text{c.c.} \right] \right\} \tag{22}$$

In the 8-band approximation, because $\left| a_{cv}^{2\omega}(\boldsymbol{k}) \right|^2$ and $\left| a_{cv}^{\omega}(\boldsymbol{k}) \right|^2$ are both even in $\boldsymbol{k}$ as discussed above, the charge currents due to 2ω only and ω only, represented by the first two terms in Eq. (22), are zero. On the other hand, $\boldsymbol{p}_{cc}(\boldsymbol{k}) \propto \boldsymbol{k}$ and $a_{cv}^{2\omega}(\boldsymbol{k})^* a_{cv}^{\omega}(\boldsymbol{k})$ are both odd functions of $\boldsymbol{k}$, so their product is even in $\boldsymbol{k}$. As an example, the current density injection rate for pathway 1 of Fig. 1(a) to lowest order in $\boldsymbol{k}$ is

$$\dot{\boldsymbol{J}} \propto q e^{i\Delta\phi} \sum_{c,v,\boldsymbol{k}} \boldsymbol{k} \left[\hat{\mathrm{e}}_{2\omega}^* \cdot \boldsymbol{p}_{vc}\left(\hat{\boldsymbol{k}}\right) \right] \left[\hat{\mathrm{e}}_{\omega} \cdot \boldsymbol{p}_{cv}\left(\hat{\boldsymbol{k}}\right) \right] (\boldsymbol{k} \cdot \hat{\mathrm{e}}_{\omega}) + \text{c.c.}, \tag{23}$$

where $\hat{\boldsymbol{k}}$ is a unit vector along $\boldsymbol{k}$. This, and the contributions from the other types of two photon transitions, lead to a coherently controllable net charge current.

Spin Current

Finally, to describe spin currents, Bhat and Sipe define the spin current pseudotensor K^{ij} as the expectation value of the velocity operator times the spin operator [35]:

[2] To find the total charge current density injection rate, including both electron and hole contributions, $\boldsymbol{p}_{cc}(\boldsymbol{k})$ can be replaced by $\boldsymbol{p}_{cc}(\boldsymbol{k}) - \boldsymbol{p}_{vv}(\boldsymbol{k})$ in Eq. (22).

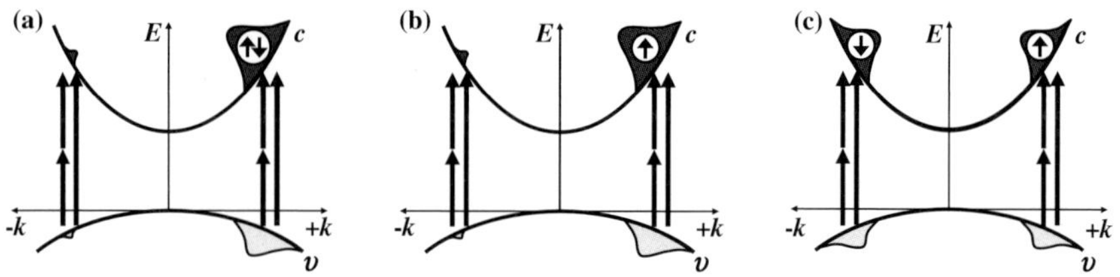

Fig. 4. Schematics of asymmetric charge and spin distributions resulting in three types of QUIC current: (a) unpolarized charge current; (b) spin-polarized charge current; (c) pure spin current.

$$K^{ij} = \left\langle \hat{\mathrm{v}}^i \hat{S}^j \right\rangle, \tag{24}$$

where the superscripts i and j represent Cartesian coordinates. In this notation, a spin current K^{ij} consists of electrons travelling in the i-direction, with average electron spin pointing along the j-direction. At the same time, holes travel in the -i-direction, with average hole spin pointing along the -j-direction [35]. For two-color excitation, the electron contribution to the spin current injection rate is [35, 50]

$$\dot{K}^{ij} \propto \sum_{c,c',v,\boldsymbol{k}} \langle \, \mathrm{v}^i S^j \left(\boldsymbol{k}\right) \, \rangle_{cc'} \, a_{c'v} \left(\boldsymbol{k}\right)^* a_{cv} \left(\boldsymbol{k}\right). \tag{25}$$

where $\langle \, \mathrm{v}^i S^j (\boldsymbol{k}) \, \rangle_{cc'}$ is the matrix element between c and c' of the product of v^i and S^j. In the 8-band parabolic approximation, for 2ω or ω acting alone, spins are distributed symmetrically in k-space, and the net spin current is zero.[3] The spin current also contains interference terms,

$$\dot{K}^{ij} \propto \sum_{c,c',v,\boldsymbol{k}} \langle \, \mathrm{v}^i S^j \left(\boldsymbol{k}\right) \, \rangle_{cc'} \left[a^{2\omega}_{c'v} \left(\boldsymbol{k}\right)^* a^{\omega}_{cv} \left(\boldsymbol{k}\right) + \mathrm{c.c.} \right], \tag{26}$$

that depend on the field polarizations and phases and the crystal symmetry and orientation.

Section 6 details how the polarizations of ω and 2ω determine whether a charge or spin current is created, and how the phases of ω and 2ω can be used to coherently control each type of current. Figure 4 sketches carrier distributions responsible for various combinations of charge and spin currents. Figure 4(a) shows a distribution that will lead to a charge current without a spin current: charge is distributed asymmetrically in k-space, but the net spin everywhere is zero. As we will discuss in Sect. 6, this occurs when ω and 2ω have parallel linear polarizations. In Fig. 4(b), both $\dot{\boldsymbol{J}}$ and $\dot{K}^{ij}$ are non-zero,

[3] Note, however, that if the inversion asymmetry of GaAs is taken into account, one-color excitation can produce charge currents and spin currents in GaAs and other semiconductors [45], as demonstrated experimentally by Laman et al. in ZnSe [31] and Ganichev et al. in off-axis-excited quantum wells [32, 33].

resulting in a spin-polarized charge current. In this case, *both* charge *and* spin are distributed asymmetrically in k-space. This occurs when ω and 2ω have the same circular polarization. Finally, Fig. 4(c) illustrates a scenario where $\dot{\boldsymbol{J}}$ is zero, but $\dot{K}^{ij}$ is non-zero: spin is distributed asymmetrically, even though the charge distribution is symmetric. This leads to a spin current *without* a net charge current; this *pure* spin current can be generated when ω and 2ω have orthogonal linear polarizations.

2.2 Symmetry Considerations

In addition to the Fermi's golden rule approach, one can also write phenomenological expressions for each control process in terms of electric field components and the appropriate crystal tensor or pseudotensor. These tensors must be invariant under all operations forming the point group of the crystal [52]. This invariance offers insight directly from the symmetry requirements. Namely, from the point group operations of the crystal, we can deduce the non-zero and independent tensor components, and determine the crystal orientations and excitation polarizations necessary for each type of control.

The charge density control signal can then be written [11–13]

$$\dot{n}_I = \sum_{ijk} \xi_I^{ijk} E_\omega^{i*} E_\omega^{j*} E_{2\omega}^k + \text{c.c.}, \tag{27}$$

where i, j, and k are dummy indices representing Cartesian components of the fields and ξ_I is a third-rank tensor that is non-zero only in crystals that lack inversion symmetry [11–13]. For a crystal with point group symmetry T_d, such as GaAs, it has six non-zero elements but only one independent component,

$$\xi_I^{abc} = \xi_I^{acb} = \xi_I^{bac} = \xi_I^{bca} = \xi_I^{cab} = \xi_I^{cba}, \tag{28}$$

where a, b, and c denote components along the cubic axes [100], [010], and [001], respectively [11–13]. Similarly, the phase-dependent injection rate for net spin aligned along the i-direction can be written [13]

$$\dot{S}_I^i = \sum_{jkl} \zeta_I^{ijkl} E_\omega^{j*} E_\omega^{k*} E_{2\omega}^l + \text{c.c.}, \tag{29}$$

where the fourth-rank pseudotensor ζ_I is also non-zero only in crystals that lack inversion symmetry. In GaAs, it has 18 non-zero elements and two independent components: $\zeta_{IA} \equiv \text{Im}\zeta_I^{abba}$ and $\zeta_{IB} \equiv \text{Im}\zeta_I^{abab}$ [13].

Therefore, when ω and 2ω are simultaneously present, the injection rates for both charge density and spin density can contain interference terms that depend on the field phases and polarizations and the crystal symmetry and orientation. Because the spin density control tensor has a higher rank than the charge density control tensor, these two processes can have different symmetries, raising the prospect of *independent* control of charge density and spin density. How these processes depend on field polarizations and phases, as well as crystal symmetry and orientation – and how they can be independently controlled – is the subject of Sect. 3.

One can also follow symmetry arguments to find the tensor expressions for charge current density injection rate along the i-direction [28, 29, 46, 50]

$$\dot{J}_I^i = \sum_{jkl} \eta_I^{ijkl} E_\omega^{j*} E_\omega^{k*} E_{2\omega}^l + c.c., \tag{30}$$

and for spin current injection rate [50],

$$\dot{K}_I^{ij} = \sum_{klm} \mu_I^{ijklm} E_\omega^{k*} E_\omega^{l*} E_{2\omega}^m + c.c. \tag{31}$$

Clearly, both $\dot{J}_I^i$ and $\dot{K}_I^{ij}$ depend on the phases and polarizations of the ω and 2ω fields. Information about the crystal symmetry and orientation are contained in the fourth-rank tensor η_I and the fifth-rank pseudotensor μ_I. Since η_I and μ_I can have different symmetries, charge currents and spin currents can be *independently* controlled. Moreover, ξ_I, ζ_I, η_I, and μ_I all have different symmetries, allowing independent control of charge density, spin density, charge currents, and spin currents.

Once the non-zero and independent tensor components are found using the symmetries of the system, one can use an appropriate microscopic model for the electronic states of the system to determine the value of these tensor components. Both electron and hole contributions can be included in such a microscopic treatment.

2.3 Summary

In this section, we have described qualitatively how quantum interference can be established between multiple excitation pathways linking an initial state in the valence band with a final state in the conduction band. By exciting this system with phase-locked, harmonically related excitation pulses with frequencies ω and 2ω, we should be able to control the amplitude and phase of the transition rate as a function of $\boldsymbol{k}$, thus controlling the form of the final excited state population.

In a Fermi's golden rule formalism, when ω and 2ω simultaneously excite the sample, QUIC arises because we must take the *sum* of the transition amplitudes for ω and 2ω *before* multiplying by the complex conjugate. This results in cross terms that contain the product of $a_{cv}^{\omega}(\boldsymbol{k})$ and $a_{cv}^{2\omega}(\boldsymbol{k})$. These cross terms can contain pieces that are even functions of $\boldsymbol{k}$ and pieces that are odd functions of $\boldsymbol{k}$. The even pieces can lead to control of the charge density, and the odd pieces can lead to a charge current. Similar criteria apply to control of spin density and spin currents. While the Fermi's golden rule approach in Sect. 2.1 focused on the electron contributions to each of these processes, the *hole* contributions have also been calculated, with qualitatively similar results [10, 35, 46]. In Sect. 3, we will review the crystal orientations and polarizations for which QUIC can be used to coherently control the charge density and spin density. Section 6 details the specific polarizations of ω and 2ω necessary to generate charge currents and spin currents.

3 Symmetry Predictions: Charge Densities and Spin Densities

In this section, we discuss all-optical injection and coherent control of both the *charge density* and the *spin density.* We review a macroscopic symmetry analysis for crystals with zincblende symmetry (such as GaAs) [13], which predicts that QUIC can be used to *independently control charge and spin densities* in GaAs by systematically varying the phases and polarizations of the two incident pulses and by rotating the sample.

3.1 Phenomenological Framework

As discussed in Sect. 2, when ω and 2ω excite a direct gap semiconductor and satisfy the relation $\hbar\omega < E_g < \hbar 2\omega$, a charge density ($n$) can be injected into this semiconductor through one-photon absorption of the 2ω field ($n_{2\omega}$) or through two photon absorption of the ω field (n_ω). In addition, carriers can be generated through QUIC (n_I). The total charge density injection rate can be written [11–13]:

$$\dot{n} = \dot{n}_{2\omega} + \dot{n}_\omega + \dot{n}_I . \tag{32}$$

Here $\dot{n}_{2\omega} = \sum\limits_{ij} \xi_1^{ij} E_{2\omega}^{i*} E_{2\omega}^{j}$ and $\dot{n}_\omega = \sum\limits_{ijkl} \xi_2^{ijkl} E_\omega^{i*} E_\omega^{j*} E_\omega^{k} E_\omega^{l}$, where i, j, k, and l are dummy indices representing Cartesian components of the fields corresponding to the crystallographic principle axes. These are variations on the well-known forms of $n_{2\omega}$ and n_ω from linear and nonlinear optics, where the tensor ξ_1 is related to $\chi^{(1)}$ and describes one photon absorption, and ξ_2 is related to $\chi^{(3)}$ and describes two photon absorption [11].

The interference component of the total charge density injection rate $\dot{n}_I$ – which we refer to as "charge control" – is given by Eq. (27), where the ξ_I tensor is related to the imaginary part of $\chi^{(2)}$ [11]. According to Eqs. (27) and (28), $\dot{n}_I$ is only non-zero if the ω field has components along at least two of the principle axes, and the 2ω field has a component along the third principle axis. Hence, it is clear why there can be no charge control for normal incidence excitation of (001)-oriented GaAs, which we use for measurements of currents in Sections 7 and 8. In this case, the ω and 2ω fields are constrained to lie in the (001)-plane, so they have components only along a and b. Since neither field can have a component along the [001] direction (the c axis), $\dot{n}_I$ must be zero.

Spin can also be optically injected into a semiconductor, since absorption of circularly polarized light creates a spin-polarized carrier population, as discussed in Sect. 2.1. Spin can be injected through one photon absorption of the 2ω pulse ($S_{2\omega}$), through two photon absorption of the ω pulse (S_ω), or through quantum interference between these two transition amplitudes (S_I). The total injection rate of spins aligned along the i direction is [13]:

$$\dot{S}^i = \dot{S}^i_{2\omega} + \dot{S}^i_\omega + \dot{S}^i_I . \tag{33}$$

In order to relate the spin density to the incident optical polarizations, while taking into account the material symmetry, we have introduced the pseudotensors ζ_1, ζ_2, and ζ_I. The components of spin injected through one- and two-photon absorption alone are $\dot{S}^i_{2\omega} = \sum_{jk} \zeta_1^{ijk} E^{j*}_{2\omega} E^k_{2\omega}$ and $\dot{S}^i_\omega = \sum_{jklm} \zeta_2^{ijklm} E^{j*}_\omega E^{k*}_\omega E^l_\omega E^m_\omega$, respectively [13]. For linear polarizations, the carriers created through one- or two-photon absorption will have no net spin-polarization along the z-direction ($\dot{S}^z_{2\omega} = \dot{S}^z_\omega = 0$) [24]. Conversely, when either pulse is circularly polarized and acting alone, $\dot{S}^z_{2\omega}$ and $\dot{S}^z_\omega$ are positive for $\hat{\sigma}^-$ polarized 2ω and ω fields, whereas $\dot{S}^z_{2\omega}$ and $\dot{S}^z_\omega$ are negative for $\hat{\sigma}^+$ polarized 2ω and ω fields [24, 35].

The interference components to spin injection $\dot{S}^i_I$ – the "spin control" terms – are given by Eq. (29). For ω and 2ω at normal incidence on bulk (001)-oriented GaAs, S_I is restricted to lie in the (001) plane [35]; for (111)-GaAs, S_I can be oriented along the propagation direction, z [13]. Since the measurement techniques described in this chapter are not sensitive to spins in the transverse (x-y) plane, only spins with components along the $\pm z$-axis are considered here.

3.2 Predictions for (111)-GaAs

To illustrate independent control of charge density and spin density, we consider 2ω and ω pulses that travel along the [111]-direction of a bulk GaAs sample. The pulses are normally incident on the (111)-oriented sample, and the sample is mounted so that it can be rotated about the propagation direction (the z-axis). For a fixed linearly or circularly polarized excitation field, the charge density injected by 2ω or ω acting independently does not depend on rotations of the sample about the z-axis [53]. The same is true for the spin density injected by 2ω or ω acting alone.

In contrast, we expect the charge control and spin control signals to depend strongly on sample angle. For the general case of ω and 2ω both linearly polarized and lying in the (111) plane, at angles α and β with respect to the $[2\bar{1}\bar{1}]$ direction, respectively, as shown in Fig. 5, Eq. (27) becomes [13]

$$\dot{n}_I = \frac{2\sqrt{2}}{\sqrt{3}} \xi_I^{abc} \left|E_{2\omega}\right| \left|E_\omega\right|^2 \cos\left(2\alpha + \beta\right) \cos\left(\Delta\phi\right) . \tag{34}$$

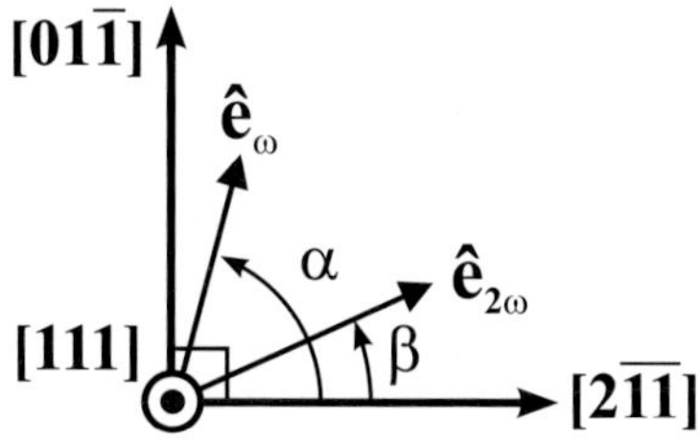

Fig. 5. Definitions of α and β. The ω and 2ω fields propagate along [111], and their polarization vectors $\hat{\mathbf{e}}_\omega$ and $\hat{\mathbf{e}}_{2\omega}$ lie in the (111) plane.

For these same polarizations, the z-component of the spin control from Eq. (29) is [13]

$$\dot{S}_I^z = -\sqrt{\frac{2}{3}}\left(\zeta_{IA} + 2\zeta_{IB}\right)|E_{2\omega}|\,|E_\omega|^2 \sin\left(2\alpha+\beta\right)\sin\left(\Delta\phi\right). \quad (35)$$

Therefore, for the sample and polarization angles that yield a (negative or positive) maximum in the charge control signal ($2\alpha + \beta = 0, \pi, \ldots$), the spin control along z will be zero. Likewise, when $\left|\dot{S}_I^z\right|$ is maximum ($2\alpha + \beta = \pi/2, 3\pi/2, \ldots$), $\dot{n}_I$ is zero. In addition, $\dot{n}_I$ for parallel linear polarizations ($\beta = \alpha$), has opposite dependence on α as $\dot{n}_I$ for orthogonal linear polarizations ($\beta = \alpha - \pi/2$). The same is true of $\dot{S}_I^z$.

By comparison, if ω and 2ω are both *circularly* polarized, the prediction is quite different. When the fields have *opposite circular* polarizations (2ω is $\hat{\sigma}^+$ and ω is $\hat{\sigma}^-$), Eqs. (27) and (29) simplify to give the charge control signal [13]

$$\dot{n}_I = \frac{4}{\sqrt{3}}\xi_I^{abc}\,|E_{2\omega}|\,|E_\omega|^2\cos\left(\Delta\phi\right) \quad (36)$$

and the z-component of the spin control signal [13]

$$\dot{S}_I^z = \frac{2}{\sqrt{3}}\left(\zeta_{IA} + 2\zeta_{IB}\right)|E_{2\omega}|\,|E_\omega|^2\cos\left(\Delta\phi\right). \quad (37)$$

Therefore, for opposite circular excitation, we should be able to control both charge and spin densities simultaneously. Conversely, when ω and 2ω have the *same* circular polarization, this macroscopic symmetry theory predicts that both $\dot{n}_I$ and $\dot{S}_I^z$ should be zero [13].

4 Experimental Coherent Control Techniques

In previous two-color coherent control experiments, Haché et al. used electrical detection to study a net charge current [28, 29], and Fraser et al. used an optical probe to study the control of the charge density [11, 12]. In addition, optical probing techniques have been used extensively to study carrier spin [1–4] and motion [26, 58–61] in semiconductors.

Here, we discuss our techniques for controlling the phases of ω and 2ω [13, 36, 37, 39]; the basis for these techniques is the coherent control experimental geometry shown in Fig. 6. The ω pulse, which is 100 - 200 fs in duration and has a central wavelength that is tuned over the range of 1.4 - 1.6 μm, is generated in an optical parametric amplifier (OPA) that is pumped by a Ti:Sapphire-laser-seeded regenerative amplifier operating at 250 kHz. Second harmonic generation in beta barium borate (BBO) produces the 2ω pulse (0.7 - 0.8 μm). The two pulses are then separated using a dichroic beamsplitter, and the relative phase between ω and 2ω is controlled with a scanning Michelson interferometer. A piezoelectric transducer attached to the retroreflector in the 2ω arm is periodically scanned to vary the phase difference between ω and 2ω, and this "piezo scan" frequency is used as the reference to a lockin amplifier. The polarization of each pulse is independently controlled with waveplates and polarizers [36]. The two beams are recombined collinearly, sent through a broadband quarter wave plate [36], and focused at normal incidence onto the semiconductor sample. In all experiments, photon energies are chosen such that $\hbar\omega < E_g < \hbar 2\omega < E_{SO}$, where E_g (E_{SO}) is the heavy hole (split off) to conduction band energy gap.

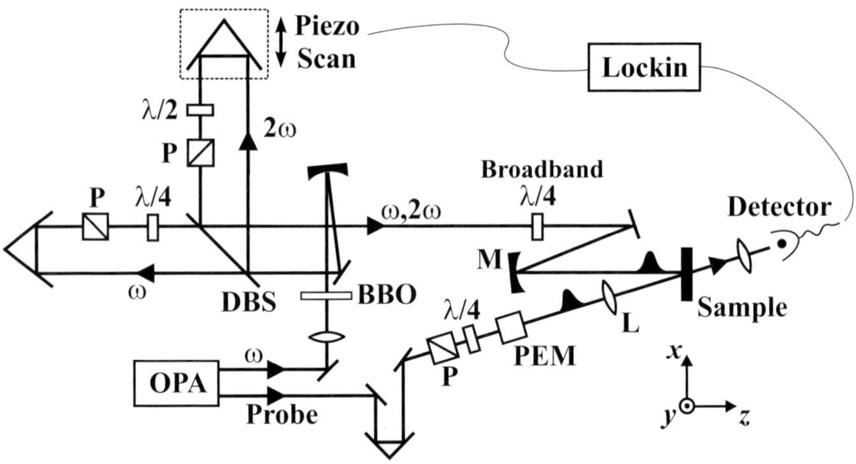

Fig. 6. Experimental geometry: BBO generates 2ω ; DBS is a dichroic beamsplitter; P, $\lambda/2$, and $\lambda/4$ are a polarizer, half wave plate, and quarter wave plate, respectively; PEM is a photoelastic modulator; L and M are a lens and a curved mirror.

The detection scheme is changed as appropriate for each type of measurement. The probe pulse shown in Fig. 6 can be used to monitor the charge density or spin density, as described in Sect. 5, or it can be used to track the transport of charge or spin, as discussed in Sect. 8. Alternatively, we can block the probe beam and use a sample with electrodes to monitor charge currents, as outlined in Sect. 7. All measurements of currents discussed in this chapter are performed for normal-incidence excitation of (001)-oriented GaAs-based samples with no applied magnetic field, where currents due to the circular photogalvanic and spin-galvanic effects are zero [32–34].

5 Control of Charge and Spin Densities: Measurements

5.1 Experimental Techniques

This section reviews our recent demonstration that the charge density and spin density *can* depend on the phases and polarizations of the ω and 2ω fields and the crystal orientation [13]. For these measurements, we use the experimental geometry in Fig. 6; here, the ω (2ω) pulse has a wavelength centered at 1.43 μm (0.715 μm). The ω and 2ω pulses are temporally overlapped, propagate collinearly, and are focused at normal incidence onto a $\sim$ 0.65 μm thick (111)-oriented bulk GaAs epilayer sample [11–13]. This sample is held at room temperature, where the two pumps each create an equal number of carriers, with excess energies up to 275 meV, at an overall density of $\sim 1.2\times10^{18}\mathrm{cm}^{-3}$. Measurements are taken for the two fixed sample orientations illustrated in Fig. 7.

The probe, which in this case is spatially centered on the pump spots, has a wavelength of 0.81 μm and monitors the density and spin of pump-injected carriers at lower energies in the band, arriving $\sim$ 0.9 ps after the pumps. This allows enough time for the carriers to thermalize, but not enough time for the electrons spins to relax. Examining either charge density or spin density requires choosing the appropriate probe polarization, as indicated by the optical selection rules in Fig. 3 [24]. The change in transmission of a $\hat{\sigma}^+$ probe is a measure of spin-down electron density, $n_\downarrow$. Likewise, a $\hat{\sigma}^-$ probe is used to study spin-up electron density, $n_\uparrow$, and a linearly polarized probe is used to monitor to the total electron density, $n = n_\downarrow + n_\uparrow$.

"Charge control" refers to a $\Delta\phi$-dependent change in charge density, Δn, regardless of spin polarization of the carriers. To measure Δn, the probe is linearly polarized, the piezoelectric transducer in the 2ω arm is periodically scanned to vary $\Delta\phi$, and the piezo scanning frequency is used as the reference to a lockin amplifier. This gives a background-free measurement of the phase-dependent differential transmission $[\Delta\mathrm{T}(\Delta\phi)]$ that is sensitive only to the change in charge density due to a change in $\Delta\phi$. We also detect the change in probe transmission with and without the pumps present $[\Delta\mathrm{T}(n_0)]$ using optical choppers and lockins with a linearly polarized probe. This gives a measure of the average background charge density, n_0, of all pump-injected carriers. For these measurements, $\Delta\mathrm{T}(n_0)$ is a linear function of n_0, so the fractional phase-dependent change in charge density is simply $\Delta n/n_0 = \Delta\mathrm{T}(\Delta\phi)/\Delta\mathrm{T}(n_0)$.

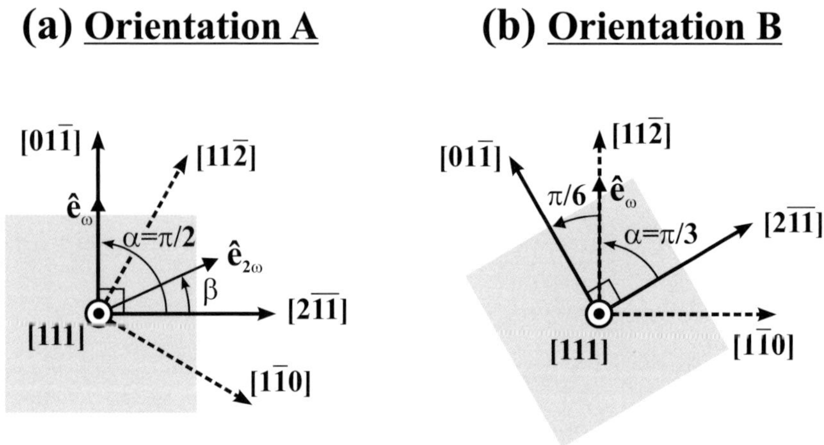

Fig. 7. Schematic of relevant crystallographic directions in (111)-oriented GaAs sample. Sample orientation B is achieved by rotating the sample $\pi/6$ about the z-axis from sample orientation A.

"Spin control" refers to a $\Delta\phi$-dependent change in net spin density along $\hat{\mathbf{z}}$, $\Delta S = \frac{\hbar}{2}(\Delta n_{\uparrow} - \Delta n_{\downarrow})$. To study ΔS, we monitor $\Delta\mathrm{T}(\Delta\phi)$ as the probe polarization is periodically modulated between right- and left-circular polarizations with a photoelastic modulator (PEM). We cascade two lockins for this measurement, using the PEM frequency as the reference to one lockin and the piezo scan frequency as the reference to another. The measured signal therefore is proportional to the difference between $\Delta\mathrm{T}(\Delta\phi)$ with a $\hat{\sigma}^-$ probe and $\Delta\mathrm{T}(\Delta\phi)$ with a $\hat{\sigma}^+$ probe, which in turn is proportional to ΔS. This quantity is normalized by a similar differential transmission measurement (using the PEM but not the piezo) of the total pump-injected spin density when both 2ω and ω are $\hat{\sigma}^-$ polarized, $S_0 = S^z_{2\omega(\hat{\sigma}^-)} + S^z_{\omega(\hat{\sigma}^-)}$.

5.2 Experimental Results

Figure 8(a)-(b) shows the results of measuring the phase-dependent change in charge density and spin density at sample orientation A. In Fig. 8(a)-(b), both 2ω and ω beams are linearly polarized, and the ω polarization is along the $[01\bar{1}]$ direction. In Fig. 8(a), the 2ω polarization is also along $[01\bar{1}]$; in this case we see a strong charge control signal but little or no modulation of the spin density. By contrast, in Fig. 8(b) the 2ω field is polarized orthogonal to the ω field and lies along the $[2\bar{1}\bar{1}]$ direction, allowing control of the spin density, but no discernable control of the charge density. [The configuration of Fig. 8(b) repeats the sample orientation and polarization combination reported by Fraser et al. [11, 12], with similar results – both qualitatively and quantitatively – for Δn.] Therefore, at orientation A, parallel linear pump

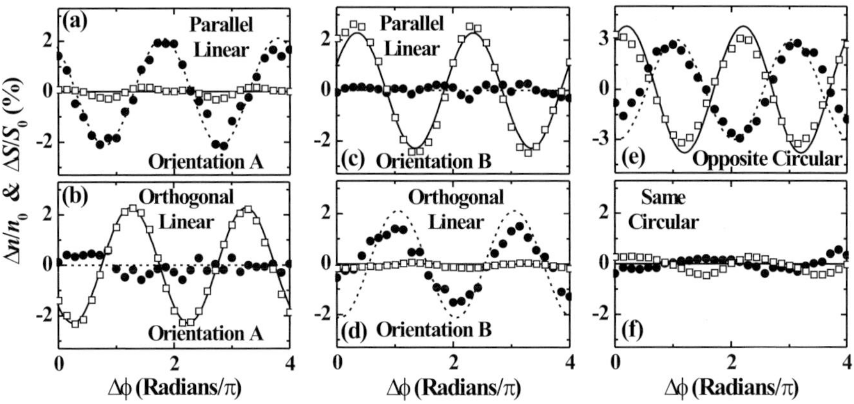

Fig. 8. Measured fractional change in spin (solid circles) and charge (open squares) densities as a function of relative phase $\Delta\phi$ for various polarization combinations and sample orientations. Also shown are simulations of charge control (solid lines) and spin control (dashed lines) using the QUIC theory discussed in Sect. 3.

polarizations allow spin control without charge control, whereas orthogonal polarizations allow charge control without spin control.

Next we rotate the sample $\pi/6$ about the z-axis to sample orientation B so ω is polarized along $[11\bar{2}]$, and repeat the parallel and orthogonal linear polarization configurations. The results are shown in Fig. 8(c)-(d). In contrast to the results at orientation A, for parallel linear excitation polarizations – now along $[11\bar{2}]$ – we observe charge control but no discernable spin control. With orthogonal linear polarizations – ω along $[11\bar{2}]$, 2ω along $[1\bar{1}0]$ – we observe spin control but little or no charge control. Therefore at sample orientation B, the dependences of the spin control and of the charge control on the linear polarization combinations are reversed from the results at sample orientation A.

Thus for fixed parallel (or orthogonal) linear pump polarizations, the sample can first be oriented so that the phase $\Delta\phi$ can be used to control the spin, but not the charge. The sample can then be rotated so that $\Delta\phi$ controls the charge, but not the spin. These results agree with the predictions of Sect. 3.

For *opposite circular* excitation, Fig. 8(e) shows that both the spin and the charge are controlled simultaneously. Since the ω and 2ω circular polarizations have opposite handedness, the carriers injected by these two pulses have opposite net spin: the $\hat{\sigma}^+$ polarized 2ω pulse injects carriers with a net spin along $-\hat{\mathbf{z}}$, while the $\hat{\sigma}^-$ polarized ω pulse injects carriers with net spin along $+\hat{\mathbf{z}}$. The magnitudes of the net spin injected per electron for one- and two-photon absorption are expected to be nearly equal [35], so that $S^z_{2\omega} + S^z_{\omega} \cong 0$, which we have verified experimentally by measuring the average background (non-phase-dependent) spin. Consequently, in the absence of spin control, there is approximately zero net spin along $\hat{\mathbf{z}}$, and the spin control signal causes the direction of the total $\dot{S}^z$ to oscillate between $+\hat{\mathbf{z}}$ and $-\hat{\mathbf{z}}$. By contrast, Fig. 8(f) shows that for *same circular* excitation, there is essentially no phase dependence to either charge density or spin density, even though a highly spin-polarized background population is produced because $\dot{S}^z_{2\omega}$ and $\dot{S}^z_{\omega}$ both point along $+\hat{\mathbf{z}}$. In addition, we have verified that the magnitudes of the $\Delta n/n_0$ and $\Delta S/S_0$ signals for same and opposite circular excitation do not depend on which sample orientation we choose (not shown), as predicted in Sect. 3.2.

5.3 Discussion

Also shown in Fig. 8 are theoretical fits to the data using the model based on the QUIC process of coherent control of carrier charge and spin densities outlined in Sect. 3. At the ω and 2ω electric field strengths that balance the number of carriers produced by one- and two-photon absorption, for linear polarizations [13]

$$\frac{\dot{n}_I}{\dot{n}_{2\omega}+\dot{n}_\omega}=\sqrt{\frac{2}{3}}\frac{\xi_I^{abc}}{\sqrt{\xi_1^{aa}\xi_2^{aaaa}}\,(1-\sigma/2)}\cos\left(2\alpha+\beta\right)\cos\left(\Delta\phi\right), \tag{38}$$

where σ is the two photon absorption anisotropy parameter [53]. For opposite circular polarizations [13],[4]

$$\frac{\dot{n}_I}{\dot{n}_{2\omega}+\dot{n}_\omega}=\frac{2}{\sqrt{3}}\frac{\xi_I^{abc}}{\sqrt{\xi_1^{aa}\xi_2^{aaaa}}\,(1-\sigma/6-\delta)}\cos\left(\Delta\phi\right), \tag{39}$$

where δ is the two photon absorption dichroism parameter [44].

The solid lines in Fig. 8 are plots of $\dot{n}_I/(\dot{n}_{2\omega}+\dot{n}_\omega)$ using Eq. (38) for (a)-(d) and using Eq. (39) for (e). These simulations have been shifted along the $\Delta\phi$-axis to line up with the data, to show that the magnitudes of the simulations are in good quantitative agreement with the data. The single value of 3.2% was chosen for $\xi_I^{abc}/\sqrt{\xi_1^{aa}\xi_2^{aaaa}}$ to give the best fit to all charge control data, while values of σ = - 0.6 and $\delta = 0$ were taken from a 14-band model calculation [44].

The dashed lines in Fig. 8 are plots of $\dot{S}^z_I\big/\dot{S}_0$ using Eq. (35) for (a)-(d) and Eq. (37) for (e). The simulations are again shifted along the $\Delta\phi$-axis, and the single value of 2.6% for $(\zeta_{IA}+2\zeta_{IB})\left|E_{2\omega}\right|\left|E_\omega\right|^2\big/\dot{S}^z_0$ has been chosen to give the best fit to all the data; again, there is good quantitative agreement between data and theory.

Our measured value of the charge control ratio with a 715 nm 2ω pulse is close to the value reported by Fraser et al. [11] with a 775 nm 2ω pulse in this

[4] Eq. (39) corrects an error in Eq. (10) of [13].

same sample. A charge control ratio of a few percent is expected, since the effect only exists as a result of the lack of inversion symmetry, which is small in GaAs [13]. Indeed, *ab initio* calculations indicate that $\xi_I^{abc}/\sqrt{\xi_1^{aa}\xi_2^{aaaa}}$ is 6.4% at these wavelengths [11]. Our experimental fit value is smaller by a factor of two, similar to the result in [11]; much of the difference between data and theory can be attributed to the use of a relatively thick sample [11]. Similar arguments could be made for the magnitude of the spin control ratio: it too relies on the lack of inversion symmetry. One can take the charge control ratio as a rough upper bound on the spin control ratio; from this point of view, the spin control ratio we have measured is in line with our expectations [13].

In addition to the sources of charge and spin control due to QUIC, there also can be contributions from cascaded second harmonic processes. In the cascaded processes, the *real* part of $\chi^{(2)}$ causes frequency conversion between the ω and 2ω pulses as they propagate through the GaAs sample. The direction of energy transfer and the relative phases and polarizations of the fields associated with that transfer are determined by the phases and polarizations of the incident ω and 2ω fields and by the crystal symmetry and orientation. If the *intensities* of the ω and 2ω fields are modified by frequency conversion, then the carrier *densities* injected through two- and one-photon absorption will be modulated. If their *polarization states* are modified, the carrier *spin* can be modulated.

In zincblende-type crystals, the cascaded and QUIC contributions to charge control have the same symmetry properties, and thus, they cannot be distinguished by varying the polarization states or crystal orientation [13]. For spin control, by contrast, the cascaded and QUIC contributions have different symmetry properties, although the difference vanishes for light normally incident on a (111) surface [13]. Therefore the qualitative predictions of whether to expect charge or spin control for specific sample orientations and polarization combinations discussed in Sect. 3 are valid for both QUIC and cascaded processes [13]. As discussed by Fraser et al. [11, 12], the cascaded and QUIC processes can, however, be separated by their different dependencies on sample thickness; for thin samples, the direct process dominates. For our sample, which is the same one used by Fraser et al. [11, 12], we expect that the cascaded and direct processes both contribute to charge control [13].

We emphasize, however, that the data in Fig. 8 clearly demonstrate independent coherent control of the carrier charge density and spin density, regardless of the underlying physical mechanism(s) responsible. Finally, we also expect to generate *currents* for normal incidence excitation of (111)-GaAs due to QUIC [10, 45] and the photogalvanic effect [32, 33]. However, spatially resolved measurements (not shown) confirm that the data in Fig. 8 are a measure of phase-dependent changes of the *overall* densities of charge and spin.

Therefore, in a (111)-oriented sample, we have shown: (i) For a selected fixed sample orientation and linearly polarized ω and 2ω pulses, parallel polarizations allow control of the charge density (i.e., the free electron-hole density) without modulating the spin density (i.e., the net spin polarization of the electrons and holes); conversely, orthogonal linear polarizations allow control of the spin, but not the charge. (ii) When the sample is rotated $\pi/6$ about the z-axis, the dependences of the spin control and of the charge control on these polarizations are reversed. (iii) For opposite circular polarizations, both the spin and the charge densities can be simultaneously controlled, regardless of sample orientation. (iv) When ω and 2ω have the same circular polarization, $\Delta\phi$ controls neither the spin nor the charge. These results agree with a theory based on a macroscopic symmetry analysis of QUIC, but the symmetries are also consistent with contributions of cascaded processes.

6 Symmetry Predictions: Charge Currents and Spin Currents

Sections 3-5 reviewed the use of QUIC to independently control the charge density and spin density in (111)-oriented GaAs, demonstrating one way that charge and spin can be decoupled. The next three sections, by contrast, discuss independent control of ballistic charge currents and spin currents in unbiased (001)-oriented GaAs, illustrating another way of decoupling charge and spin. In generating these currents, QUIC modulates the k-space *distribution* of charge and spin, without changing the *overall* charge density or spin density. As discussed in Sect. 3, neither the charge density nor the spin density can be coherently controlled through QUIC for normal-incidence excitation of (001)-oriented GaAs.

Recall that in Sect. 2, we reviewed approaches to QUIC based on Fermi's golden rule and on symmetry considerations, and that both of these approaches predict control of charge currents and spin currents, at least in principle. This section reviews the specific charge current and spin current symmetry predictions of Bhat and Sipe [35], exploring the polarization combinations of ω and 2ω necessary to generate each type of current. To find the charge current injection rate in terms of arbitrary input polarizations, Bhat and Sipe derive a specific form of the tensor expression in Eq. (30) for an isotropic material [35]:

$$\dot{\boldsymbol{J}} = -iD\frac{q}{\hbar}\left[B_1\boldsymbol{E}_\omega\left(\boldsymbol{E}_\omega\cdot\boldsymbol{E}_{2\omega}^*\right) + B_2\boldsymbol{E}_{2\omega}^*\left(\boldsymbol{E}_\omega\cdot\boldsymbol{E}_\omega\right)\right] + \text{c.c.}, \tag{40}$$

where[5] [35]:

$$D \equiv \frac{\sqrt{2}}{60\pi}\frac{q^3 E_P}{\sqrt{m}}\frac{\left(2\hbar\omega - E_g\right)^{3/2}}{\hbar^4\omega^4}. \tag{41}$$

[5] Note that $q < 0$; by consequence, $D < 0$.

E_p is the Kane energy, E_g is the band gap energy, and q and m are the electron charge and mass. Likewise, to find the spin current injection rate as a function of arbitrary input polarizations, Bhat and Sipe derive the form of Eq. (31) specific to an isotropic medium [35]:

$$\dot{K}^{ij} = \frac{d}{dt}\left\langle \hat{\mathrm{v}}^i \hat{S}^j \right\rangle = A_1 D E^i_\omega \left(\mathbf{E}^*_{2\omega} \times \mathbf{E}_\omega\right)^j + A_2 D \left(\mathbf{E}^*_{2\omega} \times \mathbf{E}_\omega\right)^i E^j_\omega$$

$$+ A_3 D \epsilon^{ijk} E^{*k}_{2\omega} \left(\mathbf{E}_\omega \cdot \mathbf{E}_\omega\right) + A_4 D \epsilon^{ijk} E^k_\omega \left(\mathbf{E}^*_{2\omega} \cdot \mathbf{E}_\omega\right) + \text{c.c.} \tag{42}$$

where ϵ^{ijk} is the Levi-Civita pseudotensor, and repeated indices are summed over. While Eqs. (40) and (42) assume an isotropic medium, Bhat and Sipe argue that they are also good approximations for GaAs, which has symmetry T_d [35]. The A and B coefficients in these equations depend on material parameters such as E_p, E_g, the effective masses of the various bands involved, and the excess energies. For GaAs, Bhat and Sipe find that $B_1 \gg B_2$ [35].

Using these general formulas for charge and spin currents, we can explore the predictions for several specific polarization configurations. Here, we take the direction of propagation of the ω and 2ω fields as the $+z$ direction, and only consider currents travelling in the transverse (x-y) plane. Figure 9 summarizes the electron velocities and spins associated with three types of transverse current for the three polarization combinations discussed below. Bhat and Sipe have also predicted that ω and 2ω can generate longitudinal currents, which travel along the z-axis [35]; since the measurement techniques described in this chapter are not sensitive to them, the longitudinal currents are not discussed here.

When ω and 2ω have parallel linear polarizations along $\hat{\mathbf{x}}$, there will be a charge current injected in the $\hat{\mathbf{x}}$ direction [28, 46],

$$\dot{\boldsymbol{J}} = 2DB_1 q\hbar^{-1} \left|E_\omega\right|^2 \left|E_{2\omega}\right| \sin\left(\Delta\phi\right) \hat{\mathbf{x}}, \tag{43}$$

but no spin current injected along $\hat{\mathbf{x}}$:

$$\dot{K}^{xz} = \frac{d}{dt}\left\langle \mathrm{v}^x S^z \right\rangle = 0. \tag{44}$$

Thus, we expect an unpolarized charge current along the x-direction, as shown in Fig. 9(a). The relative phase of the ω and 2ω fields should control the magnitude and sign of this current, and the optical polarization direction of the pump pulses should control the direction of current flow.

By contrast, when ω and 2ω have the same circular polarization, a very different transverse current is expected. When both ω and 2ω are left-hand circularly polarized ($\hat{\sigma}^- = (\hat{\mathbf{x}} - i\hat{\mathbf{y}})/\sqrt{2}$), the predicted charge current injection rate is [35]

$$\dot{\boldsymbol{J}} = \sqrt{2} DB_1 q\hbar^{-1} \left|E_\omega\right|^2 \left|E_{2\omega}\right| \; \hat{\mathbf{m}}, \tag{45}$$

where the direction $\hat{\mathbf{m}}$ is defined

$$\hat{\mathbf{m}} \equiv \hat{\mathbf{x}} \sin\left(\Delta\phi\right) - \hat{\mathbf{y}} \cos\left(\Delta\phi\right). \tag{46}$$

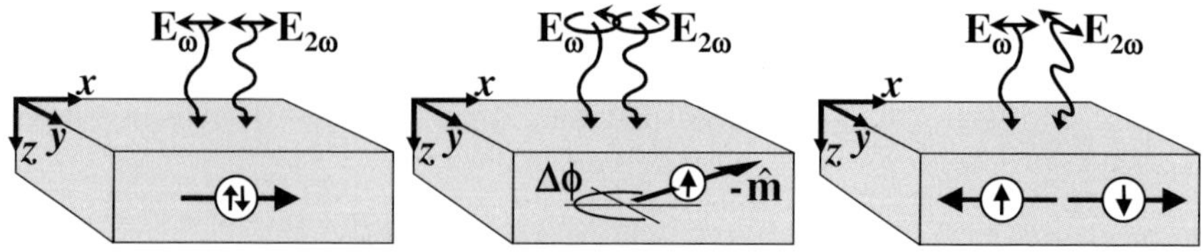

Fig. 9. Transverse currents predicted when ω and 2ω have (a) parallel linear, (b) same circular, and (c) orthogonal linear polarizations. The electron spins are illustrated by the small arrows inside each sphere, and the larger arrow intersecting each sphere shows the direction of electron travel. For clarity, holes are not shown.

The expected transverse spin current injected in this case is also along $\hat{\mathbf{m}}$ [35]:

$$\dot{K}^{mz} = \tfrac{d}{dt} \langle \mathrm{v}^m S^z \rangle = -\sqrt{2} D \left(A_1 - A_4 \right) \left| E_\omega \right|^2 \left| E_{2\omega} \right| . \tag{47}$$

Consequently, Eqs. (45) and (47) predict a spin-polarized charge current,[6] which consists of electrons travelling in the $-\hat{\mathbf{m}}$ direction with average spin along the $-\hat{\mathbf{z}}$ direction. It should be possible to *control the direction* of this current simply by adjusting the relative phase of ω and 2ω, $\Delta\phi$. However, the magnitude of this current should be independent of $\Delta\phi$. Therefore, as $\Delta\phi$ is varied, this current should rotate like a constant amplitude phasor in the x-y plane, as shown in Fig. 9(b).

Finally, if the excitation fields have orthogonal linear polarizations so that ω is along $\hat{\mathbf{x}}$ and 2ω is along $\hat{\mathbf{y}}$, Bhat and Sipe predict a charge current along $\hat{\mathbf{y}}$ [35]:

$$\dot{\boldsymbol{J}} = 2 D B_2 q \hbar^{-1} \left| E_\omega \right|^2 \left| E_{2\omega} \right| \sin \left(\Delta\phi \right) \hat{\mathbf{y}} . \tag{48}$$

The magnitude of this charge current is expected to be very small compared to the charge current produced by parallel linear polarizations [see Eq. (43)], since Bhat and Sipe predict $B_1 \gg B_2$ [35]. It is also expected to be small compared to the pure spin current discussed below [35]. Consequently, this y-directed current is not shown in Fig. 9(c) and is ignored in the remainder of our discussion in this section.

In addition, Bhat and Sipe predict a spin current along the direction of the ω polarization ($\hat{\mathbf{x}}$) [35]:

$$\dot{K}^{xz} = \tfrac{d}{dt} \langle \mathrm{v}^x S^z \rangle = -2D \left(A_1 + A_3 \right) \left| E_\omega \right|^2 \left| E_{2\omega} \right| \cos \left(\Delta\phi \right) . \tag{49}$$

Since there is no net charge current along $\hat{\mathbf{x}}$, Eq. (49) represents a *pure spin current.* Figure 9(c) shows the behavior of the two components of this pure spin current for $\Delta\phi = 0$: "spin-down" (spins along -$\hat{\mathbf{z}}$) electrons travel in the -$\hat{\mathbf{x}}$ direction, while an equal number of "spin-up" (spins along +$\hat{\mathbf{z}}$) electrons

[6] The alternative interpretation of an unpolarized charge current combined with a pure spin current, both travelling along $\hat{\mathbf{m}}$, is precluded by more detailed calculations by Bhat and Sipe [45].

travel in the $+\hat{\mathbf{x}}$ direction. When $\Delta\phi = \pi$, the directions reverse. The direction of the ω pulse polarization determines the direction of the currents in the x-y plane, while $\Delta\phi$ controls the magnitude and sign of these currents. In effect, this QUIC configuration provides a mechanism for "sorting" the spin of electrons, sending spin-up electrons in one direction and spin-down electrons in the opposite direction [35].

Therefore, by choosing the appropriate ω and 2ω polarizations, Bhat and Sipe predict three different types of coherently controllable transverse ballistic currents in unbiased semiconductors. For parallel linear polarizations, an unpolarized charge current should be injected along the direction of the incident polarizations, and we should be able to control the magnitude and sign of this current by controlling $\Delta\phi$. If ω and 2ω pules have the same circular polarization, they should generate a spin-polarized charge current whose direction depends on $\Delta\phi$, but whose magnitude does not. For orthogonal linear polarizations, we expect a pure spin current along the direction of the ω polarization, in addition to a small charge current (not shown in Fig. 9) along the direction of the 2ω polarization. Finally, when ω and 2ω have opposite circular polarizations (not shown), Bhat and Sipe predict that no charge current or spin current will be injected [35]. The next two sections review our experimental observations and comparisons of each of these currents.

7 Electrical Detection of Spin-Polarized Charge Current

This section reviews our electrical measurements of QUIC charge currents [36]. As predicted in Sect. 6, we demonstrate that when ω and 2ω have the same circular polarization they generate a charge current that is expected to be spin polarized and whose direction can be controlled by adjusting the relative phase of ω and 2ω [36]. We contrast this with the charge current for parallel linear excitation reported by Haché et al. [28, 29] that is not expected to be spin polarized, in which the optical polarization direction controls the current direction. Both currents have been investigated by collecting the charge they produce using conventional electrodes and the experimental geometry in Fig. 6 – without the probe beam [36]. The sample used for these measurements is a (001)-oriented bulk, low-temperature-grown GaAs (LT-GaAs) sample with evaporated gold electrodes on the surface [28, 29, 36].

Figure 10(a) illustrates a view of the sample along the propagation ($\hat{\mathbf{z}}$) direction. The ω and 2ω beams are focused to spot diameters of $\sim 50\mu$m and $\sim 25\mu$m, respectively, and the electrodes are spaced 10 μm apart. The sample can be rotated to measure the current along an arbitrary direction $\hat{\mathbf{r}} = \cos\theta\hat{\mathbf{x}} + \sin\theta\hat{\mathbf{y}}$, as shown in Fig. 10(a). At each fixed value of θ, the piezo in the 2ω arm of the interferometer periodically modulates $\Delta\phi$, and the voltage across the electrodes is measured using a lockin amplifier with this modulation frequency as the reference. Acting independently, the ω and 2ω

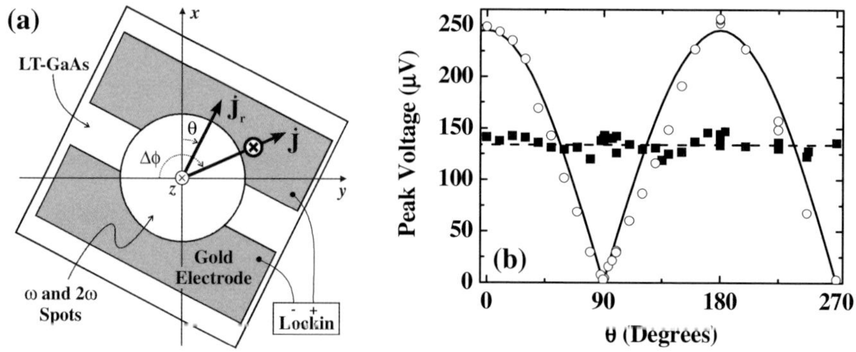

Fig. 10. (a) LT-GaAs sample with surface electrodes, showing the expected current ($\dot{\mathbf{J}}$) injected by ω and 2ω pulses that are both $\hat{\sigma}^-$ polarized. (b) Peak phase-dependent voltage as a function of sample angle, θ. The solid squares (open circles) are taken with ω and 2ω both $\hat{\sigma}^-$ ($\hat{\mathbf{x}}$) polarized.

pulses, which have wavelengths of 1.55 μm and 0.775μm, would create carrier densities of $2 \times 10^{15}\text{cm}^{-3}$ and $5 \times 10^{17}\text{cm}^{-3}$ through two- and one-photon absorption, respectively, with excess energies $(2\hbar\omega - E_g)$ of $\sim$ 180 meV.

For $\hat{\sigma}^-\hat{\sigma}^-$ excitation, (45) predicts a charge current of constant amplitude that points along the -y axis for $\Delta\phi = 0$ and rotates clockwise in the x-y plane with increasing $\Delta\phi$. The component of this current in the $\hat{\mathbf{r}}$ direction is therefore:

$$\dot{\boldsymbol{J}}_{\sigma-\sigma-} \cdot \hat{\mathbf{r}} \propto |E_\omega|^2 \, |E_{2\omega}| \, \sin(\Delta\phi - \theta) \,. \tag{50}$$

The peak amplitude of this current – and hence the peak amplitude of the measured voltage – should be independent of the orientation of the electrodes; however, the maximum voltage for each orientation should occur at a unique phase difference given by $\Delta\phi = \theta + \pi/2$.

By contrast, when both excitation pulses are linearly polarized along $\hat{\mathbf{x}}$, (43) predicts that the current should travel only along the direction of the incident polarization, and the magnitude should be controlled by the phase. The projection of this current along $\hat{\mathbf{r}}$ is proportional to the cosine of the angle of rotation:

$$\dot{\boldsymbol{J}}_{xx} \cdot \hat{\mathbf{r}} \propto |E_\omega|^2 \, |E_{2\omega}| \sin(\Delta\phi) \cos(\theta) \,. \tag{51}$$

The measured peak current amplitude as a function of θ is plotted in Fig. 10(b) for these two polarization combinations. As expected for $\hat{\sigma}^-\hat{\sigma}^-$ excitation, the peak voltage is essentially constant. For $\hat{\mathbf{x}}\hat{\mathbf{x}}$ excitation, on the other hand, the peak signal is proportional to $|\cos\theta|$, in agreement with Eq. (51); the absolute value arises because peak signal is always positive.

In addition, the dependences of these currents on the phase difference $\Delta\phi$ at two fixed sample orientations are illustrated in Fig. 11. As predicted for $\hat{\sigma}^-\hat{\sigma}^-$ excitation, Fig. 11(a) shows: (i) both x and y signals vary sinusoidally with $\Delta\phi$; (ii) the peak amplitudes are equal; and (iii) the two curves are out of phase by approximately 90^0. For $\hat{\mathbf{x}}\hat{\mathbf{x}}$ excitation, by contrast, Fig. 11(b) shows that the current in the x-direction varies sinusoidally, but there is no current along the y-direction, in agreement with Eq. (51). The results in Fig. 11(b) corroborate the earlier measurements of Haché et al. [29]. The amplitudes and relative phases of the x and y currents for $\hat{\sigma}^-\hat{\sigma}^-$ excitation can be inferred from the data in Fig. 11(a), but the absolute phases were not determined. In addition, no relative comparison should be made between the phases of curves in Fig. 11(a) and Fig. 11(b).

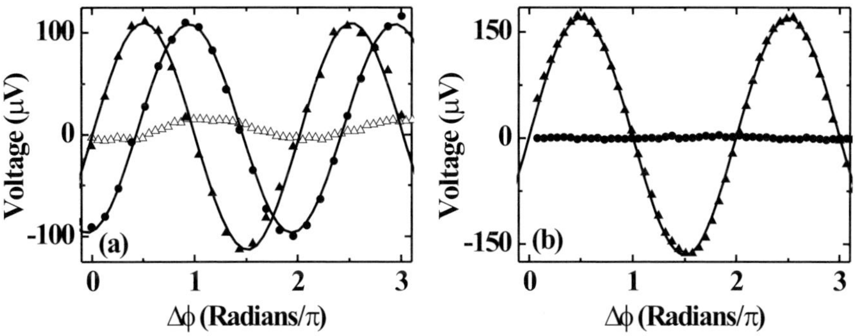

Fig. 11. Measurements of the charge current along the x (triangles) and y (circles) directions. (a) Solid triangles and circles: ω and 2ω both $\hat{\sigma}^-$ polarized; open triangles: ω field $\hat{\sigma}^-$, 2ω field $\hat{\sigma}^+$. (b) ω and 2ω both $\hat{\mathbf{x}}$. Solid curves are sinusoidal fits to the data.

Figures 10 and 11 therefore show that the interference between ω and 2ω fields with the same circular polarization produces a net in-plane transverse current that has all the salient features of the spin-polarized transverse current predicted by Bhat and Sipe [35]. Specifically, the magnitude of this current is independent of the phase difference between ω and 2ω, while the direction rotates periodically in the transverse plane with this phase difference. Although this electrical detection experiment does not measure carrier spin, it has been shown that either one- or two-photon absorption of circularly polarized light produces a spin-polarized carrier population, even when exciting well above the band gap [24, 57, 62–65], and according to the predictions of Bhat and Sipe [35], this current should be spin-polarized.

Bhat and Sipe also predict zero transverse current for opposite circular polarizations [35]. As confirmation of this prediction, the open triangles in Fig. 11(a) represent measurements of the x-component of the signal for the ω beam $\hat{\sigma}^-$ polarized and the 2ω beam $\hat{\sigma}^+$ polarized. The measured current along y (not shown) has about the same magnitude as the current along x. While not exactly zero, the current for opposite circular excitation is clearly very small compared to the current for same circular excitation.

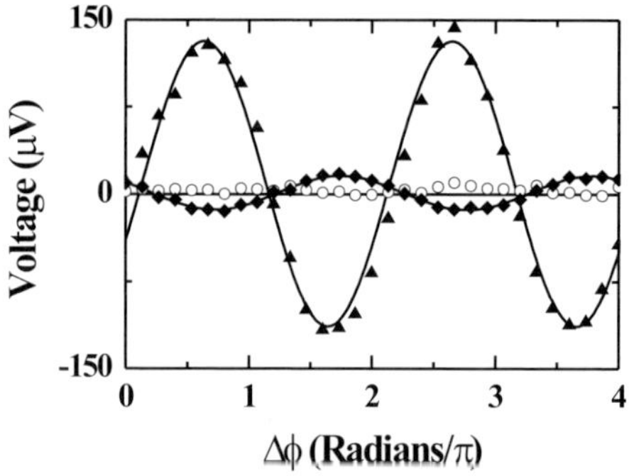

Fig. 12. Measurement of charge currents for ω (2ω) polarized along x (y). Open circles (solid diamonds) show the current in the x- (y-) direction. For comparison, solid triangles show the current in the x-direction with ω and 2ω both polarized along x.

In addition, Fig. 12 shows the measured charge current for orthogonal linear polarizations, with ω along x and 2ω along y. The solid diamonds show a small current in the y-direction, which is an order of magnitude weaker than the charge current for parallel linear excitation under the same conditions shown by the solid triangles. This confirms Bhat and Sipe's theoretical prediction discussed in Sect. 6 that $B_1 \gg B_2$ [35]. The open circles show that there is no net charge current in the x-direction; to determine whether this null net charge current along x actually consists of two counter-propagating currents having opposite spin, as Eq. (49) predicts, we use the optical probing technique discussed in the next section.

8 Optical Probing of Charge Current and Pure Spin Current

8.1 Optical Probing Technique

Another way to monitor the carrier movement resulting from the currents is to monitor the carrier position and spin using the tightly focused optical probe pulse in the apparatus shown in Fig. 6. We have performed such measurements using a probe with a wavelength of 0.81 μm that can be scanned along the x-direction across the region excited by the pumps [37, 39]. This optical probing technique can be used to determine carrier spin, and can substantiate the electrical measurements of the previous section.

The sample used for these measurements is a multiple quantum well that consists of ten periods of 14-nm-wide GaAs wells alternating with 17-nm-thick $Al_{0.3}Ga_{0.7}As$ barriers grown on a (001)-oriented GaAs substrate.[7] The sample is mounted on a glass flat, and the GaAs substrate has been removed by selective etching to permit transmission measurements. This sample is held at a temperature of 80 K, where the ω and 2ω pumps, which in this case have wavelengths 1.42 μm and 0.71 μm, generate electrons with excess energies of $\sim$ 200 meV, while the probe is resonant with the heavy-hole exciton transition. The probe pulse arrives $\sim$ 4 ps after the pumps to allow the ballistically injected hot carriers to thermalize and the spins of the holes (but not the electrons) to relax [24, 66]. The ω and 2ω pumps are focused to spot widths of $\sim 10 - 15$ μm, and the probe is $\sim$ 6 μm wide.

The probing techniques are similar to those described in Sect. 5, except that here the probe spot is moved to spatially resolve the differential transmission signals, and the PEM is not used. To monitor the *overall* local electron density at a given x position, $n(x)$, we measure $\Delta\mathrm{T}(n)$ using choppers and lockin amplifiers. By contrast, to study the *phase-dependent* local electron density, $\Delta n(x)$, we measure $\Delta\mathrm{T}(\Delta\phi)$ using the piezo scanning frequency as the lockin reference. For small changes in transmission, $\Delta\mathrm{T}(\Delta\phi)$ is proportional to the change in carrier density at a given x position, $\Delta n(x, \Delta\phi)$. Both $\Delta\mathrm{T}(n)$ and $\Delta\mathrm{T}(\Delta\phi)$ can be normalized by the transmission through the sample in the absence of the pumps, T, to give $\Delta\mathrm{T}(n)/\mathrm{T}$ and $\Delta\mathrm{T}(\Delta\phi)/\mathrm{T}$.

In addition, electron *spin* at each position is monitored by setting the probe polarization with a quarter wave plate and using the optical selection rules for excitonic transitions [24, 67, 68]. Since the probe is tuned to the heavy hole excitonic resonance, the change in transmission of a $\hat{\sigma}^+$ probe is a measure of spin-down electron density, $n_\downarrow$ (or phase-dependent change in that density, $\Delta n_\downarrow$). A $\hat{\sigma}^-$ probe monitors $n_\uparrow$ (or $\Delta n_\uparrow$), and a linearly polarized probe monitors the total electron density, $n = n_\downarrow + n_\uparrow$ (or $\Delta n = \Delta n_\downarrow + \Delta n_\uparrow$).

8.2 Expected Charge and Spin Motion

For the charge current predicted in Sect. 6 when ω and 2ω have parallel linear polarizations, Fig. 13(a) plots electron density as a function of position, showing the expected motion of the optically injected Gaussian electron distribution resulting from this current. The dashed line indicates the profile's original position, and the solid line shows its resultant position. For the case shown, electrons have a preferred ballistic momentum in the $+x$ direction; thus, initially the electron distribution will move in the $+x$ direction. As Fig. 13(b) illustrates, this movement should result in a change in electron density as a function of position, $\Delta n(x)$, that approximates the spatial derivative of

[7] The theory in Sections 2 and 6 was for bulk semiconductors, while quantum wells are used in the experiments here. Theoretical calculations for charge and spin currents in the plane of a (001)-GaAs/AlGaAs quantum well predict behavior that is qualitatively similar to that for bulk GaAs [39, 50].

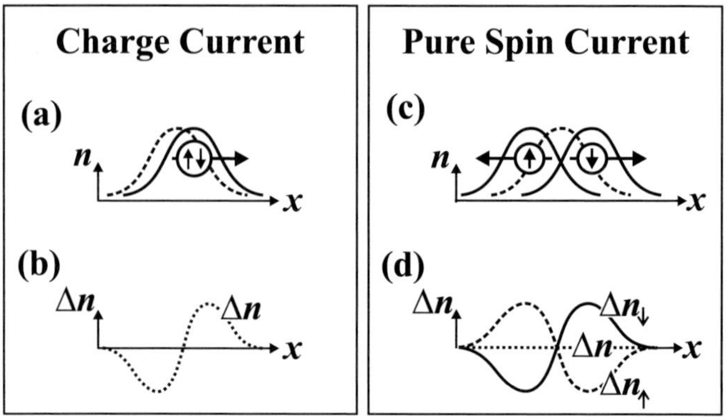

Fig. 13. Schematic motion of electron density profiles (n, $n_\uparrow$, $n_\downarrow$), and corresponding changes in local electron densities (Δn, $\Delta n_\uparrow$, $\Delta n_\downarrow$) resulting from the motion for: (a)-(b) charge current; (c)-(d) pure spin current.

the original Gaussian profile, $\partial n/\partial x$, provided the movement is small. By determining the amplitude of $\Delta n(x)$, we can use the relation

$$\Delta n\,(x) \cong \frac{\partial n}{\partial x}\Delta x \tag{52}$$

to estimate the distance carriers move, Δx.

Alternatively, for the pure spin current expected when ω and 2ω have orthogonal linear polarizations, Fig. 13(c)-(d) illustrates the expected behavior of the spin-up and spin-down carrier density profiles. Initially there are an equal number of spin-up and spin-down electrons produced at each x position; therefore, the spin-up and spin-down profiles are overlapped at the position indicated by the dashed curve in Fig. 13(c). For the case shown, however, spin-down electrons should have a preferred ballistic momentum in the $+x$ direction, while spin-up electrons should have a preferred momentum in the -x direction. Thus, before momentum is randomized, the spin-down spatial distribution should shift toward $+x$, and the spin-up distribution toward -x. When $\Delta\phi$ changes by π , the directions should reverse.

The accompanying *changes* expected in the spin-down $[\Delta n_\downarrow(x)]$, spin-up $[\Delta n_\uparrow(x)]$, and total $[\Delta n(x)]$ electron densities resulting from the pure spin current are shown in Fig. 13(d). The magnitudes of $\Delta n_\downarrow(x)$ and $\Delta n_\uparrow(x)$ are determined by the distance the spin-down and spin-up carrier profiles move. If this movement is small, the magnitude of $\Delta n_\downarrow(x)$ should be proportional to the spatial derivative of the initial Gaussian-shaped carrier profile $\partial n_\downarrow(x)/\partial x$, the magnitude of $\Delta n_\uparrow(x)$ should be proportional to $\partial n_\uparrow(x)/\partial x$, and the magnitude of $\Delta n(x) = \Delta n_\downarrow(x) + \Delta n_\uparrow(x)$ should be approximately zero.

The distance each profile in Fig. 13 moves is determined in part by the number of carriers involved in each current, their ballistic velocities, and the relaxation mechanisms [10, 29, 69, 70]. The distances in Fig. 13 are greatly exaggerated; for the carrier densities ($\sim 10^{17}\text{cm}^{-3}$), lattice temperature (80 Kelvin) and excess energy ($\sim$ 200 meV) used in our experiments, we expect electron motion on the order of a few nm.

8.3 Charge Current Measurements

Results of this spatially resolved optical probing technique when ω and 2ω pulses have parallel linear polarizations along $\hat{\mathbf{x}}$ are shown in Fig. 14. For these measurements, the ω and 2ω pulses acting independently each create roughly equal peak carrier densities (at $x = 0$) of $\sim 8 \times 10^{16}$ cm^{-3}; in a sense, the two arms of our "matter interferometer" can be considered to be balanced. We expect a charge current along $\hat{\mathbf{x}}$ in this case; as discussed in Sect. 7, this charge current has been observed using electrical detection [28, 29, 36].

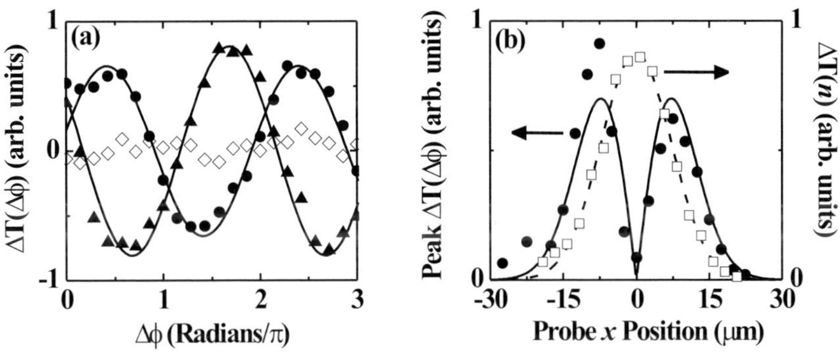

Fig. 14. (a) Phase-dependent differential transmission data at $x \simeq -7.5$ μm (solid triangles), $x \simeq 0$ (open diamonds) and $x \simeq +7.5$ μm (solid circles). Solid curves are sinusoidal fits. (b) Spatially resolved ΔT. The open squares show $\Delta\text{T}(n)$, while the solid circles show the peak value of $\Delta\text{T}(\Delta\phi)$.

Figure 14(a) shows measurements of the phase-dependent change in carrier density, $\Delta n(x, \Delta\phi)$: the phase-dependent change in probe transmission $\Delta\text{T}(\Delta\phi)$ is plotted for three different probe positions. The probe is linearly polarized, so it is sensitive to the phase-dependent change in total electron density, $\Delta n = \Delta n_\uparrow + \Delta n_\downarrow$. On the left side of the excitation region ($x \simeq -7.5$ μm), Δn varies sinusoidally with $\Delta\phi$ and is approximately out of phase with Δn on the right side ($x \simeq +7.5$ μm).[8] For $\Delta\phi \simeq \pi/2$, the carrier density increases on the right side of the excitation region while decreasing

[8] While the relative phases of Δn at different positions can be inferred from the data in Fig. 14(a), the absolute phase was not determined. The same holds for Fig. 15(a).

on the left side, showing a net motion of carriers to the right (in the $+x$-direction). When $\Delta\phi$ changes by $\sim \pi$ (to $\sim 3\pi/2$), carriers move to the left (in the -x-direction). We expect Δn at $x \simeq -7.5$ μm to be π out of phase with Δn at $x \simeq +7.5$ μm; the deviation from π in the data is reproducible and occurs because the relative phase fronts of the ω and 2ω beams are not perfectly parallel. Most importantly, there is almost no phase dependence to Δn at $x \simeq 0$, suggesting that an equal number of electrons move into and out of the probed region at $x \simeq 0$ for any fixed $\Delta\phi$.

The solid circles in Fig. 14(b) show peak values of phase-dependent oscillations like those in Fig. 14(a), plotted for a range of probe positions. These solid circles are a measure of the maximum Δn at each x position. For comparison, the open squares show $\Delta\mathrm{T}(n)$, which is a measure of the spatial profile of the background non-phase-dependent carrier density, $n(x)$. In addition, the dashed line in Fig. 14(b) is a Gaussian fit to $\Delta\mathrm{T}(n)$, and the solid line is the absolute value of the derivative of that Gaussian fit, with its overall amplitude scaled for comparison with the $\Delta\mathrm{T}(\Delta\phi)$ data. Notice that the peak value of $\Delta\mathrm{T}(\Delta\phi)$ as a function of x has approximately the same shape as $|\partial[\Delta\mathrm{T}(n)]/\partial x|$. This implies that Δn has approximately the same shape as $|\partial n/\partial x|$, which is consistent with the description surrounding Fig. 13(a)-(b), and verifies that there is carrier motion associated with a QUIC current. The absolute value arises because peak value is always taken to be positive.

Therefore, the data in Fig. 14 display the two most important features of the current previously observed using electrical detection in bulk GaAs [28, 29, 36]: (i) there is a net charge current injected along the direction of the ω and 2ω polarizations, and (ii) controlling $\Delta\phi$ allows control of the both magnitude of this current and whether it travels in the $+x$ or -x direction.

8.4 Pure Spin Current Measurements

In the next situation we study, ω and 2ω pulses have orthogonal linear polarizations such that ω is $\hat{\mathbf{x}}$ polarized and 2ω is $\hat{\mathbf{y}}$ polarized. In this case, we expect them to generate a pure spin current along $\hat{\mathbf{x}}$. The carrier densities created by ω and 2ω acting independently are again approximately equal, and here the total peak carrier density (at $x = 0$) is $\sim 4 \times 10^{17}$ cm^{-3} when both ω and 2ω excite the sample.

We again measure $\Delta\mathrm{T}(\Delta\phi)$ to study the phase-dependent local electron density, but in this case we study only the spin-up electron density at first. To monitor $\Delta n_\uparrow(x, \Delta\phi)$, the probe is $\hat{\sigma}^-$ polarized, and Fig. 15(a) plots $\Delta\mathrm{T}(\Delta\phi)$ for three probe positions along the x-axis. The phase-dependent spin-up densities $\Delta n_\uparrow$ at $x \simeq -7.5$ μm and at $x \simeq +7.5$ μm are roughly out of phase, and $\Delta n_\uparrow$ at $x \simeq 0$ shows almost no phase dependence. Thus, for a given $\Delta\phi$, spin-up electrons accumulate on one side of the excited region while being depleted on the other side. When $\Delta\phi$ changes by $\sim \pi$, the regions of accumu-

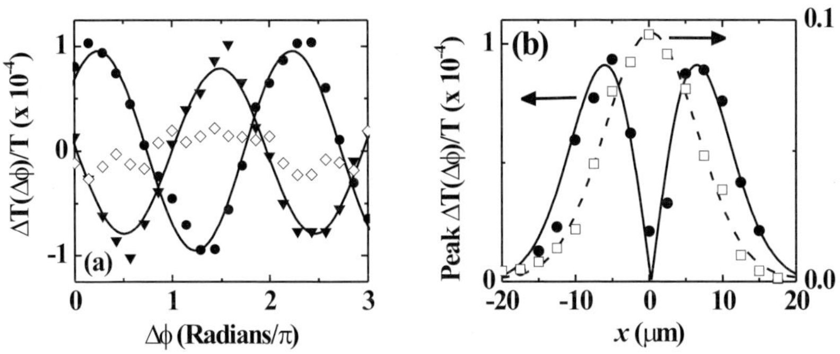

Fig. 15. (a) Phase-dependent ΔT of a $\hat{\sigma}^-$ probe at $x \simeq -7.5$ μm (solid triangles), $x \simeq 0$ (open diamonds), and $x \simeq +7.5$ μm (solid circles). (b) Spatially resolved ΔT. The open squares show $\Delta\mathrm{T}(n)/T$; the solid circles show the peak value of $\Delta\mathrm{T}(\Delta\phi)/T$. The dashed line is a Gaussian fit to $\Delta\mathrm{T}(n)/T$, and the solid line is the absolute value of the derivative of that Gaussian.

lation and depletion are reversed. In addition, Fig. 15(b) illustrates that the peak value of $\Delta\mathrm{T}(\Delta\phi)$ at each position follows $|\partial[\Delta\mathrm{T}(n)]/\partial x|$. The data in Fig. 15 therefore indicate that spin-up carriers move as part of a coherently controllable current. From the relative magnitudes of the solid and dashed curves in Fig. 15(b), we estimate that the spin-up carrier profile moves ~ 10 nm.

To verify that the spin-up current is part of a *pure spin current*, the results in Fig. 16 confirm that spin-down carriers travel in the opposite direction to spin-up carriers, yielding no net charge current. Here, the probe position is fixed at $x \simeq +7.5$ μm, and $\Delta\mathrm{T}(\Delta\phi)/\mathrm{T}$ is measured for three probe polarizations. The open triangles ($\hat{\sigma}^+$ probe) are proportional to $\Delta n_\downarrow(7.5\ \mu\mathrm{m},\ \Delta\phi)$, the solid circles ($\hat{\sigma}^-$ probe) to $\Delta n_\uparrow(7.5\ \mu\mathrm{m},\ \Delta\phi)$, and the open squares (linear probe) to $\Delta n(7.5\ \mu\mathrm{m},\ \Delta\phi)$. The phase-dependent changes in spin-up and spin-down densities have approximately equal amplitudes and are almost exactly π out of phase at this location; therefore, the spin-up and spin-down electrons do indeed move in opposite directions for a given $\Delta\phi$, as predicted by Eq. (49). Because the magnitudes of the $\hat{\sigma}^+$ probe and $\hat{\sigma}^-$ probe data are approximately equal at this and other positions (not shown), we estimate that each of the profiles sketched in Fig. 13(c) moves ~ 10 nm, resulting in a "spin separation" of ~ 20 nm. Additional measurements indicate that, in this regime, the spin separation increases at lower carrier densities, suggesting that carrier-carrier scattering is playing an important role in the relaxation of this pure spin current.

Meanwhile, the open squares in Fig. 16 demonstrate that there is no phase-dependence to the *total* carrier density at this position, verifying the absence of a net charge current. This corroborates the result of the electrode measurement in Fig. 12, which indicates an absence of net charge current along the direction of the ω polarization for orthogonal linear excitation. In addition, Hübner et al. have recently reported [38] the results of spatially resolved photoluminescence studies in ZnSe that provide independent confirmation of the existence of the pure spin current demonstrated by Figs. 15 and 16.

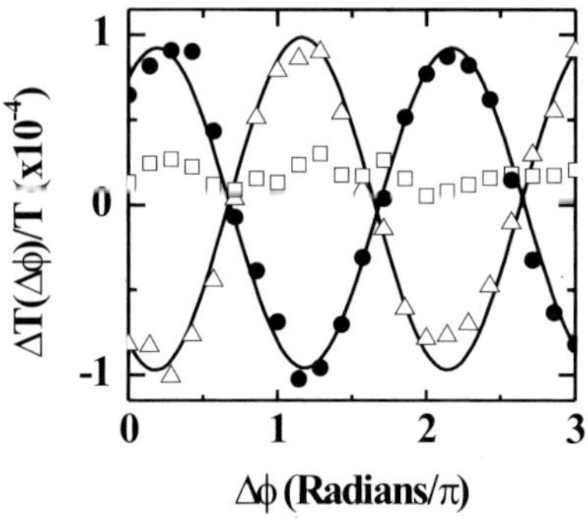

Fig. 16. Phase-dependent ΔT for three probe polarizations: $\hat{\sigma}^+$ (open triangles), $\hat{\sigma}^-$ (solid circles), and linear (open squares), all measured at $x \simeq +7.5\ \mu$m.

Finally, note that if ω and 2ω were coherently controlling the *overall* charge density [11, 12] or spin density [13], the results in Figs. 14 and 15 would be quite different. In the case of charge density control, Δn would have the same Gaussian spatial profile as n, Δn would be largest in the center, and the signals on the left and right sides would be in phase. Likewise, for spin density control, the difference between $\Delta n_\downarrow$ and $\Delta n_\uparrow$ would be largest at $x = 0$. As evidence of this, the probe is fixed at $x = 0$ for all measurements in Sect. 5; clearly, Δn and ΔS are quite large at $x = 0$ if there is control of the charge density or spin density. In addition, the ΔT($\Delta\phi$) data in Figs. 14 and 15 are not a measure of diffusion, because transport arising from diffusion – while present – will be independent of $\Delta\phi$.

8.5 Summary of Current Measurements

The data in Sects. 7 and 8 review our experimental observations of the three predicted currents illustrated in Figs. 4 and 9. For ω and 2ω having parallel linear polarizations, an unpolarized charge current is injected along the direction of the incident polarizations, and the magnitude and sign of this current is controlled by adjusting $\Delta\phi$. For ω and 2ω having the same circular polarization, we observe a charge current whose direction depends on $\Delta\phi$, but whose magnitude does not; according to the predictions of Bhat and Sipe, this charge current is spin-polarized [35]. For orthogonal linear polarizations, we observe a pure spin current along the direction of the ω polarization and a small charge current along the direction of the 2ω polarization.

9 Summary

Throughout this chapter, we have reviewed several ways that Quantum Interference Control (QUIC) of the transition amplitudes associated with a fundamental optical pulse (frequency ω) and a second harmonic pulse (2ω) can be used to coherently control both charge and spin in semiconductors. With ω and 2ω simultaneously present in a semiconductor, we have shown that these interference processes can lead to coherent control of the: (i) charge density; (ii) spin density; (iii) k-space distribution of charge, leading to a charge current; or (iv) k-space distribution of spin, leading to a spin current. Moreover, since we can choose the type of control by selecting the appropriate crystal orientations and ω and 2ω polarizations, we have demonstrated ways of controlling spin densities and spin currents without affecting the charge densities or charge currents, effectively decoupling the spin and charge degrees of freedom.

We have discussed several variations of an experimental technique that can control and detect the motion, density, and spin of optically injected carriers. Coherent control of the densities of charge and spin has been studied using an optical probe, and currents have been monitored both electrically and optically.

Acknowledgments

The experimental results reviewed here formed the basis for a dissertation for one of us (MS). The foundations of QUIC of populations and currents in semiconductors were laid by van Driel, Sipe, and their groups prior to the contributions that we review here. The numerous students and post docs who contributed to this early pioneering work over an extended period of time are acknowledged in the citations, and we are indebted to them. We also thank Eric Gansen, Scot Hawkins, Yaser Kerachian, Xinyu Pan, and Hui Zhao for many insightful conversations. Finally, we gratefully acknowledge the support of this work by the Office of Naval Research, the Army Research Office and the Defense Advanced Research Projects Agency.

References

1. S.A. Wolf, D.D. Awschalom, R.A. Buhrman, J.M. Daughton, S. von Molnar, M.L. Roukes, A.Y. Chtchelkanova, D.M. Treger: Science **294**, 1488 (2001)
2. D.D. Awschalom, M.E. Flatté, N. Samarth: Scientific American **286**, 66 (2002)
3. D.D. Awschalom, J.M. Kikkawa: Physics Today **52**, 33 (1999)
4. D.D. Awschalom, D. Loss, N. Samarth: *Semiconductor Spintronics and Quantum Computation* (Springer, Heidelberg 2002)
5. Y.H. Ahn, S.B. Choe, J.C. Woo, D.S. Kim, S.T. Cundiff, J.M. Shacklette, Y.S. Lim: Phys. Rev. Lett. **89**, 237403 (2002)

6. A.P. Heberle, J.J. Baumberg, E. Binder, T. Kuhn, K. Kohler, K. H. Ploog: IEEE J. Sel. Top. Quant. Electron. **2**, 769 (1996)
7. X. Marie, P. Lejeune, T. Amand, M. Brousseau, J. Barrau, M. Paillard, R. Planel: Phys. Rev. Lett. **79**, 3222 (1997)
8. N.H. Bonadeo, J. Erland, D. Gammon, D. Park, D.S. Katzer, D.G. Steel: Science **282**, 1473 (1998)
9. J.A. Gupta, R. Knobel, N. Samarth, D.D. Awschalom: Science **292**, 2458 (2001)
10. H.M. van Driel, J.E. Sipe: Coherence Control of Photocurents in Semiconductors. In: *Ultrafast Phenomena in Semiconductors*, ed. K.T. Tsen, (Springer, New York 2001) pp. 261-306
11. J.M. Fraser, A.I. Shkrebtii, J.E. Sipe, H.M. van Driel: Phys. Rev. Lett. **83**, 4192 (1999)
12. J.M. Fraser, H.M. van Driel: Phys. Rev. B **68**, 085208 (2003)
13. M.J. Stevens, R.D.R. Bhat, J.E. Sipe, H.M. van Driel, A.L. Smirl: Phys. Status Solidi B **238**, 568 (2003)
14. R. Fiederling, M. Keim, G. Reuscher, W. Ossau, G. Schmidt, A. Waag, L.W. Molenkamp: Nature **402**, 787 (1999)
15. M. Oestreich, J. Hübner, D. Hägele, P.J. Klar, W. Heimbrodt, W.W. Rühle, D.E. Ashenford, B. Lunn: Appl. Phys. Lett. **74**, 1251 (1999)
16. Y. Ohno, D.K. Young, B. Beschoten, F. Matsukura, H. Ohno, D.D. Awschalom: Nature **402**, 790 (1999)
17. M. Kohda, Y. Ohno, K. Takamura, F. Matsukura, H. Ohno: Jpn. J. Appl. Phys. Part 2-Letters **40**, L1274 (2001)
18. E. Johnston-Halperin, D. Lofgreen, R.K. Kawakami, D.K. Young, L. Coldren, A.C. Gossard, D.D. Awschalom: Phys. Rev. B **65**, 041306 (2002)
19. P.R. Hammar, B.R. Bennett, M.J. Yang, M. Johnson: Phys. Rev. Lett. **83**, 203 (1999)
20. S. Gardelis, C.G. Smith, C.H.W. Barnes, E.H. Linfield, D.A. Ritchie: Phys. Rev. B **60**, 7764 (1999)
21. H.J. Zhu, M. Ramsteiner, H. Kostial, M. Wassermeier, H.P. Schonherr, K.H. Ploog: Phys. Rev. Lett. **87**, 016601 (2001)
22. S.F. Alvarado, P. Renaud: Phys. Rev. Lett. **68**, 1387 (1992)
23. V.P. Labella, D.W. Bullock, Z. Ding, C. Emery, A. Venkatesan, W.F. Oliver, G.J. Salamo, P.M. Thibado, M. Mortazavi: Science **292**, 1518 (2001)
24. F. Meier, B.P. Zakharchenya eds.: *Optical Orientation: Modern Problems in Condensed Matter Sciences*, v. 8 (North-Holland, Amsterdam 1984)
25. D. Hägele, M. Oestreich, W.W. Rühle, N. Nestle, K. Eberl: Appl. Phys. Lett. **73**, 1580 (1998)
26. J.M. Kikkawa, D.D. Awschalom: Nature **397**, 139 (1999)
27. I. Malajovich, J.J. Berry, N. Samarth, D.D. Awschalom: Nature **411**, 770 (2001)
28. A. Haché, Y. Kostoulas, R. Atanasov, J.L.P. Hughes, J.E. Sipe, H.M. van Driel: Phys. Rev. Lett. **78**, 306 (1997)
29. A. Haché, J.E. Sipe, H.M. van Driel: IEEE J. Quant. Electron. **34**, 1144 (1998)
30. E. Dupont, P.B. Corkum, H.C. Liu, M. Buchanan, and Z.R. Wasilewski: Phys. Rev. Lett. **74**, 3596 (1995)
31. N. Laman, A.I. Shkrebtii, J.E. Sipe, H.M. van Driel: Appl. Phys. Lett. **75**, 2581 (1999)
32. S.D. Ganichev, E.L. Ivchenko, S.N. Danilov, J. Eroms, W. Wegscheider, D. Weiss, W. Prettl: Phys. Rev. Lett. **86**, 4358 (2001)

33. S.D. Ganichev, W. Prettl: J. Phys.: Condens. Matt. **15**, R935 (2003)
34. S.D. Ganichev, E.L. Ivchenko, V.V. Bel'kov, S.A. Tarasenko, M. Sollinger, D. Weiss, W. Wegscheider, W. Prettl: Nature **417**, 153 (2002)
35. R.D.R. Bhat, J.E. Sipe: Phys. Rev. Lett. **85**, 5432 (2000)
36. M.J. Stevens, A.L. Smirl, R.D.R. Bhat, J.E. Sipe, H.M. van Driel: J. Appl. Phys. **91**, 4382 (2002)
37. M.J. Stevens, A.L. Smirl, R.D.R. Bhat, A. Najmaie, J.E. Sipe, H.M. van Driel: Phys. Rev. Lett. **90**, 136603 (2003)
38. J. Hübner, W.W. Rühle, M. Klude, D. Hommel, R.D.R. Bhat, J.E. Sipe, H.M. van Driel: Phys. Rev. Lett. **90**, 216601 (2003)
39. M.J. Stevens, A. Najmaie, R.D.R. Bhat, J.E. Sipe, H.M. van Driel, A.L. Smirl: J. Appl. Phys. **94**, 4999 (2003)
40. P. Pfeffer, W. Zawadzki: Phys. Rev. B **53**, 12813 (1996)
41. E.O. Kane: J. Phys. Chem. Solids **1**, 249 (1957)
42. C.C. Lee, H.Y. Fan: Phys. Rev. B **9**, 3502 (1974)
43. M. Sheik-Bahae: Phys. Rev. B **60**, 11257 (1999)
44. D.C. Hutchings, B.S. Wherrett: Phys. Rev. B **49**, 2418 (1994)
45. R.D.R. Bhat, J.E. Sipe: Private Communication (2003)
46. R. Atanasov, A. Haché, J.L.P. Hughes, H.M. van Driel, J.E. Sipe: Phys. Rev. Lett. **76**, 1703 (1996)
47. D. Côté, N. Laman, A. Springthorpe, and H.M. van Driel, (unpublished)
48. D.H. Marti, M.-A. Dupertuis, and B. Deveaud, Phys. Rev. B **69**, 35335 (2004)
49. P. Kral and J.E. Sipe, Phys. Rev. B **61**, 5381 (2000)
50. A. Najmaie, R.D.R. Bhat, J.E. Sipe: Phys. Rev. B **68**, 165348 (2003)
51. C.F. Klingshirn: *Semiconductor Optics*, 2nd corr. printing (Springer, Berlin Heidelberg New York 1997)
52. M. Lax: *Symmetry principles in solid state and molecular physics* (Dover edition, Mineola, N.Y., 2001)
53. M.D. Dvorak, W.A. Schroeder, D.R. Andersen, A.L. Smirl, B.S. Wherrett: IEEE J. Quant. Electron. **30**, 256 (1994)
54. J.P. van der Ziel: Phys. Rev. B **16**, 2775 (1977)
55. E.L. Ivchenko: Sov. Phys. Solid State **14**, 2942 (1973)
56. S.B. Arifzhanov, E.L. Ivchenko: Sov. Phys. Solid State **17**, 46 (1975)
57. P. Nemec, R.D.R. Bhat, Y. Kerachian, J.E. Sipe, H.M. van Driel, A.L. Smirl: Private Communication (2003)
58. F.A. Majumder, H.-E. Swoboda, K. Kempf, C. Klingshirn: Phys. Rev. B **32**, 2407 (1985)
59. L.M. Smith, D.R. Wake, J.P. Wolfe, D. Levi, M.V. Klein, J. Klem, T. Henderson, H. Morkoç: Phys. Rev. B **38**, 5788 (1988)
60. H.F. Hess, E. Betzig, T.D. Harris, L.N. Pfeiffer, K.W. West: Science **264**, 1740 (1994)
61. J.R. Guest, T.H. Stievater, G. Chen, E.A. Tabak, B.G. Orr, D.G. Steel, D. Gammon, D.S. Katzer: Science **293**, 2224 (2001)
62. D.T. Pierce, F. Meier: Phys. Rev. B **13**, 5484 (1976)
63. J.B. Stark, W.H. Knox, D.S. Chemla: Phys. Rev. Lett. **68**, 3080 (1992)
64. A.R. Bungay, S.V. Popov, I.R. Shatwell, N.I. Zheludev: Phys. Lett. A, **234**, 379 (1997)
65. T. Matsuyama, H. Horinaka, K. Wada, T. Kondo, M. Hangyo, T. Nakanishi, S. Okumi, K. Togawa: Jpn. J. Appl. Phys. Part 2-Letters **40**, L555 (2001)

66. D.J. Hilton, C.L. Tang: Phys. Rev. Lett. **89**, 146601 (2002)
67. N.H. Bonadeo, D.G. Steel, R. Merlin: Phys. Rev. B **60**, 8970 (1999)
68. S.A. Hawkins, M.J. Stevens, A.L. Smirl: Phys. Rev. B **64**, 035302 (2001)
69. P.C. Becker, H.L. Fragnito, C.H. Brito Cruz, R.L. Fork, J.E. Cunningham, J.E. Henry, C.V. Shank: Phys. Rev. Lett. **61**, 1647 (1988)
70. D. Côté, J.M. Fraser, M. Decamp, P.H. Bucksbaum, H.M. van Driel: Appl. Phys. Lett. **75**, 3959 (1999)

Semiconductor Quantum Dots for Optoelectronic Applications

Tommy W. Berg and Jørn M. Hvam

1 Introduction

The early predictions and investigations of the optical properties of semiconductor quantum dots (QDs) focused on the improvement of already known semiconductor lasers and amplifiers by the use of low-dimensional active regions [1, 2]. Today the scope has broadened to include quantum computing [3], biophysics [4], efficient switching nodes [5], single photon emitters [6], widely tunable lasers [7], receivers and detectors of electromagnetic signals in the far infrared regime [8], and new emitters for the access and metro markets [9, 10]. This development has been driven by the desire to understand the new physics of QDs and QD-devices compared with the previously known opto-electronic semiconductor devices.

The improvement in growth and processing techniques over the past decade has allowed for the realization of high-quality QD devices suitable for the market of optical communication. Initially, most work was focused on the realization and characterization of lasers, but within recent years the QD-based semiconductor optical amplifier (SOA) has drawn a fair share of the attention [5, 11]. The QD-SOA is a natural extension of the bulk and quantum well (QW) based SOAs, offering many potential advantages as a consequence of the lower dimensionality. Among the most promising of these advantages are the potential of realizing high-quality linear SOAs, which may compare favourably with fiber amplifiers regarding saturated output power, noise figure, small signal gain, gain bandwidth, and lack of cross gain effects. Another potential application of the QD SOA is as high speed signal processing element [5].

2 Quantum Dot Fundamentals

A lot of research has been carried out on QD-based opto-electronic devices. We shall here discuss some of the advantages and disadvantages of using QDs as active material with the main focus on QD lasers, which is where most work has been carried out so far. The results specific to QD-SOAs will be given in Sect. 3.

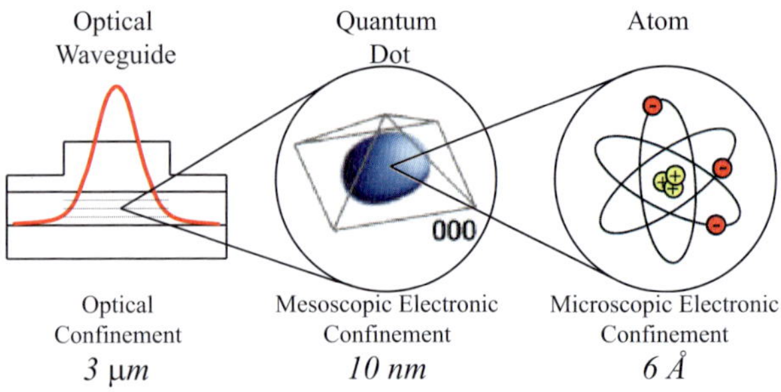

Fig. 1. Illustration of the size relationship between an optical waveguide (e.g. laser or amplifier), a quantum dot and an atom. The optical waveguide defines the profile of the optical field and the sphere inside the QD illustrates the wavefunction of the electronic ground state (from [12]). The numbers at the bottom give the typical dimension of each of the structures.

2.1 Definition of a Quantum Dot

Generally speaking a QD is a region of material with a low electronic potential compared with the surrounding material, which confines charge carriers in all three dimensions. The electronic confinement results in a complete localization of the envelope wavefunctions of the charge carriers and, consequently, QDs are often referred to as zero-dimensional. As in an optical waveguide, which must be of the size the wavelength of light in order to confine the light, the extent of the confining potential of a QD should be in the order of the wavelength of the carriers it confines. The effective wavelength of the carriers is the de Broglie wavelength, which in a typical semiconductor lattice at room temperature is about 10 nm [13] and consequently this is the typical scale of a semiconductor QD. As indicated in Fig. 1, a QD is much smaller than the dimensions of an optical waveguide, but on the other hand much larger than the lattice constant of the host semiconductor material. A QD thus typically contains several thousands of atoms allowing the use of traditional semiconductor theory (Bloch's theorem, band structure, effective masses, etc.).

The small size of the confining potential leads to the presence of discrete states, which is one of the central characteristics of QDs. They are therefore also named 'artificial atoms' . When the energy splitting between states is smaller than the homogeneous broadening, the states appear as a continuum. The nature of the active states of a given structure or device is governed by the density of states (DOS), which is the number of carrier states per unit energy and volume. Figure 2 shows schematically the DOS of bulk, QW, and QD materials in the case where no broadening mechanisms are taken into account. The bulk DOS follows the traditional square-root dependence, the QW DOS is a staircase, and the DOS for the QD material is a collection of delta-functions. The lowest state for the bulk material is at the band edge

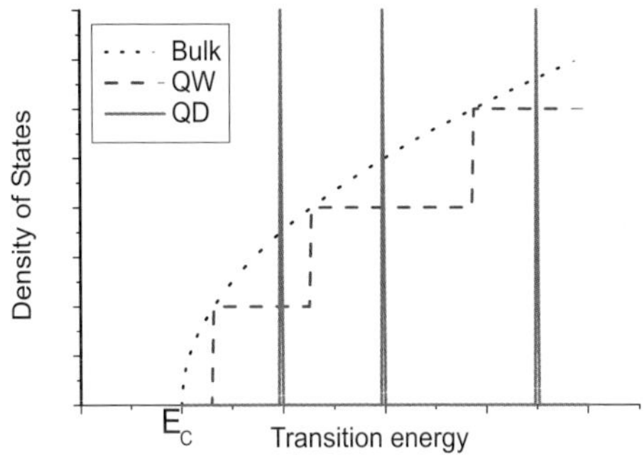

Fig. 2. Schematic DOS for a bulk, QW and QD device. E_C represents the conduction band edge.

of the material (E_C). The QW ground state (GS), however, is slightly above due to the quantization energies of electron and holes. Similarly, the ground state energy of the QD is further above the band edge due the confinement in all three dimensions.

2.2 Properties of an Ideal QD Device

An ideal QD device consists of a perfect array of identical, non-coupling QD's, where only electron and hole states of the same quantum numbers couple (sharp selection rules, the confining potential is considered to be infinitely deep). In this case the characteristics of the whole QD ensemble is a perfect match to that of a single dot, except that there is a larger overlap between the optical field and active carrier states. As a result of the simple nature of the ideal system, it is easy to make a number of predictions/observations on some of the significant characteristics of QDs. They may not in all cases materialize in real devices, but still indicate some of the potential advantages of QD-devices.

High Material Gain

One of the very first predictions on the advantages of QD devices, was published by Asada et al. [1] and regarded the material gain of low-dimensional structures. The material gain is the gain per volume of active material under the assumption of complete overlap with the optical field. The results of their calculations are seen in the left part of Fig. 3, showing the material gain of four devices made from the same active material, but with different dimensionality of the active region. The QD device material is found to exhibit significantly larger peak gain than any of the other structures and is centered at a wavelength of 1.3 μm, which is much lower than the bulk device at 1.55 μm. The difference in wavelength is caused by the quantization energy of the QD device, which is also visible in the DOS in Fig. 2. The higher material

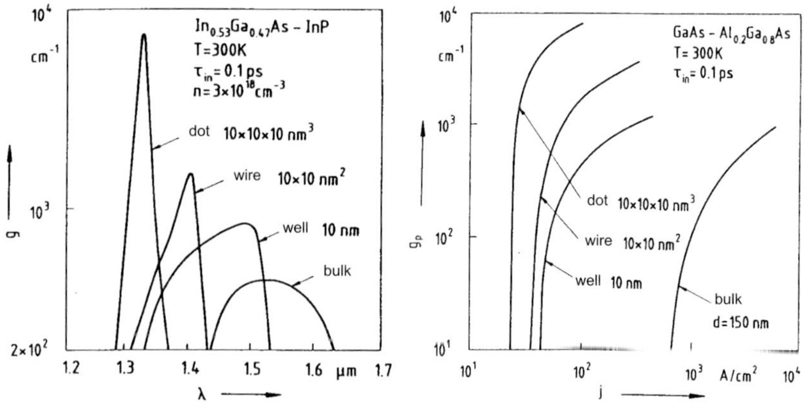

Fig. 3. Left: Material gain of bulk, well, wire, and quantum box from the same material and using equal carrier densities. Right: Peak gain for the same device species, but for a different material system, as a function of current density (from [1]).

gain is mainly caused by the smaller active volume of the QDs, which is used to normalize the material gain.

Low Threshold/Transparency Current Density

From the delta-like DOS of QD material it is clear that less carriers are required to reach a certain level of population inversion of the optically active states as compared with bulk and QW material. In devices of higher dimensionality all states at energies below the optically active states, have to be filled first before inversion is achieved at the optically resonant energies. In contrast to this the large splitting between the states of an ideal QD device means that only the optically active states are at play and these are thus readily inverted. The consequence of this is that QD lasers are expected to have a smaller threshold current density, and amplifiers to have a smaller transparency current. This effect has been demonstrated experimentally as shown in Fig. 4, displaying the record threshold current density vs. year for bulk, QW and QD lasers.

High Differential Gain

Another consequence of the easily achievable inversion is a high differential gain. Thus when every carrier enters directly into the optically active states, the gain changes more rapidly than when the carriers have to be distributed over a wide energy range as in bulk and QW devices. This is illustrated in the right half of Fig. 3, which shows the peak gain of the different species

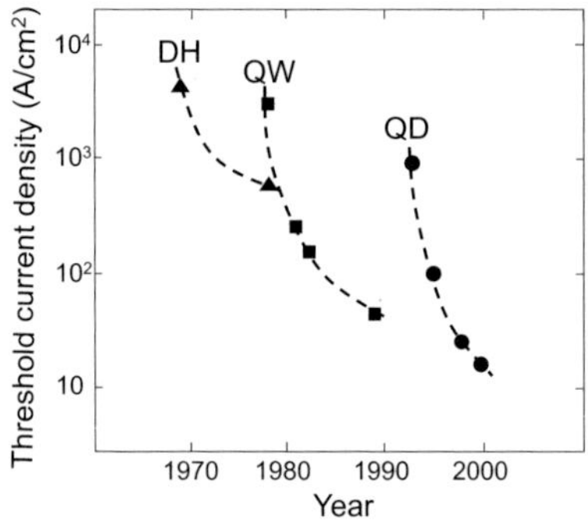

Fig. 4. The development of the threshold current density of semiconductor lasers at room temperature. Triangles mark bulk double heterostructure (DH) lasers, squares mark QW lasers, and circles mark QD lasers.

of devices as function of current density. The slope of these curves are proportional to the differential gain. Clearly the QD device has a much higher differential gain than any of the other devices. This is considered important for the potential realization of fast directly modulated devices, since only a limited number of carriers has to be moved to change the gain significantly.

High Temperature Stability

Due to the strong electronic confinement in the dots it is difficult to remove carriers from the dot by thermal excitation. In fact, the ideal QD device is completely unaffected by temperature fluctuations, in contrast to bulk and QW devices where a change in the carrier temperature results in a redistribution of carriers in the device. Since temperature control is costly, bulky, and troublesome, temperature insensitivity is a most desirable feature of QDs for device purposes.

Narrow and Symmetric Gain Spectrum

Since the gain of a device is proportional to the DOS, the delta-like DOS of the ideal QD system leads to a symmetric gain spectrum with a width determined purely by the homogeneous broadening of the system (see Fig. 3). This means that all carriers in the device interact with photons of the same wavelength.

Zero Linewidth Enhancement Factor

The linewidth enhancement factor, or α-parameter, describes the degree of coupling between gain changes and refractive index changes in a device. The size of α is proportional to the index change n_r' (varying the carrier density) divided by the differential gain, g', i.e. $\alpha \propto n_r'/g'$ [14]. The real and imaginary parts of the complex susceptibility of an optical device are related to each other through the Kramers–Kronig relation [15] and consequently gain and index changes are interdependent. There are two factors, which cause the ideal QD device to have a small linewidth enhancement factor. Firstly the differential gain is high compared to other devices. Secondly the differential refractive index is zero at the center energy for a symmetric gain spectrum [13], which causes the linewidth enhancement factor to vanish at the resonance energy. For directly modulated lasers a non-zero α-parameter causes chirp, i.e. change of emission wavelength over a pulse, and for SOAs and absorbers it causes phase changes of the signal during amplification or absorption.

2.3 Realistic QD Properties

One of the most significant differences between realistic and ideal QD devices is the presence of inhomogeneous broadening, which effectively removes the delta-like DOS. Inhomogeneous broadening is caused by size fluctuations of the individual dots within the sample and is unavoidable for large ensembles of QDs with the growth techniques used today, which are mainly based on self-assembly [16].

Low Modal Gain

Despite the fact that the material gain of QD devices is high, the modal gain is usually low due to a small overlap between the QDs and the optical field (the confinement factor). Consequently QD devices usually exhibit lower gain than bulk and QW devices. Modal gains for QD devices in the range $5-20\,\mathrm{cm}^{-1}$ have been reported [17–19]. The modal gain of QD devices can be enhanced somewhat by improving the confinement factor through an increase of the 2D density of dots in a layer and/or the number of active dot layers in the device. The small confinement factor also influences the differential gain. Still, the differential gain of QD devices is commonly reported to be higher than for bulk and QW devices and a value as high as $2.7\times10^{-14}\,\mathrm{cm}^2$ [20] has been reported at room temperature and even higher at lower temperatures [21].

Phonon Bottleneck

In higher-dimensional structures the longitudinal optical (LO) phonons interact strongly with the carriers providing an efficient way of carrier relaxation. In QD devices, however, the discrete nature of the state spectrum means that this relaxation mechanism is inhibited when the state splitting is different from the LO phonon energy (typically around 30 meV). This effect is referred to as the phonon bottleneck, the presence of which has been discussed strongly in the literature over the years. Several other relaxation mechanisms, such as multi phonon relaxation and electron hole scattering have been suggested to substitute for the LO phonon relaxation and the actual mechanism is probably a mixture of several of the suggested processes [22, 23].

Normal Threshold/Transparency Current

For bulk and QW devices, a low threshold/transparency current density is a witness of high material and device quality. For a QD device the total threshold current is of more significance. If we imagine a QD laser containing a small 2D density of QDs it would require a low current density to invert it resulting in a very low threshold current density. On the other hand the modal gain would also be extremely small and the laser would have to be long in order achieve lasing, which would result in a high total threshold current. In general QD devices do not require much less current to achieve lasing or transparency, since the advantage of a delta-like DOS is reduced by the presence of inhomogeneous broadening. One potential advantage for the QD device is that the waveguide losses are usually low in these structures, which means that less inversion is necessary in order to achieve lasing or amplification.

Difficulty in Reaching 1.55 μm Wavelength Operation

As seen in Fig. 3 the emission wavelength of QDs made from the same material as a bulk device emits at a significantly shorter wavelength due to the quantization energy. It has thus been a main obstacle for many years to reach either lasing or amplification at the important optical communication wavelength of 1.55 μm. The problems lie in avoiding strong non-radiative recombination through defects and at same time retain the essential QD properties. A natural solution would be to go towards larger QDs in order to minimize the quantization energy, however, this would also diminish the discrete nature of the states and thus remove the desired characteristics.

Recently the goal has been achieved to a certain degree by two different approaches. One approach is to grow InAs on InP substrate [24]. This results in so-called quantum dashes, which are elongated structures resembling quantum wires more than QDs. The emission wavelength in this case can be tuned all the way from 1.3 μm to around 2.0 μm simply by controlling the amount of material deposited. Another approach is based on InAs/InGaAs QDs grown on GaAs substrate applying pseudomorphic buffer layers, which has led to an emission wavelength of 1.49 μm [25]. Lasing has been demonstrated, but it is not yet clear how QD-like this structure is.

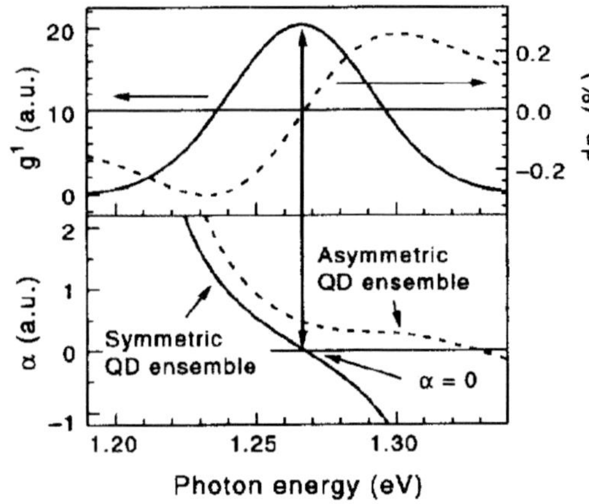

Fig. 5. The influence of an asymmetric gain spectrum on the linewidth enhancement factor (from Ref. [29]). Upper: Gain and differential refractive index for a symmetric gain spectrum. Lower: Linewidth enhancement factor as a function of photon energy for a symmetric and an asymmetric measured gain spectrum.

Finite Linewidth Enhancement Factor

Due to inhomogeneous broadening and the presence of excited QD states, wetting layer (WL), and barrier states, the gain spectrum of a QD device is not symmetric and as a consequence, the linewidth enhancement factor is not zero at the center of the GS transition peak. This is illustrated in Fig. 5. The effect of including the asymmetry in the calculations is seen to be an increase of the α-parameter across the entire spectrum. The value below 0.5 found in the proximity of the gain peak is, however, still lower than the smallest values measured for QW devices [26] and significantly lower than those of bulk devices, which typically have linewidth enhancement factors in the order of 2–10 [27]. Values as low as 0.1 and even negative values have been measured for QD devices [28] indicating that there is an advantage in using QD material in this respect.

Broad Gain Spectrum

The inhomogeneous broadening of real QD devices means that the narrow gain spectrum cannot be realized and in many cases the width of the gain spectrum is actually broader than that of bulk and QW devices. This opens up for new applications such as widely tuneable lasers and broadband SOAs. So far QD lasers with a tunability of 201 nm have been realized using InAs QDs in a $In_{0.2}Ga_{0.8}As$ QW [7]. Since there are only a finite number of states in a given QD device to deliver the gain, there is a trade-off between gain bandwidth and the amplitude of the gain at a given wavelength. Since the modal gain of QD devices is already low there is a limit to how broad the gain spectrum can be made before the modal gain becomes comparable to the waveguide loss and lasing or amplification becomes impossible.

High T_0

The change of threshold current with temperature of a laser is typically described by the characteristic temperature T_0, which is the temperature span over which the threshold current increases by a factor e [30]. A high T_0 thus indicates little temperature sensitivity of the laser. The active QD states couple with both the excited states, the WL and the barrier material, which causes a certain degree of temperature dependence of the carrier distribution. It is not only the energy separation between states, but also the ratio of the DOS of the QD to the continuum, which matters. Thus if the continuum contains a very high number of states compared with the QDs, even a small overlap of the distribution function will have severe consequences for the inversion of the QD states when the temperature is varied [31]. Another important aspect is the quality of the WL and barrier material. If the WL contains a high number of defects, which cause non-radiative recombination, the temperature sensitivity will be increased significantly [32]. At low temperatures, lasers exhibiting T_0 as high as 700 K have been realized [33]. At room temperature, the T_0 parameter is more commonly reported to be below 100 K [34], which is not better than many QW lasers [35].

High-Power Lasers

The final property of realistic QD devices is the possibility of high-power lasers. The limiting effects for the output power of semiconductor lasers is typically catastrophical facet damage caused by the high output power or heating from the carriers injected into the device in order to sustain the high rate of stimulated emission. QD materials posses some advantages reducing these problems. In bulk or QW lasers the presence of self phase modulation causes a beam filamentation with a higher than average power density, which eventually causes catastrophic optical mirror damage. The smaller linewidth enhancement factor of QDs, however, means that less self phase modulation is present and as a result filamentation of the beam is suppressed [36]. In bulk and QW lasers carriers can diffuse towards the facets where there is chance of recombining through surface states and as a result the facet is heated more than the rest of the device, which eventually leads to thermal problems. This effect is suppressed in QDs due to the confinement of carriers in all directions, which effectively limits carrier diffusion towards the facets. The final effect which may help the QD laser achieve high output power is the lower threshold current density. Output powers as high as 11.7 W in pulsed mode [37] and more than 6 W in CW mode [38] have been achieved.

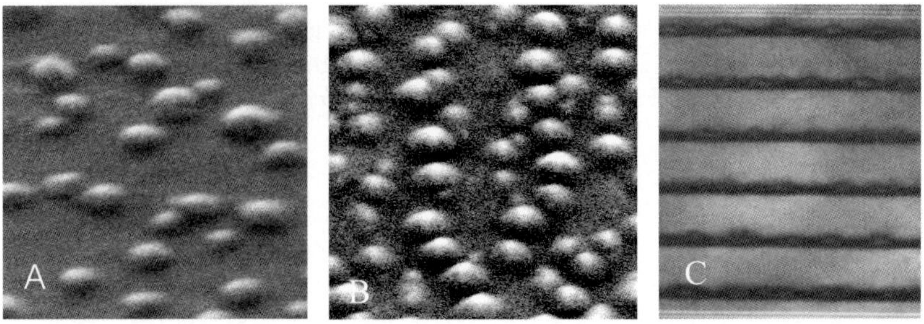

Fig. 6. (a) SEM picture of uncapped InAs QDs on a GaAs substrate with a density of $3 \times 10^{10}\,\mathrm{cm}^{-2}$. (b) Same as a) but for InAs QDs on $Ga_{0.85}In_{0.15}As$ with a surface density of $1 \times 10^{11}\,\mathrm{cm}^{-2}$. (c) XTEM from the side of 6 layers of InAs QDs with GaAs barriers (courtesy of J.P. Reithmaier).

3 Quantum Dot Semiconductor Optical Amplifiers

In this section we will focus our attention on QD SOAs by first discussing a number of elements and physical mechanisms playing a crucial role in understanding the properties of QD amplifiers. Later we will discuss some specific features of QD SOAs.

3.1 Dot Growth

It is only within the last decade that it has become possible to produce large arrays of nearly uniform dots on a regular basis. The fabrication techniques can be split into three main categories. The first technique ever used to realize QDs was by inclusion of nano-size semiconductor material in glass, dating back to 1932 [13]. The basis of the technique is to include semiconductor elements such as CdS and CdSe (and their alloys) in a supersaturated glass matrix [39]. One application of QDs produced in this matter is in colour filters. This method is, however, not suitable for electro-optic purposes, since the surrounding glass effectively isolates the semiconductor material electrically.

The second category of techniques is much broader and includes all forms of processing related production, i.e. using lithography and etching techniques. One example of the techniques of this category is dry etching of pillars from a thin film structure. In this way fairly regular dots can be produced in almost perfect arrays [13], but the lateral dimensions are limited by the lithographic processes, which at best can be in the 10 nm range [39]. The etching process, however, causes serious damage to the semiconductor material over a range of 50 nm [39], which constitutes a serious limitation for using this method at the smallest possible scales.

The third, and today most important method of making QDs is by self-assembly. The Stranski-Krastanov growth mode dominates, and the first

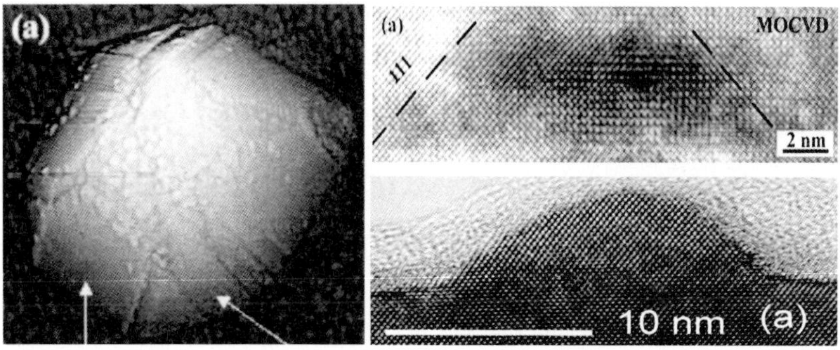

Fig. 7. Left: Scanning tunnelling microscope topography of 50×50 nm InAs QD [41]. Right: Cross-sectional high-resolution TEM micrograph of (top) an MOCVD grown $In_xGa_{1-x}As$ QD [40] and (bottom) an uncapped InAs QD [41].

demonstration of this method for the realization of self-assembled QDs was reported in 1985 by Goldstein et al. [16], growing InAs QDs on a GaAs substrate. The main idea of the technique is to deposit the dot material onto a substrate with a different lattice constant by metal-organic vapor phase epitaxy (MOVPE), molecular beam epitaxy (MBE), or similar techniques. As a result of the lattice mismatch between the two materials, the dot material will tend to cluster in small islands minimizing the surface strain, and hence forming the dots. When the dots have reached a certain size and/or density they are covered with a high band-gap (barrier) material (possibly made from the same material as the substrate). A new layer of dots can then be grown on top of a spacer layer.

Three examples of self assembled QDs can be seen in Fig. 6. The size, shape, material composition and surface density of the QDs depends strongly on the substrate material and the growth procedure, i.e. temperature, duration, and number of interrupts. The shape of self-assembled QDs is [40, 41] often found to be either lens-like or a truncated pyramid (see Fig. 7). The base length varies from a few nanometers to several tens of nanometers and the height is usually in the order of a few nanometers.

The exact nature of QDs is difficult to take accurately into account in simple device models. First of all the properties vary from device to device and secondly a single device contains millions of QDs, which among themselves vary widely. When modelling QD devices, it is essential to reduce the input to a smaller number of key parameters such as single-dot gain, dot volume, confinement energies, and capture times. These parameters are further averaged over the large ensembles of QDs within the device.

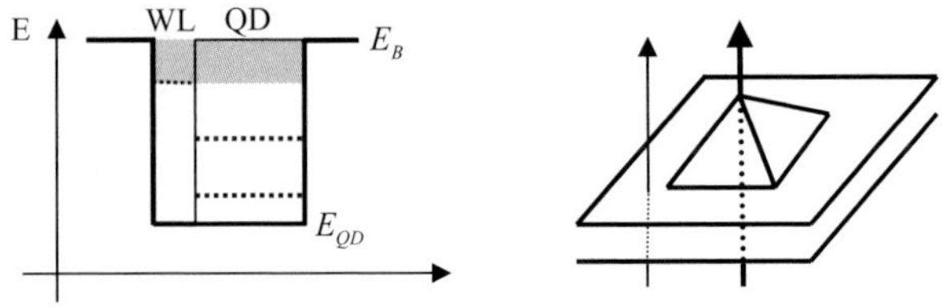

Fig. 8. Schematic variations of the band edge energy along two specific lines (arrows shown to the right). The thin line shows the case for the pure WL (between dots) and the thick line for the combined QD and WL (through the center of a dot). The horizontal dotted lines indicate the presence of confined states and the grey shaded area indicates a continuum of states.

3.2 Wetting Layer

In the Stranski-Krastanov growth mode a thin layer of the QD material, the wetting layer (WL), is present beneath each layer of QDs. The WL width is commonly between one and two monolayers of material [42], i.e. in the order of a nanometer. The presence of the WL can clearly be seen in Fig. 6 c).

Figure 8 shows schematically the confining potential of the combined QD and WL (the thick line) and for the WL alone (the thin line). The WL can be considered as a thin QW with its bound states relatively close to the barrier band edge. The QW nature of the WL means that a continuum of states is present in the WL (indicated by the area with grey shading in the figure). Since the volume of the combined QD and WL evidently is larger than the WL itself, the bound states of the QDs will always be lower than the pure WL state as indicated in the figure.

Due to the continuous nature of the WL states and the energy separation between the highest QD state and the lowest barrier states, the WL acts as a transport medium of carriers from the barrier into the QDs. Hence, the WL plays a crucial role in the dynamic properties of QD devices and acts as a nearby carrier reservoir for the QD states. The role of the WL has been sought enhanced even further by embedding each QD layer in a QW. This is often referred to as Dots-in-a-well or DWELL structures [43]. The effect of this approach is to enhance the rate and efficiency of the capture of carriers into the dots, thus enhancing the operation speed and wall plug efficiency of these devices. An alternative, but similar, approach is to grow an extra layer of material on top of the QDs, a so-called overgrowth layer, with a bandgap in-between the QD and the barrier material [44]. In this way the QDs are effectively sandwiched between two narrow QWs, which leads to the same type of behavior as for the DWELL structures.

3.3 Device Geometry and Structure

The geometry of the type of SOAs considered here is illustrated in Fig. 9. It is a ridge waveguide of length L, width W, and height H. The active region

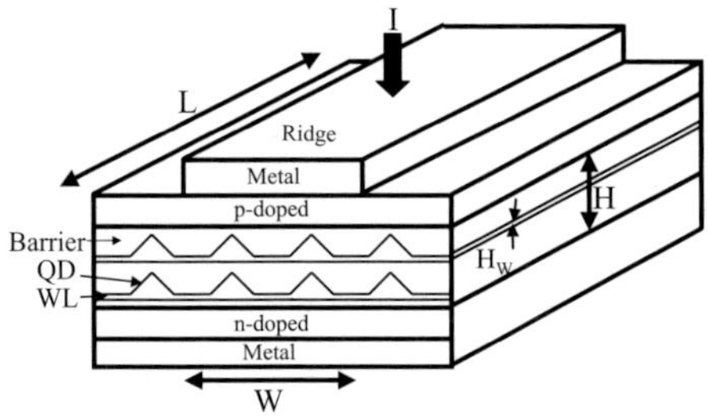

Fig. 9. Schematic illustration of the waveguide structure.

contains a number of QD layers, n_l, and beneath each layer a WL of width H_W is present. The surface density of dots is denoted D_{2D} and the volume of an average dot in the ensemble V_0.

Current is supplied through the top and bottom contacts, and carriers enter the un-doped active region by drift and diffusion through the p- and n-doped regions. Optical confinement is supplied by a separate confinement heterostructure (SCH). The design of the SCH is of importance for the carrier transport times into the active regions, the differential gain, the confinement factor and for the coupling losses in and out of the amplifier. Typical values of the confinement factor of the active region (overlapping with the waveguide) are $0.2 - 0.4$. A low reflectivity at the facets of the device is typically realized by the use of either anti-reflection (AR) coatings, tilted facets or a combination of these. In lasers the facets are either un-coated or high-reflection (HR) coated.

3.4 Density of States

The DOS of an idealized QD is very simple and consists of a set of discrete states in both the conduction and valence band and the transitions of the system obey strict selection rules. Unfortunately, this simple picture does not hold in reality. In general the conduction band contains a relatively small number of well-separated bound states. In comparison, the valence band contains, due to the larger hole masses, a higher number of closer lying states. The largest uncertainty is regarding the allowed transitions, which in some investigations take place only between states with the same quantum numbers [45], and in other investigations between states with different quantum numbers [46]. Thus, it is difficult to include the exact nature of all the individual QDs of an ensemble in device models.

Figure 10 illustrates the conduction band density of states (DOS) of a QD device. Each dot contains a spin-degenerate ground state (GS) and a four-fold degenerate excited state (ES). Due to size and composition fluctuations among the dots, these states are inhomogeneously broadened with a full width at half maximum (FWHM) of γ_G and γ_E, respectively. Above the QD states,

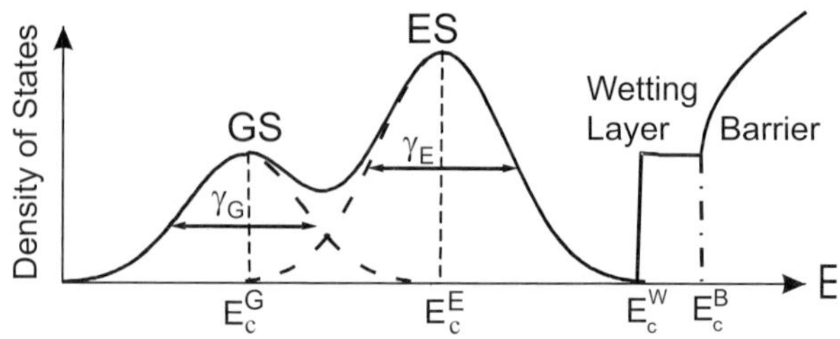

Fig. 10. Schematic illustration of the conduction band density of states for a QD device including ground state (GS), excited state (ES), wetting layer and barrier.

the step-like DOS typical of a QW is visible. We assume that due to the narrowness of the WL, typically 1–2 nm [47], only a single electronic subband exists. Finally, the barrier material is treated as a bulk-like region, leading to a square root dependence of the DOS at high energies.

Due to the vastly different volumes of the three different regions (QD, WL and barrier), the absolute number of states in the QDs is much smaller than that of the WL, which in turn is much smaller than that of the barrier. The relative number of states in these regions has profound consequences for the saturation properties of the device [31, 48]. Since the electron mass is usually smaller than the hole mass, the conduction band (electron) properties usually dominate the overall optoelectronic properties of QD devices. However, in order to calculate the true interband DOS (optical spectra) and the degree of inversion (band filling), it is necessary to take into account also the valence band DOS and to adopt some selection rules. Various more or less crude approximations have been applied in the literature [5, 49].

3.5 Band Filling Effects

One of the most fundamental properties of QD amplifiers is the filling of the active states with current in the absence of an optical signal. The degree of population inversion as a function of applied current is of importance for both the differential gain and the saturation properties of an amplifier and some of the unique characteristics of QD amplifiers, as compared to bulk or QW SOAs, can be traced back to this origin. Fig. 11 shows the occupation probability of the active states (proportional to the normalized modal gain at a certain wavelength) as a function of the total carrier density injected in devices with bulk, QW and QD active material.

For the QD case (solid line) the inversion is seen to increase nearly linearly with a high slope for small carrier (current) densities, but then quickly saturate as the active states become filled. The same fundamental behavior is observed for the QW case (dotted line), except a higher carrier density is needed to reach the same level of inversion. The main reason for this difference is the higher density of active states in the QW case compared with

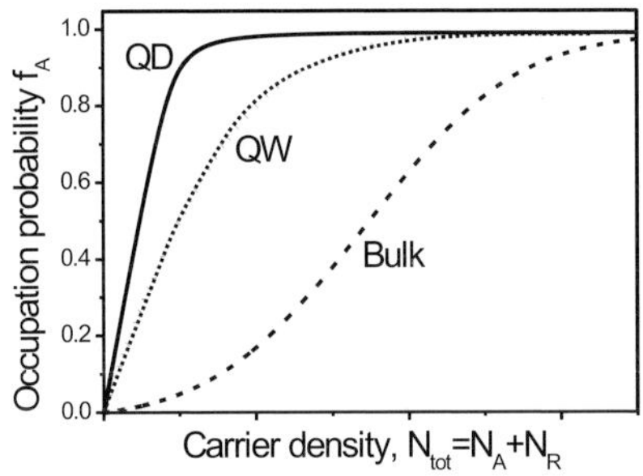

Fig. 11. Calculated occupation probability versus total carrier density for bulk, QW and QD active material. For QD, the total carrier density has a contribution also from the wetting layer states, acting as a reservoir of carriers.

the QD (when normalized with respect to the same volume). Another important factor for this difference is that the large confinement energy of the QD states with respect to the WL, results in a small overlap of the carrier distribution function with the WL states at low currents, where the quasi Fermi energy is far below the WL band edge for both electrons and holes. For bulk material (dashed line) the slow rise and saturation of the occupation probability is even more pronounced. Bearing in mind that the slope of the curves in Fig. 11 is proportional to the differential gain, one can conclude that QD active material is characterized by two well-separated regions. A low-bias region with high differential gain favorable for high-speed operation, and a high-bias region with a very low differential gain suitable for linear amplifiers featuring high saturated output power and low noise, as discussed in Sect. 3.6.

The measured inversion of QD SOAs as a function of current at room temperature reported in the literature varies substantially and ranges from the slow filling reported by Matthews et al. [31] to almost linear filling and sharp saturation reported by Borri et al. [19]. Rapid and linear filling of the active states and corresponding sharp saturation is one of the unique properties of QD SOAs compared with higher-dimensional devices. From the above it is clear that the stronger the coupling between the QD states and the WL is, the more this feature is smeared out making the QD SOA less unique.

3.6 Linear Amplification

Linear amplification of optical signals is a key application of SOAs. So far semiconductor based amplifiers have played only a minor role on the commercial markets for linear optical amplification purposes. The main reasons for this is the higher noise figures and lower saturated output powers of bulk

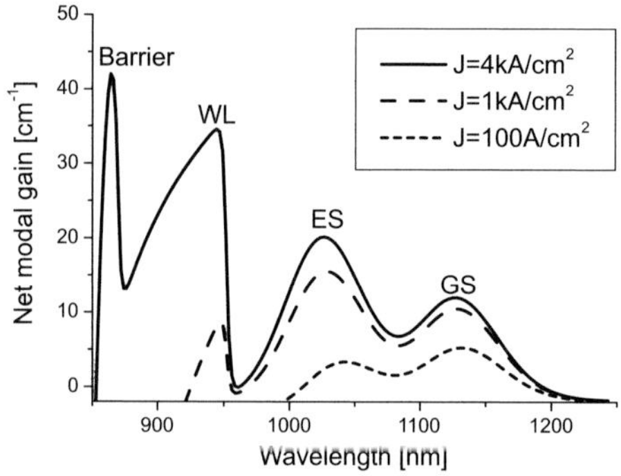

Fig. 12. Modal gain of QD SOA for three different current densities. No propagation is included in this case, corresponding to an absence of ASE.

and QW SOAs compared with fiber amplifiers, which are today the standard choice for high-end amplifiers. On the other hand, the compactness, flexibility, efficiency, and potential low price of SOAs render them interesting candidates for future communication networks if the performance issues mentioned above are improved. It is the hope that QD SOAs may overcome at least some of the problems of conventional SOAs.

Amplified Spontaneous Emission

Due to the small modal gain of present-day QD devices it is necessary to use long devices in order to reach a significant device gain. Reports of QD SOAs as long as 25 mm have thus been given in the literature [50]. At the same time, it is often desired to have a high degree of inversion, which means that also the WL and barrier are significantly populated and as a result exhibit gain. The combination of long devices and high modal gain can be expected to produce a large amount of amplified spontaneous emission (ASE), and it is interesting to investigate how this influences the properties of the device.

Another aspect of the high modal gain in the upper levels is that it could be expected to lead to device gains of many times the gain from the QD states, which could mean that all carriers would be consumed in the WL through ASE and that an unrealistically good AR coating would be necessary in order to avoid lasing at these upper levels. Neither of these predictions hold true. Actually, it is the ASE itself which limits the device gain of the WL and barrier.

Figure 12 shows the net modal gain spectrum of a QD SOA, i.e. including waveguide loss, for three different current densities. In the spectrum, the different carrier populations are clearly visible. The increase and saturation of the GS gain with current density corresponds to the behavior seen in Fig. 11 and similarly the ES gain saturates, but at larger current densities. For a current density of $4\,\mathrm{kA\,cm^{-2}}$ the WL and barrier are seen to exhibit signifi-

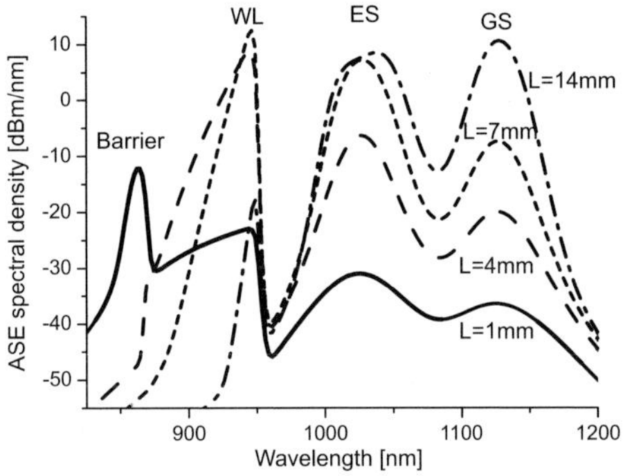

Fig. 13. Spectral density of ASE at the output of a QD amplifier for four different device lengths. The current density is $4\,\mathrm{kA\,cm^{-2}}$ for all devices.

cantly higher gains than the QD transitions. The maximum gain coefficient of the WL under full inversion is thus $56\,\mathrm{cm}^{-1}$ and at realistic current densities the barrier gain can reach significantly higher values.

It could be expected that the ASE output spectrum of an SOA on a logarithmic scale would look similar to the modal gain spectrum. However, due to saturation effects when the ASE is propagated through the amplifier, this is only the case for very short devices. The ASE output spectrum of a QD SOA for four different device lengths for a current density of $4\,\mathrm{kA\,cm^{-2}}$ is shown in Fig. 13. For the shortest device shown with $L = 1\,\mathrm{mm}$ the ASE spectral density is seen to resemble the gain spectrum. For the 4 mm long device, however, the barrier peak has disappeared, whereas the WL contribution has increased substantially and is much higher than the QD contributions. For the even longer devices of $L = 7\,\mathrm{mm}$ and $L = 14\,\mathrm{mm}$ the QD contributions are seen to eventually become the dominant.

The gradual redshift of the ASE peak wavelength with increasing device length is a result of gain saturation at all wavelengths. Since the upper transitions from the barrier and WL have a higher modal gain, these transitions have the shortest saturation length. For long devices, the QD states consume increasingly more carriers through stimulated emission, which reduces the inversion of the WL and barrier states and eventually decreases the ASE power generated from these transitions. The energetically lower QD states will thus always win the competition for carriers over the reservoir states, ensuring the shift towards the low energy side. This phenomenon is also found in bulk SOAs where the width of the gain peak is reduced with increasing device length.

Saturated Output Power

The saturated output power (SOP) is an important parameter for an amplifier and the realization of SOAs with high SOP is important in order to compete

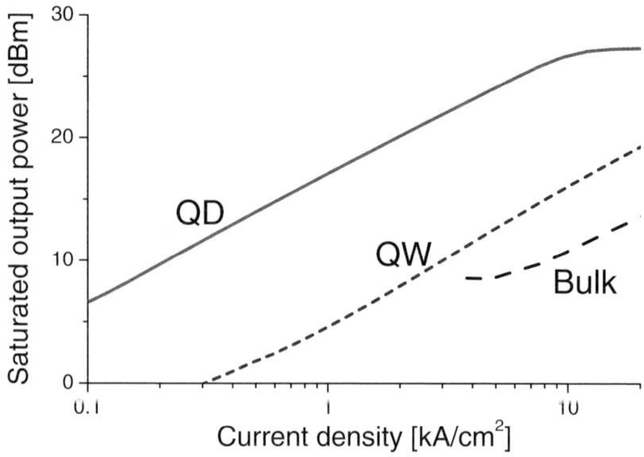

Fig. 14. Saturated output power versus current density for QD, QW, and bulk SOAs of length $L = 0.25\,\mathrm{mm}$.

with fiber amplifiers. The SOP is defined as the output power at which the saturated gain is $G_{\mathrm{sat}}^{\mathrm{out}} = (G_0 + A)/2$, where G_0 is the unsaturated gain and A is the fully saturated gain, i.e. the total waveguide loss.

Calculated SOP for QD, QW and bulk amplifiers are shown in Fig. 14 as a function of current density. For the QD SOA the SOP scales linearly with current until it saturates at a very high value (27 dBm) for current densities above $10\,\mathrm{kA\,cm^{-2}}$. The increase with current is caused by the increased degree of band filling of the ES, WL and barriers, and these carriers can be consumed by the energetically lower GS transition without significant reduction of the gain. Eventually, the inversion becomes sufficiently high that the limiting mechanism for the SOP becomes the capture of carriers into the active QD states. The maximum SOP is in this case limited by the minimum capture time (2 ps) of electrons from the WL to the ES.

In comparison with the QD device, the QW SOA (dotted line in Fig. 14) starts out at a smaller value of 0 dBm for the SOP around transparency and increases steadily with current to a value slightly below 20 dBm for a current density of $20\,\mathrm{kA\,cm^{-2}}$. The increase with current is in this case caused by a combination of increased inversion of both the QW states, as well as the barrier and a decrease of the effective carrier lifetime in the device. The decrease of the carrier lifetime is caused by the polynomial dependence of the carrier lifetime on carrier density, describing the influence of radiative and Auger assisted recombination. The small value close to transparency is oppositely determined by a long carrier lifetime and a high differential gain at this point. The same mechanisms cause a slight increase of the SOP of the bulk device from a value of 9 dBm at transparency to 14 dBm at $20\,\mathrm{kA\,cm^{-2}}$. When compared at the same current density, it is clear that the QD SOA outperforms both the bulk and the QW SOAs regarding saturated output power. The result in Fig. 14 are all for short devices ($L = 0.25\,\mathrm{mm}$). However, the SOP for QD devices remain superior also for long devices with correspondingly higher device gain.

Device Gain

The magnitude of the small-signal gain is of course a most important parameter for amplifiers. Since the modal gain of QD SOAs is usually significantly smaller than for QW and bulk devices, it is necessary to use longer devices in order to get a reasonable device gain. The modelled (see [49] for details) dependence of the small-signal peak gain on device length is shown in Fig. 15 (a). For the bulk and QW devices, the gain is seen to increase linearly with length for up to 1 mm in the bulk case and 2 mm in the QW case, and then saturate at a value of 40 dB for longer devices. Note that since the gain peak shifts with length, different signal wavelengths are used for the different lengths. The smaller modal gain of the QD device leads to a slow increase of the device gain with length. For the QD SOA the optimum signal wavelength is nearly independent of device length, due to the nature of band filling in the QDs. For devices longer than 6 mm, the gain increases sub-linearly with length, and for devices longer than the 14 mm shown here, the gain does eventually saturate completely as for the bulk and QW case.

In the absence of a saturating input signal, gain saturation is caused by the presence of strong ASE intensity in long devices [51]. The optical power of the ASE at any given wavelength is proportional to the gain at that wavelength. Consequently, the ASE power increases with length until the point where it becomes comparable to the saturation power of the device and the gain, as a result, starts to saturate. The maximum gain obtainable thus scales linearly with the saturation power of the device. However, other parameters such as the waveguide dimensions and the detailed assumptions in the simulation [49] also influence the maximum gain value. These points apply to all three types of amplifiers in similar ways, so that the high saturation power of QD SOAs should lead to higher saturated gains of QD SOAs compared to bulk and QW amplifiers at the same current density.

Noise Properties

An important figure of merit of optical amplifiers is the noise figure. Fig. 15 (b) shows the calculated noise figures (for non-saturated signals) corresponding to the gain curves in Fig. 15 (a), including shot noise and signal-spontaneous beat noise, but neglecting spontaneous-spontaneous beat noise. For the QD device the noise figure is seen to increase from below 3 dB for devices shorter than 2 mm and stabilize slightly above 3 dB with a small increase for the longest devices shown. For the short QD SOAs the gain is very low and the noise figure is consequently below 3 dB. The reason for the very low noise figure even for devices longer than 2 mm is that the active electronic QD

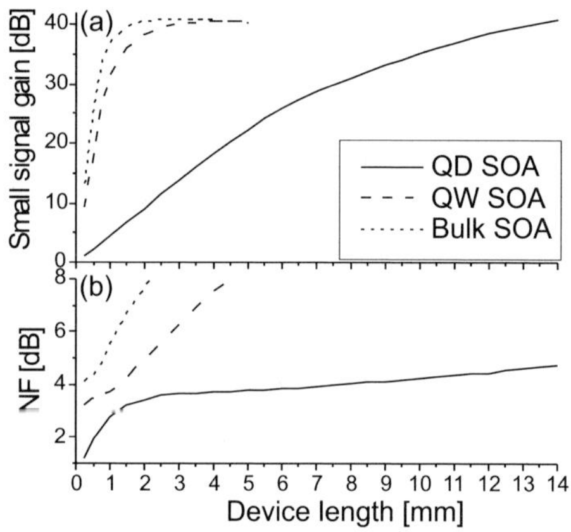

Fig. 15. Small signal gain (a) and noise figure at the signal wavelength (b) vs. device length. The current densities are: $1\,\mathrm{kA\,cm^{-2}}$ (QD SOA), $2\,\mathrm{kA\,cm^{-2}}$ (QW SOA), and $10\,\mathrm{kA\,cm^{-2}}$ (bulk SOA).

states are nearly completely inverted. For long devices, the ASE starts to saturate the device, i.e. the inversion decreases, and the noise figure starts to increase. Similar behavior is seen for the bulk and QW devices, except that the increase in noise figure occurs at much shorter device lengths due to the higher gain of these devices and consequently faster ASE saturation. The QW device is seen to have a noise figure close to 3 dB for devices shorter than 1 mm, indicating that the active electronic states of this device are also nearly completely inverted. In contrast, the bulk device remains far from full inversion at the signal wavelengths, despite the higher bias current density, resulting in higher noise figures for all device lengths. It should be noted that in order to perform a fair comparison of the amplifier types, the noise figure should be compared at the same device gain. If we thus compare for a device gain of 20 dB corresponding to $L = 0.4\,\mathrm{mm}$ for the QW device and $L = 4\,\mathrm{mm}$ for the QD, we see that the QW amplifier actually has a slightly smaller noise figure of 3.4 compared to 3.7 for the QD SOA.

3.7 Gain Recovery

The process of gain recovery after amplification of a single short pulse is unique for QD amplifiers compared with bulk and QW devices. In the higher dimensional devices it is common to see an initial fast recovery on a sub-picosecond timescale, followed by a slow recovery in the order of the carrier lifetime of the device, i.e. in the order of hundreds of picoseconds. It is this slow component of the recovery, that limits the bit rate at which processes like cross gain modulation can be performed without significant pattern dependence.

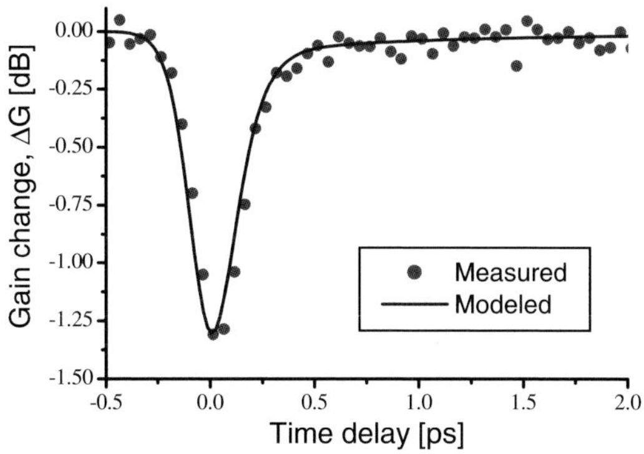

Fig. 16. Measured [19] and simulated gain recovery of a QD amplifier following the amplification of a 150 fs pulse.

A very common way to measure the gain recovery of a device is through pump-probe measurements [52]. In these experiments a short, high intensity, pump pulse is injected into the device together with a weaker probe pulse. At the output the amplified pump and probe are separated and the amplitude and phase of the probe is measured. By repeating this process for varying time delay between the two pulses the time resolved gain and index changes induced by the pump pulse can be measured. The time resolution is determined by the width of the pulses and is typically in the order of 100 fs.

The measured gain recovery of a QD amplifier after the injection of 150 fs pump pulse is shown in Fig. 16 [19]. Following the shape of the pump pulse closely the gain change is seen to be nearly symmetric and the gain recovers almost completely within less than 0.5 ps. At longer time-scales a slow component of the recovery is seen, but it is very weak compared with the fast component. The maximum gain change induced by the pump pulse is only -1.25 dB, which reflects the combination of small modal gain of 12 cm^{-1} and a device length of 0.475 mm giving a small signal device gain of only 2 dB [19].

The observed behavior is very different from the gain recovery process in bulk and QW devices due to the negligible slow component and, therefore, these measurements stirred great hopes for the use of QD amplifiers for ultrafast signal processing [5]. The major question in this context is: How can such fast gain recovery take place and why is the slow component so small? In order to answer these questions, the measurements were simulated (see [53] for details) and the result is shown with the solid line in Fig. 16. The good agreement between measurement and model seen in the figure is achieved by fitting the modal gain and two-photon absorption coefficients to the measurements [53]. The remaining parameters have been supplied by the grower [19, 44] or have been estimated from other types of measurements on the same device.

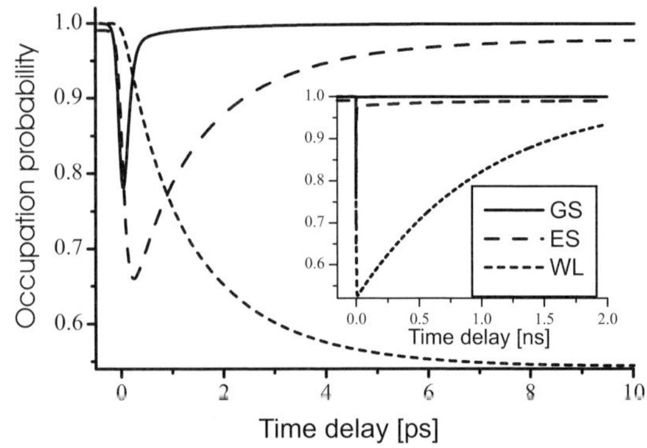

Fig. 17. Variation of the carrier densities of the three different levels in a QD SOA during the amplification of a short pulse. The GS and ES carrier densities are normalized with respect to their respective DOS and for the WL the initial carrier density (before the arrival of the pulse) has been used for the normalization. The insert shows the same, but on a longer time-scale.

The explanation for the fast gain recovery can be seen in Fig. 17, which shows the variations of the carrier densities of the three different levels during the amplification of the strong pump pulse. Before the pulse arrives the two dot levels are seen to be nearly completely inverted and furthermore a large amount of carriers is present in the WL. When the pulse arrives the GS level experiences a rapid depletion through stimulated emission. However, at the same time ES-level carriers start to relax into the empty states at the GS level on a time-scale of the duration of the pulse. This fast relaxation is a result of two features: the large energy splitting between the dot levels, which ensures slow thermal re-excitation of carriers, and secondly a high WL carrier density, resulting in fast Auger assisted relaxation. The ES level thus acts as a nearby carrier reservoir for the GS level enabling ultrafast gain recovery. Since the process of carrier capture is slower than intra-dot relaxation, the ES level recovers on a longer time-scale of several picoseconds. Finally, the WL is in the insert of Fig. 17 seen to recover on a nanosecond time-scale. The rate of refilling of this upper level is essentially determined by the injection current and the spontaneous recombination rate of the WL.

These results predict that a gain change should be observed at the ES transition following the passing of the pump pulse. This effect has been verified also experimentally by Kim et al. [54]. It is clear that despite the fast and complete recovery of the gain, the effect of the pump pulse is present in the device for much longer time and the complete recovery of a QD SOA is governed by the effective carrier lifetime of the device as the case is for bulk and QW devices.

4 Conclusions

The use of semiconductor QDs in optoelectronics is still in the infant stage of research and development. Some of the early predictions of QD properties for optoelectronic applications have turned out to be overly optimistic and based on the expected behavior of ideal (and isolated) QD systems rather than on QDs as they are realized in a true solid-state environment with imperfections and strong many-body interactions. Still, semiconductor QD devices have been realized with superior properties over devices based on conventional bulk or QW semiconductors.

QD semiconductor lasers have shown record-low threshold current densities and they seem very promising for high-power lasers due to better mode-control with less tendency of damaging beam filamentation and facet burns. Less feed-back sensitivity is also a striking and advantageous feature. Wavelength control and inhomogeneous broadening due to size and alloy fluctuations are issues that in some cases are limiting factors and in other cases can be exploited to their advantage.

We have paid particular attention to QDs as active material in SOAs. For low bias QD SOAs feature high differential gain as desired for high speed applications and signal processing. For high bias, on the other hand, there is low differential gain, which can be exploited in linear amplifiers featuring high saturated output power and low-noise performance. We have also investigated the ultrafast gain recovery after a short-pulse excitation in QD SOAs. It is explained by a fast carrier capture into the dots and ultrafast intra-dot relaxation. For fast pulse sequences, however, the carrier lifetime in the wetting layer may still limit the application of QD SOAs for high bit-rate operation.

In the present article, we have only discussed QD devices as replacements for conventional optoelectronic devices based on bulk and QW semiconductors. Semiconductor QDs may also provide completely new functionalities based on their ultimate quantum nature. Single-photon emitters for quantum communication and quantum logic elements, or qubits, for quantum communication have been mentioned. It is clear, however, that serious technological challenges lie ahead before the QD formation is sufficiently under control for these applications to materialize.

Acknowledgment

The authors want to thank Jesper Mørk for many enlightening and fruitful discussions.

References

1. M. Asada, Y. Miyamoto, Y. Suematsu: IEEE J. Quantum Electron. **22**, 1915 (1986)
2. R. Dettmer: IEE Review **34**, 395 (1988)
3. M. Bayer: Nature **418**, 597 (2002)
4. M. Bruchez Jr., M. Moronne, P. Gin, S. Weiss, A.P. Alivisatos: Science **281**, 2013 (1998)
5. M. Sugawara, N. Hatori, T. Akiyama, Y. Nakata, H. Ishikawa: Jpn. J. Appl. Phys. **40**, L488 (2001)
6. Z. Yuan, B.E. Kardynal, R.M. Stevenson, A.J. Shields, C.J. Lobo, K. Cooper, N.S. Beattie, D.A. Ritchie, M. Pepper: Science **295**, 102 (2002)
7. P.M. Varangis, H. Li, G.T. Liu, T.C. Newell, A. Stintz, B. Fuchs, K.J. Malloy, L.F. Lester: Electron. Lett. **36**, 1544 (2000)
8. S. Komiyama, O. Astafiev, V. Antonov, T. Kutsuwa, H. Hirai: Nature **403**, 405 (2000)
9. J.A. Lott, N.N. Ledentsov, V.M. Ustinov, N.A. Maleev, A.E. Zhukov, A.R. Kovsh, M.V. Maximov, B.V. Volovik, Zh.I. Alferov, D. Bimberg: Electron. Lett. **36**, 1384 (2000)
10. A. Fiore, J.X. Chen, M. Ilegems: Appl. Phys. Lett. **81**, 1756 (2002)
11. P. Borri, W. Langbein, J.M. Hvam, F. Heinrichsdorff, M.-H. Mao, D. Bimberg: IEEE J. Sel. Top. Quantum Electron. **6**, 544 (2000)
12. O. Stier, M. Grundmann, D. Bimberg: Phys. Rev. B **59**, 5688 (1999)
13. D. Bimberg, M. Grundmann, N.N. Ledentsov: *Quantum Dot Heterostructures* (John Wiley & Sons 1999)
14. C.H. Henry: IEEE J. Quantum Electron. **18**, 259 (1982)
15. H. Haug, S.W. Koch: *Quantum Theory of the Optical and Electronic Properties of Semiconductors* (World Scientific, Singapore 1990)
16. L. Goldstein, F. Glas, J.Y. Marzin, M.N. Charasse, G. Le Roux: Appl. Phys. Lett. **47**, 1099 (1985)
17. A. Bilenca, R. Alizon, V. Mikhelashvili, G. Eisenstein, R. Schwertberger, D. Gold, J.P. Reithmaier, A. Forchel: Electron. Lett. **38**, 1350 (2002)
18. L.V. Asryan, M. Grundmann, N.N. Ledentsov, O. Stier, R.A. Suris, D. Bimberg: J. Appl. Phys. **90**, 1666 (2001)
19. P. Borri, W. Langbein, J.M. Hvam, F. Heinrichsdorff, M.-H. Mao, D. Bimberg: IEEE J. Photon. Technol. Lett. **12**, 594 (2000)
20. P. Bhattacharya, S. Ghosh, S. Pradhan, J. Singh, Z.-K. Wu, J. Urayama, K. Kim, T.B. Norris: IEEE J. Quantum Electron. **39**, 952 (2003)
21. N. Kirstaedter, O.G. Schmidt, N.N. Ledentsov, D. Bimberg, V.M. Ustinov, A.Yu. Egorov, A.E. Zhukov, M.V. Maximov, P.S. Kop'ev, Zh.I. Alferov: Appl. Phys. Lett. **69**, 1226 (1996)
22. I. Magnusdottir, A.V. Uskov, S. Bischoff, B. Tromborg, J. Mork: J. Appl. Phys. **92**, 5982 (2002)
23. A.V. Uskov, F. Adler, H. Schweizer, M.H. Pilkuhn: J. Appl. Phys. **81**, 7895 (1997)
24. R. Schwertberger, D. Gold, J.P. Reithmaier, A. Forchel: IEEE J. Photon. Technol. Lett. **14**, 735 (2002)
25. N.N. Ledentsov, A.R. Kovsh, A.E. Zhukov, N.A. Maleev, S.S. Mikhrin, A.P. Vasil'ev, E.S. Semenova, M.V. Maximov, Yu.M. Shernyakov, N.V. Kryzhanovskaya, V.M. Ustinov, D. Bimberg: Electron. Lett. **39**, 1126 (2003)

26. M.P. Mullane, J.G. McInerney: IEEE J. Photon. Technol. Lett. **12**, 1147 (2000)
27. M. Osinski, J. Buus: IEEE J. Quantum Electron. **23**, 9 (1987)
28. T.C. Newell, D.J. Bossert, A. Stintz, A. Fuchs, K.J. Malloy, L.F. Lester: IEEE J. Photon. Technol. Lett. **11**, 1527 (1999)
29. D. Bimberg, N. Kirstaedter, N.N. Ledentsov, Zh.I. Alferov, P.S. Kop'ev, V.M. Ustinov: IEEE J. Sel. Top. Quantum Electron. **3**, 196 (1997)
30. G.P. Agrawal: *Fiber-Optic Communication Systems* (John Wiley and Sons 1997)
31. D.R. Matthews, H.D. Summers, P.M. Smowton, M. Hopkinson: Appl. Phys. Lett. **81**, 4904 (2002)
32. L.V. Asryan, R.A. Suris: IEEE J. Quantum Electron. **34**, 841 (1998)
33. G. Park, O.B. Shchekin, D.L. Huffaker, D.G. Deppe: Electron. Lett. **36**, 1283 (2000)
34. M. Sugawara, K. Mukai, N. Yoshiaki: Appl. Phys. Lett. **75**, 656 (1999)
35. D. Klotzkin, P. Bhattacharya: J. Lightwave Technol. **17**, 1634 (1999)
36. Ch. Ribbat, R.L. Sellin, I. Kaiander, F. Hopfer, N.N. Ledentsov, D. Bimberg, A.R. Kovsh, V.M. Ustinov, A.E. Zhukov, M.V. Maximov: Appl. Phys. Lett. **82**, 952 (2003)
37. R.L. Sellin, C. Ribbat, D. Bimberg, F. Rinner, H. Konstanzer, M.T. Kelemen, M. Mikulla: Electron. Lett. **38**, 883 (2002)
38. G. Erbert, J. Fricke, J.P. Reithmaier, A. Forchel, R. Staske, B. Sumpf, S. Deubert, G. Trankle: Electron. Lett. **39**, 1655 (2003)
39. C. Weisbuch, B. Vinter: *Quantum Semiconductor Structures: Fundamentals and Applications* (Academic press 1991)
40. A. Passaseo, M. De Vittorio, M.T. Todaro, I. Tarantini, M. De Giorgi, R. Cingolani, A. Taurino, M. Catalano, A. Fiore, A. Markus, J.X. Chen, C. Paranthoen, U. Oesterle, M. Ilegems: Appl. Phys. Lett. **82**, 3632 (2003)
41. G. Costantini, C. Manzano, R. Songmuang, O.G. Schmidt, K. Kern: Appl. Phys. Lett. **82**, 3194 (2003)
42. H. Lee, S.M. Kim, Y.J. Park, E.K. Kim: J. Appl. Phys. **90**, 2290 (2001)
43. G.T. Liu, A. Stintz, H. Li, T.C. Newell, A.L. Gray, P.M. Varangis, K.J. Malloy, L.F. Lester: IEEE J. Quantum Electron. **36**, 1272 (2000)
44. F. Heinrichsdorff, M.-H. Mao, N. Kirstaedter, A. Krost, D. Bimberg, A.O. Kosogov, P. Werner: Appl. Phys. Lett. **71**, 22 (1997)
45. K.H. Schmidt, G. Medeiros-Ribeiro, M. Oestreich, P.M. Petroff, G.H. Dohler: Phys. Rev. B **54**, 11346 (1996)
46. M. Grundmann, O. Stier, D. Bimberg: Phys. Rev. B **52**, 11969 (1995)
47. A.T. Winzer, R. Goldhahn, G. Gobsch, H. Heidemeyer, O.G. Schmidt, K. Eberl: Physica E **13**, 289 (2002)
48. D.G. Deppe, D.L. Huffaker: Appl. Phys. Lett. **77**, 3325 (2000)
49. T.W. Berg, J. Mørk: Appl. Phys. Lett. **82**, 3083 (2003)
50. T. Akiyama, N. Hatori, Y. Nakata, H. Ebe, M. Sugawara: Electron. Lett. **38**, 1139 (2002)
51. T. Liu, K. Obermann, K. Petermann, F. Girardin, G. Guekos: Electron. Lett. **33**, 2042 (1997)
52. K.L. Hall, G. Lenz, A.M. Darwish, E.P. Ippen: Opt. Commun. **111**, 589 (1994)
53. T.W. Berg, S. Bischoff, I. Magnusdottir, J. Mørk: IEEE J. Photon. Technol. Lett. **13**, 541 (2001)
54. K. Kim, J. Urayama, T.B. Norris, J. Singh, J. Phillips, P. Bhattacharya: Appl. Phys. Lett. **81**, 670 (2002)

GaInNAs: Fundamentals of a New Material System for Near-Infrared Optoelectronics

Michael Hetterich

1 Introduction to GaInNAs-Based Laser Structures

In recent years, dilute nitrides such as GaAsN and GaInNAs with typical nitrogen concentrations in the range of a few percent have become the subject of intense research, because they are promising materials for the realization of optoelectronic devices operating in the near infrared (NIR), including semiconductor lasers, resonant cavity enhanced photodetectors [1] and tandem solar cells [2]. Heterojunction bipolar transistors have also been investigated [3]. Of particular technological interest are quantum well structures based on the quaternary compound GaInNAs. As first suggested by Kondow et al. [4] they can form the active region of laser diodes operating at 1.3 or even 1.55 µm, the wavelengths used in present optical fibre communication networks. Up to now, InP-based InGaAsP devices are commonly applied for that purpose. However, this approach has several disadvantages [4, 5]: The electron confinement in InGaAsP structures is quite poor, leading to a high temperature sensitivity of the lasing threshold (i.e. a low T_0). As a practical consequence, thermoelectric cooling is often required. Furthermore, the development of InP-based vertical-cavity surface-emitting lasers (VCSELs) has turned out to be quite problematic, although the latter would have many advantages (e.g. single mode operation and improved coupling to optical fibres). This is due to the lack of materials lattice-matched to InP which comprise a sufficiently large refractive index contrast for the realization of high-reflectivity distributed Bragg reflectors (DBRs) and also have a sufficient thermal conductivity.

In GaInNAs-based laser diodes many of the problems typical for InGaAsP devices can be avoided by taking advantage of the unusual physical properties of this material system: As shown in Fig. 1, alloying of GaAs with small amounts of nitrogen results in a strongly reduced band gap while the lattice parameter decreases. Since adding indium to GaAs lowers the band gap as well but increases the lattice parameter, quaternary $Ga_{1-x}In_xN_yAs_{1-y}$ with a band gap corresponding to an emission wavelength of 1.3 µm can be grown coherently on GaAs. One of the main advantages of this approach is the fact that VCSEL structures based on high refractive index contrast AlGaAs-DBRs become possible. Indeed, optically pumped VCSELs could be demonstrated by a number of groups, e.g. [5–8]. However, first attempts to

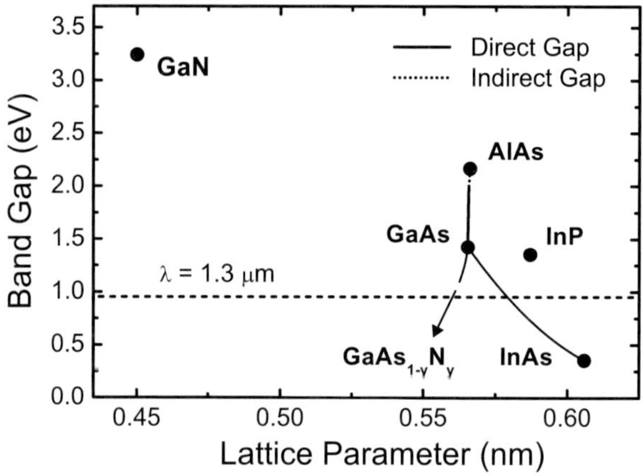

Fig. 1. Room-temperature band gap versus lattice parameter for some common binary zinc blende III-V compounds and their ternary alloys [9]. The curve for $GaAs_{1-y}N_y$ has been calculated using the band anticrossing model.

realize p-i-n-type surface-emitting lasers with electrical injection turned out to be unsuccessful. Among other reasons this could be traced back to the high losses through inter- and intra-band free carrier absorption, particularly in the p-doped Bragg mirror [8]. Therefore, tunnel junctions or intra-cavity contacts have been used in improved designs to minimize this effect, enabling the first demonstrations of electrically pumped GaInNAs-VCSELs with $\approx 1.3\,\mu\mathrm{m}$ emission wavelength [10, 11] and – after further optimization – even the commercialization of these devices.

Equally impressive is the performance of edge-emitting lasers achieved already years ago, particular in terms of lifetime and output power (see e.g. [12, 13]). Significantly improved T_0 values compared to InGaAsP devices have also been reported by many groups, e.g. [14, 15]. This improvement was expected due to the strong electron confinement caused by the large GaInNAs/GaAs conduction band offset as well as the increased effective electron mass and conduction band non-parabolicity (see e.g. [16–18] and Sect. 2) which should lead to a reduced Auger recombination. However, only little quantitative information about the recombination processes and loss mechanisms in GaInNAs laser diodes is available to correctly interpret and maybe further improve these results. In [19] the authors analyze among other things the spontaneous emission from a window in the laser substrate as a function of injection current and temperature in order to determine the temperature dependence of the monomolecular, radiative and Auger-related contributions to the threshold current in edge-emitting laser diodes. They conclude that monomolecular (defect-related) recombination plays a very important role in GaInNAs devices at low temperatures, in contrast to InGaAsP structures. Even at room-temperature a contribution of about 50% to the threshold current is found, thus suggesting the possibility of further technological improvement. Interestingly, a strong monomolecular recombination can also lead to anomalously high T_0 values. Indeed, high performance GaInNAs lasers in particular often tend to have an unexpectedly *low* T_0 (70–110 K), which may be related to an insufficient hole confinement [20]. However, a better understanding of the electronic structure of GaInNAs-based quantum well structures is required to clarify this point.

2 Band Structure of GaAsN and GaInNAs

From a band structure point of view, the properties of Ga(In)NAs are quite unusual: As shown in Fig. 1, incorporation of only a few percent of nitrogen into GaAs (or GaInAs) leads to an enormous band gap reduction. Furthermore, the so-called E_+ band appears above the band edge, the effective electron mass increases and a significant conduction band non-parabolicity can be observed [16–18, 21–24]. On the other hand, the valence band structure is hardly affected. For example, the spin-orbit splitting energy Δ_0 has been reported to be essentially independent of the nitrogen content [25]. In the so-called band anticrossing (BAC) model [22] the described effects are interpreted to be the result of a repulsive interaction between the conduction band and a (strongly localized) nitrogen-related energy level E_N above the conduction band edge. For bulk material the corresponding Hamiltonian reads in its simplest form:

$$\begin{pmatrix} E_\mathrm{M}(\boldsymbol{k}) & V_\mathrm{NM} \\ V_\mathrm{NM} & E_\mathrm{N} \end{pmatrix} \tag{1}$$

where $E_\mathrm{M}(\boldsymbol{k})$ is the conduction band dispersion of the host $\mathrm{Ga}_{1-x}\mathrm{In}_x\mathrm{As}$ matrix M and V_NM represents the coupling of the latter with the nitrogen level. An extended 10-band $\boldsymbol{k}\cdot\boldsymbol{p}$ model version has also been suggested [26]. However, presently the material parameters in this Hamiltonian are unfortunately not yet known well enough. To determine the $\boldsymbol{k}$-dependent eigenvalues of (1) the characteristic equation

$$\begin{vmatrix} E_\mathrm{M}(\boldsymbol{k}) - E & V_\mathrm{NM} \\ V_\mathrm{NM} & E_\mathrm{N} - E \end{vmatrix} = 0 \tag{2}$$

has to be solved. This leads to

$$E_\pm(\boldsymbol{k}) = \frac{E_\mathrm{N} + E_\mathrm{M}(\boldsymbol{k})}{2} \pm \frac{1}{2}\left\{[E_\mathrm{N} - E_\mathrm{M}(\boldsymbol{k})]^2 + 4V_\mathrm{NM}^2\right\}^{1/2} , \tag{3}$$

where $E_+(\boldsymbol{k})$ and $E_-(\boldsymbol{k})$ correspond to the experimentally observed upper and lower branch of the $\mathrm{Ga}_{1-x}\mathrm{In}_x\mathrm{N}_y\mathrm{As}_{1-y}$ conduction band, respectively. V_NM depends on the nitrogen concentration y as [22, 24]

$$V_{NM} = C_{NM}\sqrt{y} , \tag{4}$$

where C_{NM} is a constant coupling parameter. As an example for the resulting dispersion curves, Fig. 2 shows $E_+(\boldsymbol{k})$ and $E_-(\boldsymbol{k})$ for $GaAs_{0.99}N_{0.01}$ assuming $C_{NM} = 2.7\,\text{eV}$ and $E_N = 1.65\,\text{eV}$ [22]. Finally, it should also be mentioned that E_N and E_M are themselves weakly y-dependent [24]. Therefore, $E_M(\boldsymbol{k})$ actually represents the conduction band dispersion of $Ga_{1-x}In_xN_yAs_{1-y}$ neglecting the interaction with E_N.

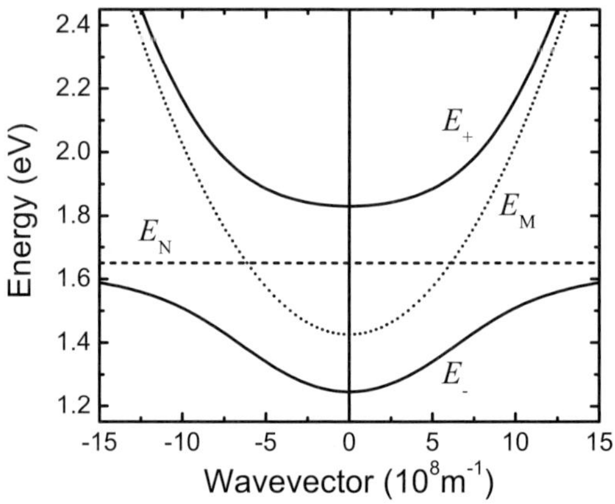

Fig. 2. Dispersion relation of the upper ($E_+(\boldsymbol{k})$) and lower ($E_-(\boldsymbol{k})$) branch of the conduction band in $GaAs_{0.99}N_{0.01}$ at room-temperature according to the band anticrossing model. Also shown are the unperturbed conduction band dispersion of the GaAs matrix material $E_M(\boldsymbol{k})$ and the nitrogen level E_N. All energies are given relative to the valence band edge.

It is quite nontrivial, that an approach as simple as the model just presented should be able to describe the complex effects of nitrogen on the $Ga_{1-x}In_xN_yAs_{1-y}$ band structure. Indeed, the results of empirical pseudopotential calculations [27] suggest, that the observed anticrossing in $GaAs_{1-y}N_y$ can in general not be ascribed to the interaction of the conduction band with a single nitrogen level. The incorporation of nitrogen into GaAs removes the translational symmetry of the crystal, thus splitting e.g. the L_{1c} valleys into $a_1(L_{1c}) + t_2(L_{1c})$ states and X_{1c} into $a_1(X_{1c}) + e(X_{1c})$. The conduction band valley Γ_{1c} remains as $a_1(\Gamma_{1c})$. Higher-energy host crystal bands are also perturbed, producing, among others, a_1 states of their own. In particular, in the dilute nitrogen regime, an additional nitrogen localized state built from many high-energy host bands forms above the conduction band minimum of GaAs. However, *all* states of the same symmetry representation can be expected to interact under the perturbation due to the presence of N, i.e. we should have a coupling and anticrossing of *all* a_1-like states (Γ, X, L band edges etc.) instead of the two level BAC approach where only the interaction of the conduction band with the localized nitrogen level E_{N} is taken into account. Applying the BAC model therefore actually means, that we replace the complex interaction with higher energy states by the coupling to only *one* "effective" nitrogen level E_{N} (which in general may not correspond to a real state) to get a simple description of the Ga(In)NAs band structure. Indeed, there is some theoretical support for this initially semi-empirical approach: Using a Green's function model it has been shown that the two-level BAC model (with an "effective" E_{N}) can be derived explicitely, if only the influence of the onsite change in orbital energies due to the N impurity is considered [26]. In further numerical calculations the authors also find evidence, that the model should stay appropriate when treating the full Hamiltonian.

From an experimental point of view this result is confirmed by the fact that the BAC model has been applied very successfully so far. It has proven its ability to accurately describe not only the electronic states in bulk material but also in quantum well (QW) structures many times and can therefore be regarded as well established (see e.g. [17, 18, 21–23, 28–32] and references therein).

3 Boundary Conditions for the BAC Model Wave Function in GaInNAs-Based Heterostructures

When the band anticrossing model is applied to a quantum well or any other structure with a spatially varying composition profile along the growth direction z, all material parameters in the BAC Hamiltonian become position-dependent. Furthermore, $E_{\mathrm{M}}(\boldsymbol{k})$ has to be replaced by a suitable Hermitian operator. For the latter we choose

$$E_{\mathrm{M}}(z) - \frac{\hbar^2}{2}\frac{\partial}{\partial z}\frac{1}{m^*(z)}\frac{\partial}{\partial z} \tag{5}$$

which has already been used quite successfully in the past, also for quantum structures based on $\mathrm{Ga}_{1-x}\mathrm{In}_x\mathrm{N}_y\mathrm{As}_{1-y}$ [28]. $E_{\mathrm{M}}(z)$ and $m^*(z)$ are the z-dependent conduction band edge and effective mass, respectively, that would be obtained if no anticrossing with E_{N} was present. In practice, their values should essentially coincide with those of the corresponding nitrogen-free material. The kinetic energy operator chosen in (5) implies that we assume a parabolic dispersion for the unperturbed conduction band of the matrix material. For the E_- branch ($\mathrm{Ga}_{1-x}\mathrm{In}_x\mathrm{N}_y\mathrm{As}_{1-y}$ conduction band) which we are primarily interested in, this approximation is usually sufficient, because a nonparabolicity in $E_{\mathrm{M}}(\boldsymbol{k})$ would occur for high k values where $E_-(\boldsymbol{k})$ flattens out anyway due to the strong anticrossing with the nitrogen level. With the changes just discussed the total BAC Hamiltonian for the one-dimensional case now reads

$$H^{\mathrm{BAC}} = \begin{pmatrix} E_{\mathrm{M}}(z) - \frac{\hbar^2}{2}\frac{\partial}{\partial z}\frac{1}{m^*(z)}\frac{\partial}{\partial z} & V_{\mathrm{NM}}(z) \\ V_{\mathrm{NM}}(z) & E_{\mathrm{N}}(z) \end{pmatrix} \tag{6}$$

and the corresponding time-dependent Schrödinger equation becomes

$$H^{\mathrm{BAC}} \begin{pmatrix} \Psi_{\mathrm{M}}(z,t) \\ \Psi_{\mathrm{N}}(z,t) \end{pmatrix} = \mathrm{i}\hbar \frac{\partial}{\partial t} \begin{pmatrix} \Psi_{\mathrm{M}}(z,t) \\ \Psi_{\mathrm{N}}(z,t) \end{pmatrix} . \tag{7}$$

Let us now turn to the probability current density $j(z,t)$ of the BAC model. The latter can be deduced by calculating the time derivative of the probability density

$$\varrho(z,t) = \Psi_{\mathrm{M}}(z,t)\Psi_{\mathrm{M}}^*(z,t) + \Psi_{\mathrm{N}}(z,t)\Psi_{\mathrm{N}}^*(z,t) \tag{8}$$

using (7). After some mathematics one finally obtains the continuity equation

$$\frac{\partial \varrho(z,t)}{\partial t} + \frac{\partial j(z,t)}{\partial z} = 0 \tag{9}$$

with the probability current density

$$j(z,t) = \frac{1}{2m^*(z)} \left[\Psi_{\mathrm{M}}^*(z,t) \frac{\hbar}{\mathrm{i}} \frac{\partial}{\partial z} \Psi_{\mathrm{M}}(z,t) - \Psi_{\mathrm{M}}(z,t) \frac{\hbar}{\mathrm{i}} \frac{\partial}{\partial z} \Psi_{\mathrm{M}}^*(z,t) \right] . \tag{10}$$

It is interesting to note that there is no contribution of $\Psi_{\mathrm{N}}(z,t)$ to j. This is caused by the fact that E_{N} is dispersionless and reflects the localized nature of the nitrogen states.

Based on the equations presented so far we are now able to investigate the boundary conditions for the solutions of the stationary Schrödinger equation

$$H^{\mathrm{BAC}} \begin{pmatrix} \psi_{\mathrm{M}}(z) \\ \psi_{\mathrm{N}}(z) \end{pmatrix} = E \begin{pmatrix} \psi_{\mathrm{M}}(z) \\ \psi_{\mathrm{N}}(z) \end{pmatrix} \tag{11}$$

at the interface between two homogeneous $\mathrm{Ga}_{1-x}\mathrm{In}_x\mathrm{N}_y\mathrm{As}_{1-y}$ layers A and B with different compositions. The first point to make here is that the lower component $\psi_{\mathrm{N}}(z)$ of the wave function is in general *discontinuous* at the interface. Let us first discuss this for the (somewhat unphysical) special case $V_{\mathrm{NM}}(z) = 0$: From the Schrödinger equation (11) we obtain $E_{\mathrm{N}}(z)\psi_{\mathrm{N}}(z) = E\psi_{\mathrm{N}}(z)$. This equation only leads to a nonvanishing $\psi_{\mathrm{N}}(z)$ if

$$E = E_{\mathrm{N}}(z) . \tag{12}$$

For a single homogeneous $\mathrm{Ga}_{1-x}\mathrm{In}_x\mathrm{N}_y\mathrm{As}_{1-y}$ layer we would have $E_{\mathrm{N}}(z) = E_{\mathrm{N}}$. Therefore, all nitrogen-like eigenfunctions would correspond to the same eigenvalue

$$E = E_{\mathrm{N}} , \tag{13}$$

as expected. However, the eigenfunctions themselves could have arbitrary shape. This is related to the fact already mentioned above that there is no contribution of the nitrogen states to the probability current density. As a consequence the values of $\psi_N(z)$ at different positions are completely decoupled and will always result in a stationary solution of the Schrödinger equation. (This changes for $V_{NM} \neq 0$, because different sites become indirectly coupled via the E_M states.)

If we now repeat the procedure described in the last paragraph for the case of two interfacing layers A and B, (13) holds for both sides of the interface separately, i.e. there are states with $E = E_N^A$ and $E = E_N^B$, respectively. Since the energetic position of E_N depends on both, the composition and microscopic structure of the layer under consideration (see e.g. [31]), we will generally have $E_N^A \neq E_N^B$. For that reason, (12) can only be satisfied on both sides simultaneously if $\psi_N(z)$ vanishes identically in material B for states with $E = E_N^A$ and vice versa, in other words, each type of nitrogen wave function is confined to its own material (where it can have arbitrary shape, see above). However, this implies that $\psi_N(z)$ is in general *discontinuous* at the interface.

For the realistic case $V_{NM} \neq 0$ the situation gets more complicated. From the Schrödinger equation (11) we now obtain

$$\psi_N(z) = \frac{V_{NM}(z)}{E - E_N(z)} \psi_M(z) , \tag{14}$$

i.e. $\psi_N(z)$ and $\psi_M(z)$ become coupled. As shown below, $\psi_M(z)$ is continuous at the interface. However, $V_{NM}(z)$ and $E_N(z)$ are of course both discontinuous which means $\psi_N(z)$ can not be continuous for all possible energy values E either.

Let us now proceed with the more important boundary conditions for the upper component $\psi_M(z)$ of the BAC wave function. For a heterostructure containing two $Ga_{1-x}In_xN_yAs_{1-y}$ layers with different compositions interfacing at $z = 0$ we integrate (11) over an infinitesimal interval $[-\varepsilon, +\varepsilon]$:

$$\int_{-\varepsilon}^{+\varepsilon} \left\{ E_M(z)\psi_M(z) - \frac{\hbar^2}{2} \frac{\partial}{\partial z} \left[\frac{1}{m^*(z)} \frac{\partial}{\partial z} \psi_M(z) \right] + V_{NM}(z)\psi_N(z) \right\} \mathrm{d}z = E \int_{-\varepsilon}^{+\varepsilon} \psi_M(z) \mathrm{d}z . \tag{15}$$

In the limit $\varepsilon \to 0$ all terms apart from the second vanish. Carrying out the integration therefore finally leads to the boundary condition

$$\frac{1}{m^*(-\varepsilon)} \frac{\partial}{\partial z} \psi_M(-\varepsilon) = \frac{1}{m^*(+\varepsilon)} \frac{\partial}{\partial z} \psi_M(+\varepsilon) , \tag{16}$$

which means $\psi_M'(z)/m^*(z)$ has to be continuous at the interface. This expression looks very similar to the commonly used condition for quantum wells with different effective masses in the well and barrier material, respectively (see e.g. [33]). However, the important difference to note is, that for nitrogen-containing material m^* does *not* correspond to the actual effective electron mass itself but rather to the value that would be obtained if there was no anticrossing interaction of E_M with E_N, in other words, m^* is essentially the effective mass of the $\mathrm{Ga}_{1-x}\mathrm{In}_x\mathrm{As}$ matrix.

The second boundary condition for $\psi_\mathrm{M}(z)$ follows from the continuity equation (9). For the stationary case $\partial\varrho/\partial t = 0$ the latter implies that the probability current density $j(z)$ is the same at both sides of the interface. Using (10) and (16) it follows that for $\varepsilon \to 0$

$$\psi_\mathrm{M}(-\varepsilon) = \psi_\mathrm{M}(+\varepsilon) \tag{17}$$

i.e. $\psi_\mathrm{M}(z)$ must be continuous at the interface.

4 BAC Model Calculation of Bound States in GaInNAs Quantum Wells

Based on the derived boundary conditions (16) and (17) we can now quite easily calculate the bound states in $\mathrm{Ga}_{1-x}\mathrm{In}_x\mathrm{N}_y\mathrm{As}_{1-y}$-based quantum well structures, avoiding the otherwise required numerical diagonalization of the BAC Hamiltonian. It is interesting to note that this is possible without explicitly treating $\psi_\mathrm{N}(z)$, since neither (16) nor (17) contains this component of the wave function. We start by using (14) to eliminate $\psi_\mathrm{N}(z)$ from the Schrödinger equation (11). One obtains:

$$-\frac{\hbar^2}{2}\frac{\partial}{\partial z}\frac{1}{m^*(z)}\frac{\partial}{\partial z}\psi_\mathrm{M}(z) + \left[E_\mathrm{M}(z) + \frac{V_\mathrm{NM}^2(z)}{E - E_\mathrm{N}(z)}\right]\psi_\mathrm{M}(z) = E\psi_\mathrm{M}(z)\,. \tag{18}$$

This is the ordinary Schrödinger equation for systems with a position-dependent effective mass [33], apart from an additional energy-dependent contribution $V_\mathrm{NM}^2(z)/[E - E_\mathrm{N}(z)]$ to the potential energy caused by the interaction with the nitrogen states. (18) can be solved using the standard ansatz

$$\psi_\mathrm{M}(z) = \begin{cases} A\exp(ikz) + B\exp(-ikz) & \text{(well)} \\ C\exp(\kappa z) + D\exp(-\kappa z) & \text{(barrier)} \end{cases} \tag{19}$$

which leads to:

$$k = \left\{\frac{2m^*(\mathrm{A})}{\hbar^2}\left[E - E_\mathrm{M}(\mathrm{A}) + \frac{V_\mathrm{NM}^2(\mathrm{A})}{E_\mathrm{N}(\mathrm{A}) - E}\right]\right\}^{1/2} \quad \text{(well material A)}$$

$$\kappa = \left\{\frac{2m^*(\mathrm{B})}{\hbar^2}\left[E_\mathrm{M}(\mathrm{B}) - E - \frac{V_\mathrm{NM}^2(\mathrm{B})}{E_\mathrm{N}(\mathrm{B}) - E}\right]\right\}^{1/2} \quad \text{(barrier material B)} \tag{20}$$

The latter equations clearly reflect the strongly nonparabolic dispersion of both the E_- and the E_+ band. Since the Schrödinger equation (18) as well as the boundary conditions (16) and (17) for $\psi_\mathrm{M}(z)$ in the BAC model are formally identical to those of the textbook quantum well (apart from the additional contribution to the potential energy, the different meaning of m^* and the changed dispersion relation) the energies of the bound eigenstates can be found by numerically solving the well-known implicit equations [33]:

$$\cos\left(\frac{1}{2}kL\right) - \frac{m^*(\mathrm{B})k}{m^*(\mathrm{A})\kappa}\sin\left(\frac{1}{2}kL\right) = 0 \qquad \text{(even states)}$$

$$\cos\left(\frac{1}{2}kL\right) + \frac{m^*(\mathrm{A})\kappa}{m^*(\mathrm{B})k}\sin\left(\frac{1}{2}kL\right) = 0 \qquad \text{(odd states)} \tag{21}$$

where L is the thickness of the quantum well. Please note, that the derived equations also cover quantized states arising from the E_+ branch. The latter could in principle occur in structures where a suitable barrier material leads to a sufficiently high conduction band offset, however, in practice this is normally not the case. To determine the quantization energies of the E_- states, E is scanned from the bottom to the top of the quantum well (or the end of the E_- branch), k and κ are calculated for each E using (20) and the energy eigenvalues are then found by checking if one of the equations (21) is satisfied. The described procedure is much simpler, faster and more transparent than the usual numerical diagonalization of the BAC Hamiltonian and therefore ideally suited for the fitting of experimental data.

5 Influence of the Local N Environment on the BAC Hamiltonian Parameters

For the BAC model to be useful in practice the parameters E_N and C_NM entering the Hamiltonian have to be known. Obviously, the latter will depend on the composition of the $Ga_{1-x}In_xN_yAs_{1-y}$ layer under consideration. However, as discussed e.g. in [34], the macroscopic composition alone does not fully characterize the material in case of a quaternary compound, because it does not uniquely specify the statistical distribution of the different possible group-III/group-V bond types. This holds in particular for the local configuration of the nitrogen atoms in $Ga_{1-x}In_xN_yAs_{1-y}$. As shown in Fig. 3, nitrogen has a tetrahedral coordination in the zinc blende lattice. The same fixed finite In and N concentrations can therefore lead to different local configurations N–In_nGa_{4-n} of the nitrogen atoms with $0 \leq n \leq 4$ In and $4-n$ Ga nearest neighbors. It has been shown by Fourier transform infrared (FTIR) absorption measurements [35] and Raman spectroscopy [36] that in as-grown $Ga_{1-x}In_xN_yAs_{1-y}$ nitrogen is preferentially surrounded by Ga, i.e. the N–Ga_4 configuration (Fig. 3 (a)) is dominant. However, the thermal annealing usually performed to improve the optical quality of $Ga_{1-x}In_xN_yAs_{1-y}$ structures leads to configuration changes in which In–N bonds are preferentially formed. This happens most likely through a hopping of the nitrogen atoms to more In-rich environments via As vacancies [31, 35]. As a result, N configurations with In nearest neighbors (in particular N–$InGa_3$, see Fig. 3 (b)) are favored in annealed material [31, 34–37].

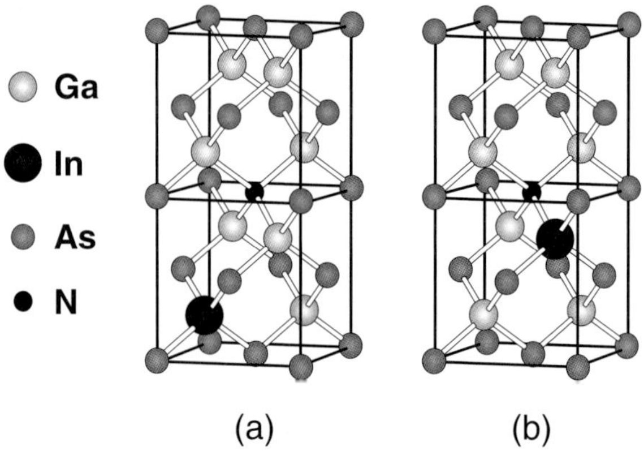

Fig. 3. Two possible nearest neighbor configurations of a nitrogen atom in $Ga_{1-x}In_xN_yAs_{1-y}$. (a) N–Ga_4 configuration, dominant in as-grown material. (b) N–$InGa_3$ configuration, formed through thermal annealing.

To explain the described effects it has been argued, that during growth nitrogen incorporation at the surface is mainly controlled by chemical bonding aspects. Due to the larger cohesive energy of the N–Ga bond (2.24 eV [38]) compared to the N–In bond (1.93 eV [38]) the N–Ga_4 configuration should therefore be preferred. In bulk material, on the other hand, the formation of N–In bonds leads to a reduction of local strain, thus favoring configurations like N–$InGa_3$ in thermal equilibrium. Experimental evidence, that the model just discussed is indeed reasonable, comes from the investigation of another quaternary dilute nitride material, AlGaAsN. Since the cohesive energy of the N–Al bond (2.88 eV [38]) is even larger than that of the N–Ga bond, a complete dominance of chemical bonding aspects should be expected in this case. Indeed, Raman investigations strongly indicate that a preferential formation of N–Al bonds takes place in this material [36], thus supporting the model discussed above.

It has been shown both experimentally and theoretically, that the nearest neighbor environment of nitrogen has a strong influence on the electronic and optical properties of $Ga_{1-x}In_xN_yAs_{1-y}$ [31, 34].[1] This is quite plausible, because the nitrogen level E_N has a strongly localized nature. sp^3s^* tight-binding supercell calculations supported by experimental data [31] suggest E_N to shift to higher energies by an amount proportional to the number of nearest In neighbors, the total shift between the N–Ga_4 and the N–In_4 configuration being about 220 meV. (More distant neighbors are found to have only a minor influence.) At the same time, the coupling parameter C_{NM} is reported to decrease significantly with the number of N–In bonds. On the one hand, this should imply that C_{NM} decreases when the In concen-

[1] Please see also [39] for a discussion of the ability of hydrogen to passivate the electronic activity of N in $Ga_{1-x}In_xN_yAs_{1-y}$ due to the formation of N–H bonds.

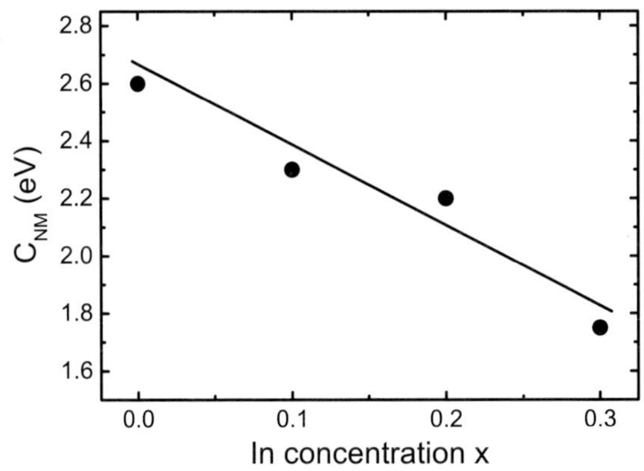

Fig. 4. BAC model coupling parameter C_{NM} as a function of In concentration x in thermally annealed $\mathrm{Ga}_{1-x}\mathrm{In}_x\mathrm{N}_y\mathrm{As}_{1-y}$ [28]. The solid line is just to guide the eye.

tration in $\mathrm{Ga}_{1-x}\mathrm{In}_x\mathrm{N}_y\mathrm{As}_{1-y}$ increases. Such a trend could indeed be confirmed by systematic room-temperature photoreflectance measurements of thermally annealed $\mathrm{Ga}_{1-x}\mathrm{In}_x\mathrm{N}_y\mathrm{As}_{1-y}$/GaAs quantum well structures with various In concentrations [28], leading to the results summarized in Fig. 4. However, even more interesting is the fact that the band gap of quaternary $\mathrm{Ga}_{1-x}\mathrm{In}_x\mathrm{N}_y\mathrm{As}_{1-y}$ with *fixed* In and N concentrations should increase (as a result of the changed E_{N} and C_{NM} values), when the nearest neighbor environment of the nitrogen atoms changes from predominantly Ga-rich to more In-rich configurations. As suggested in [31], this could explain the frequently observed blue-shift in the effective band gap of $\mathrm{Ga}_{1-x}\mathrm{In}_x\mathrm{N}_y\mathrm{As}_{1-y}$/GaAs quantum wells after thermal annealing [21]. However, further investigations are needed to clarify the possible contribution of additional mechanisms like interdiffusion to this effect. This will require the utilization of techniques like quantitative transmission electron microscopy [40], which are able to simultaneously evaluate the In and N concentration profiles in $\mathrm{Ga}_{1-x}\mathrm{In}_x\mathrm{N}_y\mathrm{As}_{1-y}$-based quantum well structures on a nanometer scale.

6 Conclusion

This chapter could only give a brief introduction to the rapidly developing topic of dilute arsenide nitrides. Although the interest in this material system was originally triggered mainly by its potential for the realization of near-infrared optoelectronic devices (in particular laser diodes for telecom applications), the many unusual physical properties of $\mathrm{Ga}_{1-x}\mathrm{In}_x\mathrm{N}_y\mathrm{As}_{1-y}$ have actually opened up a new and fascinating area of semiconductor research. The band anticrossing model discussed in detail above provides an important tool for the description of electronic states in arsenide nitride structures. However, there is still a long way to go, before a detailed understanding of this material system will have been reached.

Acknowledgments

I would like to thank my present and former colleagues, students and collaborators for their support and their contributions to our common work concerning $Ga_{1-x}In_xN_yAs_{1-y}$, in particular J. Kvietkova, A. Grau, D. Saez de Jauregui, K.C. Agarwal, T. Passow, M. Grün, C. Klingshirn, D. Litvinov, A. Rosenauer, and D. Gerthsen (Universität Karlsruhe), H. Riechert (Infineon Technologies, Munich), A.Yu. Egorov (A.F. Ioffe Physico-Technical Institute, St. Petersburg), P. Gilet and L. Grenouillet (CEA-DRT-LETI/DOPT-CEA/GRE, Grenoble), as well as M.D. Dawson (University of Strathclyde, Glasgow). Financial support by the Deutsche Forschungsgemeinschaft (DFG) within the Emmy Noether program and project A2 of the Center for Functional Nanostructures (CFN) at the University of Karlsruhe is also gratefully acknowledged.

Finally, I would like to dedicate this contribution to Claus Klingshirn on his 60th birthday and take this opportunity to thank him for his continuous support in all respects.

References

1. J.B. Heroux, X. Yang, W.I. Wang: Appl. Phys. Lett. **75**, 2716 (1999)
2. S.R. Kurtz, A.A. Allerman, E.D. Jones, J.M. Gee, J.J. Banas, B.E. Hammons: Appl. Phys. Lett. **74**, 729 (1999)
3. H.P. Xin, C.W. Tu, M. Geva: Appl. Phys. Lett. **75**, 1416 (1999)
4. M. Kondow, K. Uomi, A. Niwa, T. Kitatani, S. Watahiki, Y. Yazawa: Jpn. J. Appl. Phys. **35**, 1273 (1996)
5. M.C. Larson, M. Kondow, T. Kitatani, K. Tamura, Y. Yazawa, M. Okai: IEEE Photonics Technol. Lett. **9**, 1549 (1997)
6. C. Ellmers, F. Höhnsdorf, J. Koch, C. Agert, S. Leu, D. Karaiskaj, M. Hofmann, W. Stolz, W.W. Rühle: Appl. Phys. Lett. **74**, 2271 (1999)
7. M. Fischer, M. Reinhardt, A. Forchel: IEEE Photon. Technol. Lett. **12**, 1313 (2000)
8. M. Hetterich, M.D. Dawson, A.Yu. Egorov, H. Riechert: *Proc. 25th Int. Conf. Phys. Semicond.* (Osaka 2000), part I, ed by N. Miura, T. Ando (Springer proceedings in physics **87**, Springer, Berlin Heidelberg New York 2001) pp 693–694
9. I. Vurgaftman, J.R. Meyer, L.R. Ram-Mohan: J. Appl. Phys. **89**, 5815 (2001)
10. K.D. Choquette, J.F. Klem, A.J. Fischer, O. Blum, A.A. Allerman, I.J. Fritz, S.R. Kurtz, W.G. Breiland, R. Sieg, K.M. Geib, J.W. Scott, R.L. Naone: Electron. Lett. **36**, 1388 (2000)
11. G. Steinle, H. Riechert, A.Yu. Egorov: Electron. Lett. **37**, 93 (2001)

12. M. Kondow, T. Kitatani, K. Nakahara, T. Tanaka: Jpn. J. Appl. Phys. Part 2 **38**, L1355 (1999)
13. A. Livshits, A.Yu. Egorov, H. Riechert: Electron. Lett. **36**, 1381 (2000)
14. M. Kondow, S. Nakatsuka, T. Kitatani, Y. Yazawa, M. Okai: Jpn. J. Appl. Phys. Part 1 **35**, 5711 (1996)
15. M. Fischer, D. Gollub, A. Forchel: Jpn. J. Appl. Phys. Part 1 **41**, 1162 (2002)
16. M. Hetterich, M.D. Dawson, A.Yu. Egorov, D. Bernklau, H. Riechert: Appl. Phys. Lett. **76**, 1030 (2000)
17. C. Skierbiszewski, P. Perlin, P. Wisniewski, T. Suski, J.F. Geisz, K. Hingerl, W. Jantsch, D.E. Mars, W. Walukiewicz: Phys. Rev. B **65**, 035207 (2001)
18. C. Skierbiszewski: Semicond. Sci. Technol. **17**, 803 (2002)
19. R. Fehse, S. Tomić, A.R. Adams, S.J. Sweeney, E.P. O'Reilly, A. Andreev, H. Riechert: IEEE J. Select. Topics Quantum Electron. **8**, 801 (2002)
20. N. Tansu, L.J. Mawst: Appl. Phys. Lett. **82**, 1500 (2003)
21. I.A. Buyanova, W.M. Chen, B. Monemar: MRS Internet J. Nitride Semicond. Res. **6**, 2 (2001) and references therein
22. W. Shan, W. Walukiewicz, K.M. Yu, J.W. Ager III, E.E. Haller, J.F. Geisz, D.J. Friedman, J.M. Olson, S.R. Kurtz, H.P. Xin, C.W. Tu: Phys. Status Solidi B **223**, 75 (2001)
23. P.J. Klar, H. Grüning, W. Heimbrodt, G. Weiser, J. Koch, K. Volz, W. Stolz, S.W. Koch, S. Tomić, S.A. Choulis, T.J.C. Hosea, E.P. O'Reilly, M. Hofmann, J. Hader, J.V. Moloney: Semicond. Sci. Technol. **17**, 830 (2002)
24. A. Lindsay, E.P. O'Reilly: Solid State Commun. **112**, 443 (1999)
25. W. Shan, W. Walukiewicz, J.W. Ager III, E.E. Haller, J.F. Geisz, D.J. Friedman, J.M. Olson, S.R. Kurtz: Phys. Rev. Lett. **82**, 1221 (1999)
26. E.P. O'Reilly, A. Lindsay, S. Tomić, M. Kamal-Saadi: Semicond. Sci. Technol. **17**, 870 (2002)
27. P.R.C. Kent, A. Zunger: Phys. Rev. B **64**, 115208 (2001)
28. M. Hetterich, A. Grau, A.Yu. Egorov, H. Riechert: J. Appl. Phys. **94**, 1810 (2003)
29. I. Suemune, K. Uesugi, W. Walukiewicz: Appl. Phys. Lett. **77**, 3021 (2000)
30. R.J. Potter, N. Balkan, X. Marie, H. Carrère, E. Bedel, G. Lacoste: Phys. Status Solidi (a) **187**, 623 (2001)
31. P.J. Klar, H. Grüning, J. Koch, S. Schäfer, K. Volz, W. Stolz, W. Heimbrodt, A.M. Kamal Saadi, A. Lindsay, E.P. O'Reilly: Phys. Rev. B **64**, 121203(R) (2001)
32. A. Polimeni, M. Capizzi, M. Geddo, M. Fischer, M. Reinhardt, A. Forchel: Phys. Rev. B **63**, 195320 (2001)
33. G. Bastard: *Wave Mechanics Applied to Semiconductor Heterostructures* (Les Editions de Physique, Les Ulis 1988) pp 74–76, 112, 113
34. K. Kim, A. Zunger: Phys. Rev. Lett. **86**, 2609 (2001)
35. S. Kurtz, J. Webb, L. Gedvilas, D. Friedman, J. Geisz, J. Olson, R. King, D. Joslin, N. Karam: Appl. Phys. Lett. **78**, 748 (2001)
36. T. Geppert, J. Wagner, K. Köhler, P. Ganser, M. Maier: Appl. Phys. Lett. **80**, 2081 (2002)
37. M. Hetterich, A. Grau, D. Saez de Jauregui, A.Yu. Egorov, H. Riechert: *Physics of Semiconductors 2002, Proc. 26th Int. Conf. Phys. Semicond.*, (Edinburgh 2002), ed by A.R. Long, J.H. Davies (Institute of Physics Conference Series Number **171**, Bristol, Philadelphia 2003) paper D80

38. W.A. Harrison: *Electronic Structure and Properties of Solids* (Dover, New York 1989) p 176
39. A. Polimeni, G. Baldassarri Höger von Högersthal, F. Masia, A. Frova, M. Capizzi, S. Sanna, V. Fiorentini, P.J. Klar, W. Stolz: Phys. Rev. B **69**, 041201(R) (2004) and references therein
40. V. Grillo, M. Albrecht, T. Remmele, H.P. Strunk, A.Yu. Egorov, H. Riechert: J. Appl. Phys. **90**, 3792 (2001)

Nitride-Based Light Emitting Diodes and Laser Diodes: Optical Properties and Applications

Martin Kuball

1 Introduction

Nitride-based semiconductor products emerged onto the scene in the last decade, and have become a major player in the opto-electronics market with their wide wavelength tunability from the ultraviolet (UV) to the visible spectral region illustrated in Fig. 1. Revenues from the use of blue and green light emitting diodes (LEDs) in home and commercial lightening business alone are predicted to reach almost three billion dollars by 2009. Furthermore, the rapid development of high-brightness LEDs has opened applications in traffic signals consuming less than a tenths of the power of the standard filtered incandescent solution. The energy saving results in a short (<1 year) payback time on the initial investment of the LED solution. Automotive applications of LEDs include center-high-mount-stop-lamps (CHMLs), rear combination lamps (RCLs) and turn lights. The main driving force for LED application in automotive at present is styling, where the small size of LEDs allows flexible layout. Other applications of LEDs include backlights for liquid crystal displays (LCD) displays and full color displays.

Fig. 1. Wavelength tunability of nitride-based LEDs (Image courtesy Deutsche Post AG, Bundesministerium für Finanzen (Germany), Professor P. Krüll).

White lightening applications of nitride-based LEDs, although currently still limited due to their low optical output power, might be in general use within the next decade. LED-based daylight running lights of vehicles might appear soon, while vehicle forward lighting using white LEDs will not appear before 2006 because of the long automotive design cycle. UV nitride-based LEDs (<360 nm) are of great potential for biological and chemical sensing applications. Most biological molecules have a strong optical resonance within the optical range of approximately 240–340 nm. Compact light sources are therefore desirable for portable fluorescence based bioagent detection systems. Other applications include UV sources for short-range non-line-of-sight (NLOS) covert communication due to the increased Rayleigh scattering ($\sim \lambda^{-4}$) by the atmosphere in the UV or UV LEDs for sterilization, for example, in air-conditioning units.

Although the blue nitride laser diode (LD) optical storage market is presently negligible, it is predicted to reach almost three billion dollars by 2009. While current ultra-high storage density media such as DVDs provide a capacity of 4.5 GBytes using a 650 nm red laser diode, blue nitride-based LDs have the promise of achieving almost 20 GBytes. Other applications of blue LDs include high-density high-speed printing and full-color projection displays. Applications of nitrides, however, are not restricted to the opto-electronics market, but are also significant in the electronics market. Examples include high-speed high-power transistors for radars, satellite communication and mobile communication base station applications. This review will restrict to opto-electronic devices.

2 Historical Background

The first attempts to synthesize GaN were made by Johnson et al [1] and Juza and Hahn [2] in 1932 and 1938, respectively, and a decade later by Grimmeiss et al [3] by passing ammonia over hot gallium, resulting in small GaN needles and platelets. First epitaxial growth of GaN, on sapphire substrates, was demonstrated by Maruska and Tietjen [4] at RCA (now David Sarnoff Research Center) on 13 May 1968, a crucial milestone on the pathway to nitride-based devices. The method relied on the reaction between GaCl vapor and NH_3 in a hot-wall-reactor, a halide vapor phase epitaxy (HVPE) system [5]. HVPE as well as metalorganic vapor phase epitaxy (MOVPE) and molecular beam epitaxy (MBE) are employed today for nitride growth. All of the GaN made initially was n-type, although not intentionally doped, due to oxygen contamination [6]. Oxygen with its six valence electrons on a nitrogen site is a single donor in GaN. The possibility to epitaxially grow GaN sparked off intense research activities, which led to the first blue LED by Pankove et al [7] in 1972. This device was a metal-i-n device, since p-doping was not possible at that time, with light emission related to donor-acceptor transitions in Zn-doped GaN. The LED could either emit blue, green, yellow or red light dependent on the Zn concentration in the light-emitting region. Maruska et al [8] utilized alternatively Mg as luminescence center in metal-i-n GaN LED structures in 1973. Many unsuccessful attempts were made to achieve p-doping of GaN at this time. This key problem hampered the further development of nitride-based devices and possible commercial applications. Akasaki et al [9] and Amano et al [10] overcame this challenge by applying Low-Energy Electron Beam Interaction (LEEBI) to activate p-doping in Mg-doped GaN films in 1988. This discovery was an accidental discovery resulting from the observation of several orders of magnitude increased catholuminescence intensity in Mg-doped GaN after rastering an electron beam over the sample surface. The energy of the electron beam releases hydrogen from a Mg-H complex present after GaN growth resulting in a shallow acceptor about 0.16 eV above the valence band [11]. Although LEEBI is a feasible p-activation process on a small laboratory scale, p-activation of most devices today is the result of a subsequent discovery by Nakamura et al [12]. This process is based on the annealing of Mg-doped GaN above 750°C in N_2 or vacuum. It also converts the material into conducting p-type, making large-scale industrial device production feasible.

The success of p-doping in the nitrides intensified research activities in nitride-based devices. First pn-junction nitride-based LEDs were demonstrated, although, still based on donor-acceptor transitions such as InGaN co-doped with Zn and Si [13], i.e., a light emission mechanism similar to the early nitride metal-i-n structures. Until 1992, the brightness of nitride-based LEDs therefore reached only the milli-cd range and was unable to surpass SiC-based blue LEDs available at that time. LED brightness surpassed the 1-cd-range only in 1994 [14]. Luminous efficiency advanced significantly when introducing InGaN/GaN quantum wells (QWs) into the LED device structure [15]. The subsequent rapid evolution of the luminous efficiency of the nitride-based LEDs is illustrated in Fig. 2. In a rather short time-span of a few years luminous efficiency of nitride-based LEDs became comparable to other material systems. First nitride-based electrically pumped LDs were reported in 1995/96 shortly thereafter [16, 17] with further developments resulting into devices with lifetimes in excess of 15000 hours reported today [18].

In the following, optical properties of the components of nitride-based devices are discussed, and the current status of nitride-based LEDs and LD is reviewed. We will restrict on the wurtzite form of the nitrides. This is their most stable structural form and commonly used in opto-electronic devices, while information on the meta-stable form of zincblende nitrides and their properties can be found elsewhere in the literature (for example [19]). The wurtzite structure of the nitrides has a hexagonal unit cell with 6 atoms of each type. Its symmetry belongs to the space group C_{6v}^4. We denote in the following as c-axis the [0001] axis, i.e., the direction of the six-fold symmetry axis of the wurtzite structure unit cell.

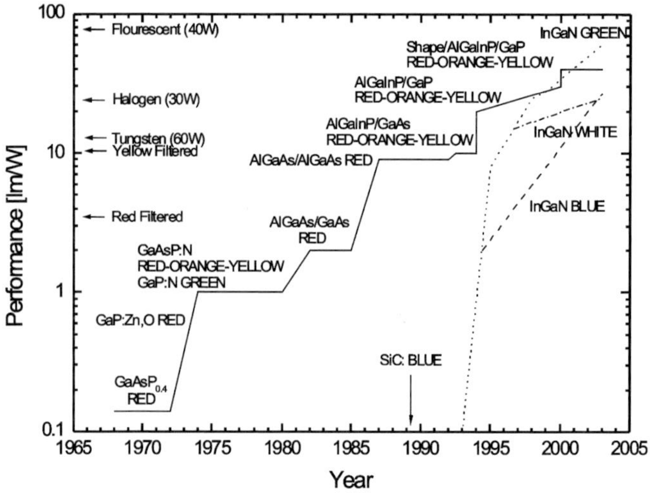

Fig. 2. Development of nitride-based LED luminous efficiency in comparison to other material systems.

3 Optical Properties of InN, GaN, AlN and Their Alloys

The band gap of the binary InN, GaN and AlN span the energy range from 0.7 eV to 6.2 eV covering the infrared, visible and UV spectral range. While the band gap of GaN with 3.4 eV and of AlN with 6.2 eV are well established, there is currently significant controversy regarding the band gap of InN with estimates ranging from 0.7 eV to 1.9 eV. Figure 3 illustrates the fundamental band gap of the nitrides versus basal-plane lattice constant assuming a band gap of 0.7 eV for InN, illustrating the large band gap tunability of the nitride material system. Full and dashed lines indicate the band gap of the ternary alloys $In_xGa_{1-x}N$, $Al_xGa_{1-x}N$ and $In_xAl_{1-x}N$. Most nitride material is characterized by a high dislocation density due to the need for heteroepitaxy commonly on sapphire or SiC substrates due to the current lack of large size GaN or AlN substrates. Dislocation densities in the 10^8–$10^{10}\,\mathrm{cm}^{-2}$ range [20] are typical due to the large lattice mismatch, for example, of 13% for the GaN/sapphire and of 3.5% for the GaN/SiC material system. Although GaN and AlN substrates have been demonstrated and been shown to result in a significant reduction of defect density for nitride growth, their current size and the currently available quantities are not adequate for wide-range use at present.

Photoluminescence of GaN is dominated at low temperature by contributions from free and bound exciton transitions [21] with excitonic transitions preserved up to room temperature [22]. Degeneracy of the top of the valence band of the wurtzite-structure nitrides is lifted by crystal field and spin-orbit interaction at the Γ point. The emerging three states are termed Γ_9^v, upper

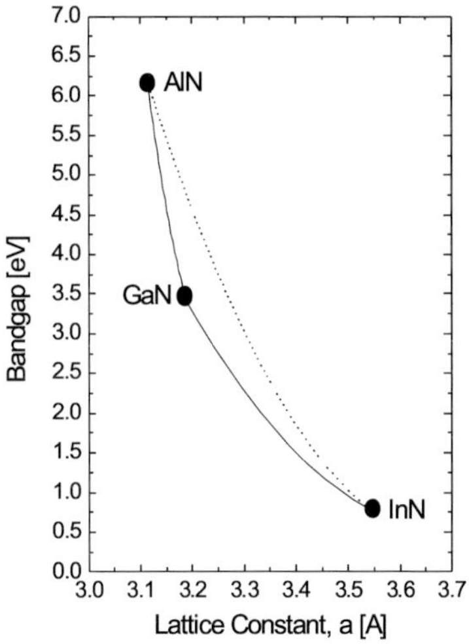

Fig. 3. Fundamental band gap of nitrides versus basal-plane lattice constant using experimental data for the InN/GaN [24] and AlN/GaN material system [25], and theoretical data on the bowing parameter for the AlN/InN material system [26]. Full and doted lines present the ternary alloys.

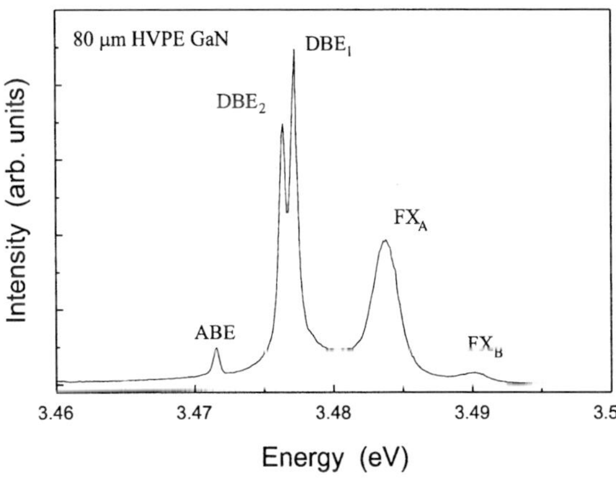

Fig. 4. Photoluminescence spectrum of a thick HVPE GaN layer (from [28]).

Γ_7^v and lower Γ_7^v denoting their symmetry. The conduction band has Γ_7^c symmetry. Three free excitons, named A ($\Gamma_7^c \rightarrow \Gamma_9^v$), B ($\Gamma_7^c \rightarrow$ upper Γ_7^v) and C ($\Gamma_7^c \rightarrow$ lower Γ_7^v) excitons result. For unstrained GaN energies of 3.4776 eV, 3.4827 eV and 3.5015 eV have been reported for the A, B and C exciton, respectively, at 10 K. Viswanath et al [23] have summarized free exciton energies reported for GaN by different groups. Figure 4 illustrates an example of a low-temperature photoluminescence spectrum of a thick HVPE-grown GaN layer. A and B free exciton transitions are clearly visible. A, B and C excitons can be typically observed in α polarization ($E \perp c$ and $k \parallel c$), A and B exciton in σ polarization ($E \perp c$ and $k \perp c$) with the C exciton weak, and A and C exciton in π polarization ($E \parallel c$ and $k \perp c$) with the B exciton weak. Apart from free exciton transitions, bound exciton transitions are commonly observed in GaN, bound to a neutral shallow donor (donor bound exciton = DBE) or neutral shallow acceptor (acceptor bound exciton = ABE). Excitons bound to charged defects are also possible, but are rarely observed to be prominent in semiconductors. Both ABE and DBE are visible in Figure 4. Energies of DBE are in the range of 3.470 eV at low temperatures for strain-free GaN [21, 27]. Note that in Fig. 4 transitions are shifted to higher energies by about 8 meV due to the presence of a small compressive strain in the HVPE GaN. Different donors typically result in slight differences in DBE energies as illustrated in Fig. 4. This enables to identify the nature of a specific donor in a GaN sample, for example, silicon or oxygen donors have different binding energies of 29 and 33 meV, respectively [28–30]. The most prominent ABE in strain-free GaN is observed around 3.466 eV [28], tentatively assigned to Mg on a nitrogen site [31, 32]. The Mg acceptor state has a binding energy of 155–165 meV [11]. Further optical transitions are present well below the band edge transitions. Examples include donor-acceptor pair (DAP) transitions at 3.27 eV [33] in doped GaN. The involved deep acceptor state (binding energy of 225–230 meV) is thought to be either carbon on a nitrogen site or a gallium vacancy [21]. Broad photoluminescence bands around 2.8–3 eV are observed in p-type GaN:Mg [21, 34]. Phonon replicas of optical transitions can often be observed in luminescence experiments due to the large ionicity of GaN providing the basis for a strong coupling to longitudinal optical (LO) phonons (phonon energy of 92 meV).

Optical transitions related to excitons are also characteristic for AlN. Luminescence in AlN is due to free and bound exciton recombination up to room temperature, similar to GaN. Exciton energies between 6.06 and 6.21 eV [35–37] have been reported at 10 K. Three types of excitons exist, but due to material quality issues only A excitons could be observed so far [36, 37]. Since Al has a large affinity to oxygen, it is almost impossible to eliminate oxygen contamination in AlN. This results in additional deep-level luminescence from AlN. Youngman and Harris [38] as well as Slack et al [39] have identified a broad luminescence feature around 3.5 eV to be related to oxygen in AlN, namely to clustered aluminium impurity oxygen complexes. Their luminescence energy decreases with increasing oxygen contamination. Additional broad luminescence features are known to occur around 2.8 eV and have been attributed to nitrogen vacancies (or excess aluminum).

A number of groups have reported optical measurements on InN. Tansley and Foley [40] were probably one of the first to accurately measure the band gap of an InN film and determined a band gap of 1.89 eV. In contrast recent experiments by Davydov et al [41] and Wu et al [24] suggest a band gap of about 0.7 eV. Several mechanisms have been proposed to explain the different band gaps measured for different samples. These include Burstein-Moss shifts of optical transitions due to the presence of a high free carrier concentration in present InN samples up to 10^{19}–$10^{20}\,\mathrm{m}^{-3}$ possibly due to oxygen impurities and the presence of oxynitrides at grain boundaries [24, 41, 42]. Photoluminescence observed at 0.7 eV has been attributed to both sub-band gap transitions as well as band-to-band transitions. Further research is clearly needed to clarify the band gap of InN.

The ternary nitride alloys $\mathrm{In}_x\mathrm{Ga}_{1-x}\mathrm{N}$ and $\mathrm{Al}_x\mathrm{Ga}_{1-x}\mathrm{N}$ are key components of visible and UV LEDs/LDs, respectively. They provide the basis for their wide wavelength tunability as illustrated in Fig. 1. The less explored quarternary alloy $\mathrm{In}_x\mathrm{Al}_y\mathrm{Ga}_{1-x-y}\mathrm{N}$ alloy has become an attractive alternative to $\mathrm{Al}_x\mathrm{Ga}_{1-x}\mathrm{N}$ for blue/UV opto-electronic devices due to its higher luminous efficiency than the ternary alloy $\mathrm{Al}_x\mathrm{Ga}_{1-x}\mathrm{N}$. $\mathrm{In}_x\mathrm{Ga}_{1-x}\mathrm{N}$ covers the infrared/visible spectral range with a band gap of $E_g(x) = (1-x)E_g(\mathrm{InN}) + xE_g(\mathrm{GaN}) - bx(1-x)$ over $0 \le x \le 1$, where $E_g(\mathrm{GaN}) = 3.4\,\mathrm{eV}$, $E_g(\mathrm{InN}) = 0.7\,\mathrm{eV}$, and values for b reported between 1.4 and 2.5 eV [24, 41]. Properties of InGaN are subject to two important effects, (i) In clustering and (ii) polarization effects. Theoretically it has been predicted by Saito et al [43] that the alloy has a strong tendency to segregate for compositions $x > 0.1$. Nurukawa et al [44], and O'Donnell et al [45] provide experimental evidence using high spatial resolution imaging techniques for the formation of high indium composition dots in InGaN layers and QWs. The extend of In clustering might depend on growth conditions, and in some cases may be avoided [28]. The InGaN alloy therefore shows compositional fluctuations that go well beyond compositional disorder known for other semiconductor alloys. Carrier capture in the region of higher indium composition is responsible for the high luminous efficiency of InGaN despite the high dislocation density present in this material system, keeping carriers away from dislocations. Carrier localization is also responsible for the long carrier recombination times characteristic for time-resolved photoluminescence from this ternary alloy [46–48]. We note that carrier localization, namely hole localization at the site of indium atoms, is also predicted to occur in InGaN even without any In clustering [49].

Due to in-plane strain significant piezoelectric field polarizations are present in $\mathrm{In}_x\mathrm{Ga}_{1-x}\mathrm{N}$ heterostructures. The wurtzite structure, however, not only allows piezoelectric polarization, but also spontaneous polarization, due to the difference in bond length in different directions. Calculations on spontaneous polarization in this material system have been reported by Bernadini et al [50]. While macroscopic fields are screened by surface charges, microscopic fields are present in heterostructures. When the potential in a heterostructure such as a QW is tilted by an electric field, electrons and holes move in opposite directions of the QW. As result the effective band gap is reduced by as much as 0.5 eV. In addition the separation of electrons and holes significantly reduces the their wavefunction overlap. This so-called quantum confined Stark effect (QCS) therefore reduces radiative efficiency and increases recombination times. QCS is responsible for the decreasing luminous efficiency of InGaN towards the UV spectral range where carrier localization due to indium clustering decreases. Electric fields as high as 2.5 MW/cm have been measured in $In_{0.2}Ga_{0.8}N$ heterostructures at $x = 0.2$ by Grandjean et al [51].

The ternary alloy $Al_xGa_{1-x}N$ spans the wavelength range from the blue to the UV spectral range with a band gap of $E_g(x) = (1-x)E_g(\mathrm{GaN}) + xE_g(\mathrm{AlN}) - bx(1-x)$ over $0 \leq x \leq 1$, where $E_g(\mathrm{GaN}) = 3.4\,\mathrm{eV}$, $E_g(\mathrm{AlN}) = 6.2\,\mathrm{eV}$ with a bowing parameter b between 0.4 and 1 eV [25, 52]. Electric fields due to piezoelectric and spontaneous polarization reduce overlap between electron and hole wavefunction in heterostructures containing AlGaN. In contrast to InGaN no significant localization effects occur in AlGaN to counter this effect. This is of disadvantage for the suitability of $Al_xGa_{1-x}N$ for blue/UV light emitting devices structures as active light emitting medium and has initiated research into the quarternary alloy $In_xAl_yGa_{1-x-y}N$. Several groups [53–57] have shown enhanced photoluminescence efficiencies of $In_xAl_yGa_{1-x-y}N$ alloys in the UV spectra region due to carrier localization related to the presence of indium. Also heterostructures can then be engineered with minimal piezoelectric polarization benefiting from the possibility of lattice matched growth. Teisseyre et al [58] recently illustrated that electric fields can be reduced dramatically on an $In_{0.04}Al_{0.12}Ga_{0.84}N/In_{0.01}Al_{0.30}Ga_{0.69}N$ QW system. Growth of $In_xAl_yGa_{1-x-y}N$ quarternary alloys, however, is challenging due to the different binding energies of the binary compounds, the different mobility and desorption temperature of the growing species.

4 Nitride-Based Light Emitting Diodes

4.1 Blue-Green-Yellow InGaN-Based Light Emitting Diodes

Nitride-based LEDs for the visible spectral range contain InGaN single or multiple QWs or InGaN/GaN/AlGaN double heterostructures (DH) as active medium. Figure 5 shows a schematic of the device structure. Indium composition is varied to control the emission wavelength. The AlGaN layer serves as electron blocking layer. Due to the rather low hole mobility, the spreading resistance of the top p-doped GaN (GaN:Mg) layer is rather poor. A large semitransparent metal contact resolves this problem. Only a small portion of the p-layer contains a thick metal contact where the bond wire is attached. The device shown in Fig. 5 is grown on sapphire. Since sapphire is an insulator this requires fabrication of the n-contact by etching the structure down to the n-doped GaN (GaN:Si) as illustrated. Alternatively use of electrically conductive substrates such as SiC allows a vertical device structure with p-contact on top and n-contact on the back of the substrate. Metal contacts such as Ti/Al/Ti/Au for ohmic contacts to n-type GaN and annealed Ni/Au

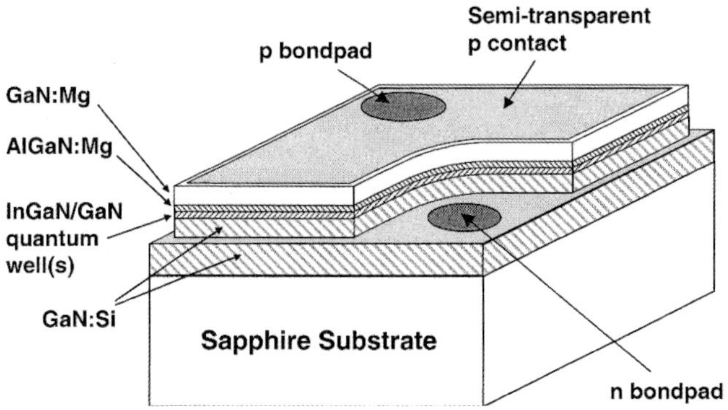

Fig. 5. Schematic of nitride-based LED (Image courtesy Lumileds Lighting, from [59]).

contacts to p-type GaN are employed with other contact schemes also available. Early LEDs contained a GaN/InGaN/AlGaN DH with a thick InGaN layer co-doped with Zn and Si. Due to the need for a large enough InGaN layer thickness, the InGaN contained only a small indium composition. Light emission was due to donor-acceptor pair transitions. This limited both the accessible wavelength range as well as the LED brightness. A broad emission spectrum of 70 nm width for an LED operating at 450 nm [13] was also characteristic. Nichia Chemical Industries commercialized devices based on this concept in 1993 with a luminous intensity of 2 cd with external quantum efficiencies up to 5% at 450 nm. This was possible despite the high dislocation density of 10^8–10^{10} cm^{-2} present in those devices. Similar dislocation densities would prevent in other material systems such as GaAs any significant light emission from an LED. This caused significant surprise at that time and is effectively the consequence of the luminescence mechanism in InGaN via localized states. Lifetimes in excess of 10000 hours were achieved even in these early days of nitride-based LED development. A minimum of 10000 hours is required by the display society (CIE).

Nowadays nitride-based LEDs are predominantly based on single or multiple InGaN QW structures. External quantum efficiencies as high as 20-30% in the blue-green spectral range are possible, with emitted optical power up to 21 mW in the blue and 7 mW in the green for a standard 5 mm LED lamp [60]. Also yellow and amber light emitting nitride-based LEDs are available. Figure 6 shows electroluminescence spectra of blue, green and yellow LEDs. Increasing the In composition shifts the emission spectrum to longer wavelengths due to the decreasing band gap (Figure 3). Figure 7 shows a cross section of a standard 5 mm LED lamp. The LED chip is placed onto one of the leads of the lamp with clear epoxy used as encapulant, which also provides the mechanical stability. The LED chip is connected by means of a thin Au ($<25\,\mu$m) wire. Lifetimes of current LEDs reach more than 50000 hours.

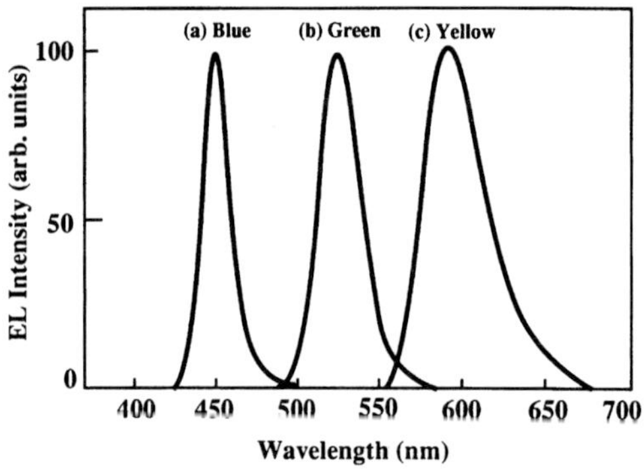

Fig. 6. Electroluminescence of blue, green and yellow InGaN-based LEDs (from [66]).

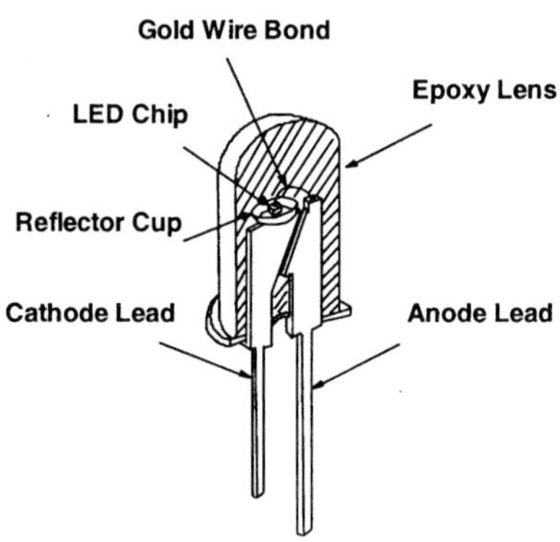

Fig. 7. Cross section of a standard 5 mm LED lamp (Image courtesy Lumileds Lighting, from [59]).

Elimination of the Zn- and Si-centers and using InGaN QWs instead led to a significant narrowing of the emission spectrum to 20 nm and 30 nm for blue and green LEDs, respectively [61]. Figure 8 illustrates how the external quantum efficiency varies with emission wavelength. External quantum efficiency decreases towards the UV spectral region, since then piezoelectric and spontaneous polarization reduce electron and hole wavefunction overlap while carrier localization due to indium clustering decreases. It is most efficient in the 400–450 nm spectral window. Towards the red spectral range, phase separation effects in the InGaN increase, reducing material quality, resulting in a decrease in the quantum efficiency. Figure 9 shows the chromatic diagram indicating the location of the green and blue InGaN-based LEDs. The outer periphery corresponds to saturated colors. The center corresponds to various grades of white. As a electroluminescence spectrum get wider, such as the

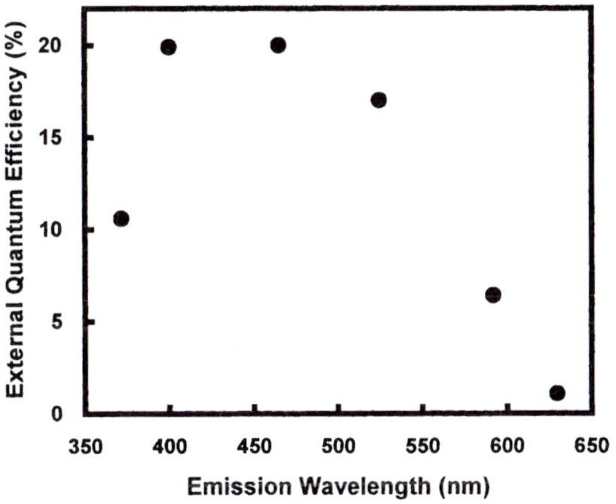

Fig. 8. External quantum efficiency of InGaN-based LED as function of emission wavelength (from [67]).

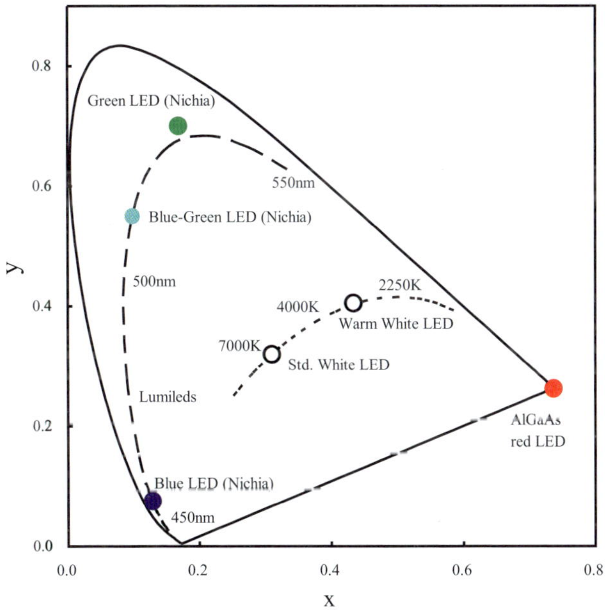

Fig. 9. CIE chromaticity diagram. Nitride-based blue, green LEDs and red AlGaAs LEDs as well as recent nitride-based white LEDs are indicated. The central area represents various grades of white, the outer line saturated colors. The line through the white-light region indicates the color diagram for white light with the accompanying color temperature (Image courtesy Lumileds Lighting and Nichia Chemical Industries).

case for the early blue LEDs, the corresponding color in the CIE chromaticity diagram gets pulled to the center reducing the accessible range of colors, illustrating one of the key disadvantages of the early LEDs. The color of current InGaN based blue LEDs is basically saturated while this is not fully the case for the InGaN based green LEDs which can be attributed to the wider spectrum due to the larger indium compositional clustering in the active device region. Nevertheless InGaN based green and blue LEDs present two important legs of the triad. Together with red AlGaAs-based LEDs full color displays (and white lighting discussed in section 4.3 by combining these three colors) can therefore be fabricated. Also micro-displays based on nitride LEDs have been demonstrated, although only monochromatic at present [62, 63].

Several groups have investigated degradation mechanisms of blue-green LEDs. Osinski et al [64] studied metal migration along defect tubes under high electric stress and identified it as the main failure mechanism on early Nichia LEDs, while Meneghesso et al [65] identified on more recent devices generation of non-radiative recombination centers as the main degradation mechanism. Degradation of the encapsulation epoxy of 5 mm lamps was also observed [59].

Challenges for blue-green LEDs are mainly concerned with the goal of achieving higher optical output power. Although the quantum efficiency of current blue and green LEDs is rather high, it is important to realize that an important criterion for an LED for lightening applications is total emitted flux. Conventional 5 mm LED lamps emit typically one or at best a few lumens of light whereas a light bulb or a halogen lamp can emit hundred of even thousands of lumens. For applications where high flux is required, hundreds of conventional LEDs therefore need to be combined to generate the required amount of light. This is possible, for example, in traffic lights since the LEDs can be arranged over a rather large area of an 8 or 12 inch traffic light, however, for applications that require a high-flux point source, use of the 5 mm LED lamps is not adequate. High optical output power LEDs are here needed. Several key issues need to be addressed for high total flux devices [59]: (i) improved thermal management of the devices to remove Joule's heat generated in the devices; (ii) remove epoxy encapsulant since it photochemically reacts with light of wavelength shorter than 500 nm, i.e., it degrades (this effect worsens for wavelengths below 380 nm); (iii) increase external quantum efficiency. In a 5 mm LED lamp (Fig. 7) the glass transition temperature of the epoxy is between 100 and 150°C. At temperatures larger than this temperature the epoxy's coefficient of linear expansion triples, resulting in stress on the Au wire connecting the LED chip rapidly causing the device to fail. Device temperature needs to be therefore limited to 10–20°C below this temperature restricting the possible electrical input and subsequently the optical output power of the LED. We note that temperature rise of the LED chip needs not only be minimized due to this packaging issue, but also since a temperature rise can directly result in a decrease in optical output of the LED chip, although this effect is not dominant for the InGaN material system, but of great importance to the AlGaN or InAlGaN materials systems needed for UV LEDs discussed later in section 4.2. Heat generated in the LED can be extracted from a 5 mm lamp only via the two thin leads (Figure 7) resulting in a thermal resistance of the packaging as high as >200°C/W. This limits the electric power input into the device to <0.1 W. Therefore alternative packaging approaches are needed. Figure 10 illustrates an example of a package where devices can be operated up to much higher electrical input power. This is possible due to improved heat sinking with devices directly mounted on a metal slug. Devices are mounted flip-chip, i.e., the active device region is directly mounted on the heat sink slug and light is extract through the sapphire substrate. Heat then does not need to be extracted via the thermally highly resistive sapphire substrate. This also reduces the need for a semi-transparent top p-contact metal layer, reducing absorption losses. Thick metal contacts can then be used for the p-side metallization of the devices. They reflect light emitted downward up again increasing the external quantum efficiency of the LED. To avoid degradation of the epoxy encapsulant, silicone is used as encapsulant. This package reduces thermal package resistance by more than an order of magnitude to between 8 and 15°C/W.

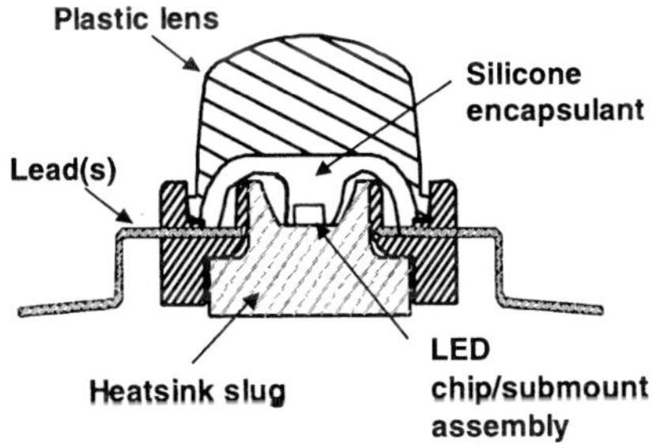

Fig. 10. Cross section of a high-power LED package (Image courtesy Lumileds Lighting, from [59]).

Apart from optimisation of device packaging, luminous efficiency of the LED chip needs to be improved. Efficiency of InGaN-based LEDs is dependent on drive current. This is in contrast to for example AlInGaP LEDs where LED efficiency is virtually independent of drive current [59]. Efficiency of InGaN-based LEDs decreases with increasing drive current, also a small blue shift of the luminescence peak occurs. Responsible is band filling with increasing injection current, resulting in a transition from localized to delocalised states [68]. Carriers can then reach non-radiative recombination centers. This is an inherent property of InGaN, i.e., cannot be prevented easily unless defect density is reduced dramatically. The solution for current nitride-based LEDs lies in increasing chip size rather then current density to increase luminous efficiency and to limit the maximum current density. Commercial manufacturers limit current density in LED chips to typically about $50\,\mathrm{A/cm^2}$. Necessary steps need to be taken in terms of metallization to ensure a homogenous current distribution when using large chip sizes to prevent current crowding. This is no significant difficulty when flip-chip mounting the devices since then a thick p-metallization can be employed, while for standard mounting interdigited finger contacts can be used [59]. Typical device sizes for 5 mm LED lamps are around 200–350 μm square. Output powers greater than 250 mW at 428 nm at $50\,\mathrm{A/cm^2}$ have been illustrated for chip sizes of 2 mm square in a high power device package such as the one shown in Fig. 10 [59]. Other concepts to increase external quantum efficiency have also been explored. For device structures grown on SiC substrates shaping of the SiC substrate was shown to increase light extraction efficiency significantly [69]. Light emitted towards the SiC substrate is then reflected back into the desired light emission direction.

4.2 Ultraviolet Light Emitting Diodes

There have been significant advances in UV nitride-based LEDs (λ<360 nm) within the last year or two. Early attempts focused on GaN as light emitting medium in DH structures [70] and QW structures [71, 72] with emission wavelengths at around 330–350 nm and external quantum efficiency of about 0.1%. Although high external quantum efficiencies are possible with the InGaN material system in blue-green LEDs, the lack of significant carrier localization in the AlGaN/GaN material system coupled with the existence of piezoelectric and spontaneous polarization, poses a difficulty for AlGaN/GaN device structures for achieving high external quantum efficiencies. Apart from high dislocation densities, primary evidence further suggests a very high density of point defects in $Al_xGa_{1-x}N$ ($x > 0.3$) with luminescence properties similar to LT-GaAs, in which the excited carriers are rapidly consumed by deep levels [73]. Improvements in material quality of the AlGaN are therefore needed for higher efficiency UV-LED based on AlGaN/GaN DH and QW structures.

Significant advancements have been made in this area recently by growing device structures on novel substrates. Nishida et al [74] reported on 352 nm LEDs with an optical output as high as 10 mW. The active medium was a $Al_{0.04}Ga_{0.96}N/Al_{0.1}Ga_{0.9}N$ single QW grown on GaN grown on a bulk GaN substrate instead of the commonly used sapphire or SiC substrates with external quantum efficiencies of 1%. Alternatively, to reduce defect density in AlGaN layers grown on sapphire substrates two approaches have been employed. A combination of faceted growth and low-temperature AlN interlayer allowed Kamiyama et al [75] to reduce the defect density for LEDs in the 323 nm to 363 nm spectral range. Improved external quantum efficiencies were demonstrated. Chitnis et al [76] introduced AlN(20Å)/$Al_{0.2}Ga_{0.8}N$(300Å) superlattices (SLs) on sapphire substrates to reduce defect density in thick

$Al_{0.2}Ga_{0.8}N$ layers used as buffer for $Al_{0.18}Ga_{0.82}N/Al_{0.12}Ga_{0.88}N$ multiple QW device structures. 325 nm LEDs with optical output as high as 4 mW under pulsed conditions were achieved for a 300 µm square device. Improvements in the growth of the strain-relief AlN/AlGaN SL by employing pulsed-atomic-layer epitaxy (PALE) instead of MOCVD resulted in a significant increase in optical output power to 10.2 mW at 325 nm with external quantum efficiency up to 0.45% under pulsed conditions for a 200 µm square device [77]. Wavelengths as short as 278 nm have been reported for UV nitride-based LEDs using AlN/AlGaN SL buffer layers with optical output as high as 3 mW with external quantum efficiencies of 0.07% for a 200 µm square device [78]. A typical output spectrum is shown in Figure 11. A strong emission peak is visible at 278 nm. Weaker luminescence is visible towards longer wavelengths at 320 nm, possibly due to deep levels in the AlGaN buffer layer of the device structure. Hanlon et al [79] demonstrated 292 nm emitting LEDs with continuous wave optical output of 2–3 µW grown on an AlN base.

Many UV LED chips are flip-chip mounted on a high-thermal conductivity carrier such as SiC or AlN to extract heat generated in the devices efficiently. Despite flip-chip mounting the devices thermal management is still an issue of significant concern for UV nitride-based LEDs. This is illustrated in Fig. 12. The optical output power of devices operated in continuous wave and pulsed mode is compared. Continuous mode operation with its larger device heating has detrimental effect on the device optical output resulting not only in a significant drop in optical output power but also in the inability to operate the device at high input currents. Sarua et al [80] measured temperature rises as high as 70°C in 325 nm flip-chip mounted LEDs at 50 mA (8 V) using micro-Raman spectroscopy. Optical output in UV LED depends significantly on temperature. Background heating of a device operated under pulsed condition (where device self-heating can be neglected to first approximation) of 70°C resulted in a drop in optical output by a factor of 2–3 [81]. Reduction in device forward voltage, i.e., better electrical properties of the AlGaN layers to reduce self-heating, and improved device packaging for better heat extraction are important components for achieving higher efficiency UV LED devices in future.

It has to be noted that conductivity control of AlGaN is quite complex. Good mobility and high carrier concentrations are difficult to achieve considering the high density of point defects in AlGaN, although there has been some recent progress. Taniyasu et al [82] illustrated free carrier concentrations up to $1\times10^{19}\,cm^{-3}$ for $Al_xGa_{1-x}N$ with x=0.42 and for AlN up to $1\times10^{17}\,cm^{-3}$. Pophristic et al [83] reported good conductive n-type $Al_{0.6}Ga_{0.4}N$ films with carrier concentrations up to $3.5\times10^{18}\,cm^{-3}$ and a mobility between 25 and $35\,cm^2/V\,s$. AlN/AlGaN SLs similar to the one used by Chitnis et al [76] reduced the defect density in the AlGaN layers. We note that DX-centers have been detected in some Si-doped AlN and higher Al composition AlGaN films [84], however, apparently did not prevent achieving

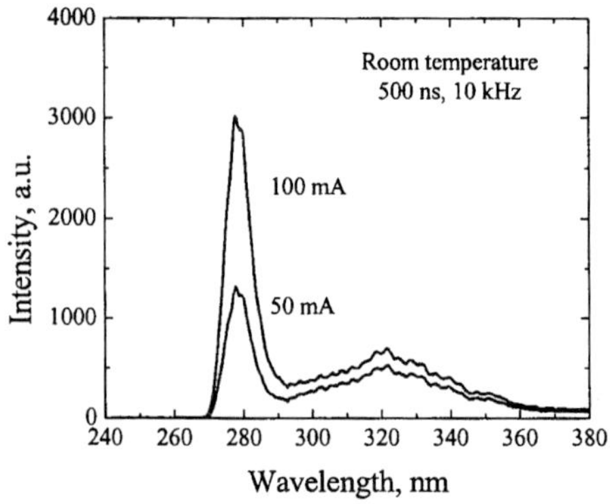

Fig. 11. Spectrum of UV LED at 278 nm operated at two different input currents under pulsed condition (from [78]).

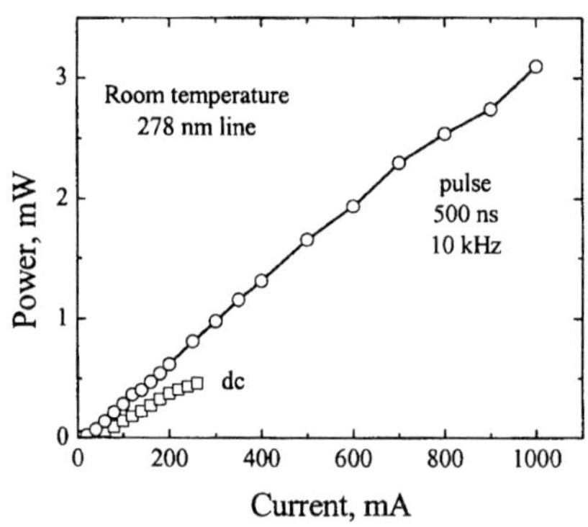

Fig. 12. Optical output power of 278 nm UV LED in continuous wave and pulsed mode as function of current (from [78]).

n-type conductivity in AlN by Taniyasu et al [82]. The question of the role of DX-centers in AlGaN still needs to be resolved. Mg remains to date the most robust approach for p-type conductivity in AlGaN in particular for low Al composition films while several other acceptor candidates such as beryllium and carbon have also been suggested for p-doping of AlGaN. P-doping of higher composition AlGaN films is difficult due to the deep acceptor nature of Mg. Many of the device structures reported for UV LEDs contain a higher composition p-type AlGaN layers, however, no precise information on carrier concentration and mobility of this layer in the device structure is typically given. Recently AlGaN/GaN SL have shown promise of achieving high free hole concentrations utilizing piezoelectric and spontaneous polarization effects. Carrier concentrations as high as $2.5–4.2\times10^{18}\,\mathrm{cm}^{-3}$ have been reported by Kozodoy et al [85] and by Yasan et al [86]. The rather high resistivity of AlGaN layers in the device structure makes apart from issues such as device self-heating current crowding an issue of concern for UV LEDs [87]. Interdigited finger geometry designs have been explored successfully to reduce current crowding effects and have resulted in improvement in optical output power from UV LED devices.

The second component for increasing optical output of UV LEDs is to introduce carrier localization into the AlGaN material system. Carrier localization is known to be responsible for the high luminous efficiency of InGaN-based LEDs. The introduction of In into the AlGaN material system is therefore currently being explored, i.e., the use of the quarternary alloy $In_xAl_yGa_{1-x-y}N$. Khan et al [88] reported AlInGaN UV LEDs with 305 nm emission with 1 mW output for a device of 20 μm×1000 μm stripe geometry. PALE growth of InAlGaN allowed the growth of InAlGaN layers with luminescence efficiencies one order of magnitude higher than AlGaN layers of similar band gap [53]. This illustrates the great potential In localization in the InAlGaN material system offers for UV LEDs. LEDs with optical output ranging from 355 to 340 nm with InAlGaN QWs and barriers of different composition with optical output power ranging from 0.23 mW (315 nm) to 1.1 mW (340 nm) were demonstrated for devices between 400 μm×400 μm and 400 μm×20 μm size [53]. Hirayama et al [54] and Wang et al [89] also demonstrated quarternary UV nitride-based LEDs.

4.3 White Nitride-Based Light Emitting Diodes

There are basically two concepts to generate a white light spectrum using LEDs. The first method is the red, green and red (RGB) approach and involves the combination of three or more emitters of different wavelength to generate a spectrum that is perceived by the human eye as white. Although this is the most efficient approach, the requirement of color mixing optics, variation in lumen output and wavelength for nominally identical LEDs and the change in these parameters with ambient temperature and time, makes the RGB approach also the most challenging. These issues can be overcome with suitable feedback control schemes that can be practically implemented at present with current LEDs [90], however, is less attractive from a cost viewpoint.

The second technique involves a single blue InGaN-based LED with wavelength between 410 and 460 nm for the excitation of a luminescent material such as yellow emitting phosphor Ce-YAG (Ce-doped Yttrium Aluminum Garnet) [91]. A schematic of such a luminescence conversion (LUCO) LED is depicted in Figure 13 for a 5 mm package. The luminescence converting material is either deposited onto the LED chip as illustrated or can be mixed into the epoxy lens. In these LUCO LEDs a certain portion of the blue emission is allowed to leak through the yellow phosphor, so that the combined spectrum of blue LED and yellow phosphor emission is perceived as white light by the human eye in the CIE chromaticity diagram (Figure 9). Alternatively

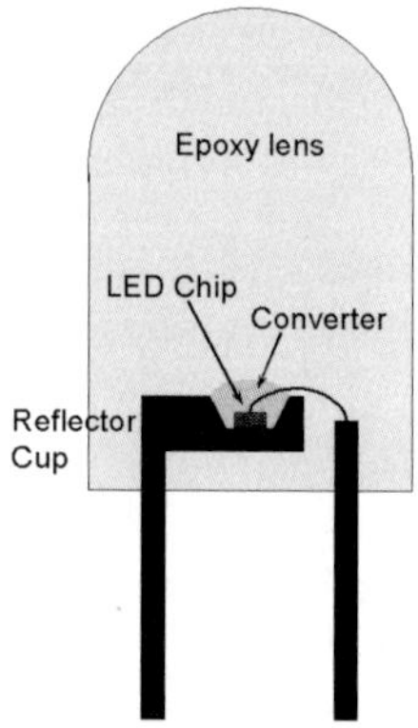

Fig. 13. Schematic of a white light emitting LUCO LED.

organic dyes can be employed for color mixing. These LEDs quickly found applications in various low/medium brightness illumination systems such as flashlights, dashboard back illumination and car interior lighting, commercially available. One problem can be that conventionally the reflector cup is partially filled with the phosphor as shown in Fig. 13 resulting in color variations in the radiation pattern caused by variations in the thickness of the phosphor layer, however, processing ways have been found which minimize this problem of the past where bluish or yellow rings in the far-field pattern where not uncommon. The advantage of LUCO LEDs is their rather good luminous efficiency between 20 and 32 lm/W [59, 91], however, their weakness is the white (hue) toning and color rendering. Since the white light is generated by mixing two complementary colors hue toning is restricted to points on a straight line in the CIE chromaticity diagram, connecting the blue pump with the yellow converter, and to high color temperatures, usually above 5500 K corresponding to a cold white. Due to the inability of such a light source to illuminate colors in the blue-green and red wavelength regime the color rendering index (CRI) is typically no more than 75, lower than the 80 or greater that is normally required for indoor illumination. CRI is a numerical system that rates the color rendering ability of light in comparison with natural daylight, which has a CRI of 100. An LED with a CRI for example of 90 shows colors more naturally than a lamp with a CRI of 75.

More than two wavelengths are required to achieve white light emitters with a high CRI. The problem of a low CRI in LUCO LEDs can be overcome by using the RGB principle, i.e. instead of using only yellow phosphor a RGB phosphor blend can be employed similar to a luminescent lighting tube. It requires a near-UV/UV LED for excitation (<400 nm) of the phosphor blends, for example a 380 nm InGaN-based LED. For example, the phosphor blends used by Kauffmann et al [92] are rare-earth doped oxide and sulfide materials which under UV excitation emit broad bands at 460 nm, 550 nm and 610 nm.

White LEDs with color temperatures between 4000 and 4300 K and CRI of 78 were achieved. Conversion efficiency of the triphosphors, in particular of the blue and red phosphors is the limiting factors in terms of brightness. Luminous efficiencies of 15 lm/W have been reported [93]. Alternatively, white LEDs have been demonstrated by Krames et al [94] combining a blue LED with Ce-YAG (yellow) phosphor and a broadband red phosphor with luminous efficiencies of 20–25 lm/W, color rendering above 85 and color temperature between 2800 and 4000 K for a 1 mm square chip. Good color rendering is therefore possible when using the two or three phosphor solution, however, at the expense of a somewhat reduced luminous efficiency. Improved luminescence conversion schemes will be needed to enable use of white nitride-based LEDs for general lightening applications.

5 Nitride-Based Laser Diodes

The rapid advancement of nitride-based LEDs ultimately led to the development of electrically pumped nitride-based LDs. Dingle et al [95] reported on optically pumped stimulated emission in GaN, namely GaN needles in 1971, Amano et al [96] and Khan et al [97] on epitaxially grown GaN and Amano et al [98] subsequently on GaN/AlGaN DH structures in 1991 to 1993. Optically pumped stimulated emission on an InGaN/GaN material system was achieved by Amano et al [99] and by Khan et al [100] in 1994. Akasaki et al [16] and Nakamura et al [17] reported the first electrically pumped nitride-based LDs in 1995/96. The latter one contained as active medium a 26 period $In_{0.2}Ga_{0.8}N(25Å)/In_{0.05}Ga_{0.95}N(50Å)$ multiple QW, with $Al_{0.15}Ga_{0.85}N$ cladding layers grown on c-face sapphire substrates with emission at 417 nm and a threshold current density of $4\,kA/cm^2$. Since the cleavage plane of the nitride device structure and the c-face sapphire substrate are rotated with respect to each other, laser facets were fabricated by reactive ion etching (RIE). The device operated at room temperature under pulsed condition to prevent device failure by overheating, considering the large threshold voltage of about 30 V. Improvements were subsequently made by using a ridge geometry InGaN/GaN/AlGaN separate confinement heterostructure (SCH) combined with a reduction of the number of QWs from 26 to 5 with laser emission at 411 nm, lower threshold current density of $3\,kA/cm^2$ and lower threshold forward voltage of 22 V [101]. Since grown on a-plane sapphire substrates laser facets could also be cleaved. Continuous wave operation was finally achieved in late 1996 with the device operating at a temperature of 233 K [102]. Room temperature continuous wave operation was demonstrated shortly thereafter [103]. LD lifetime initially 40 min increased rapidly to 27 hours [104].

Defect density in the LDs was high since grown on sapphire substrates. Although acceptable for nitride-based LEDs, this high defect density was affecting LD lifetime since LDs operate at significantly higher current densities

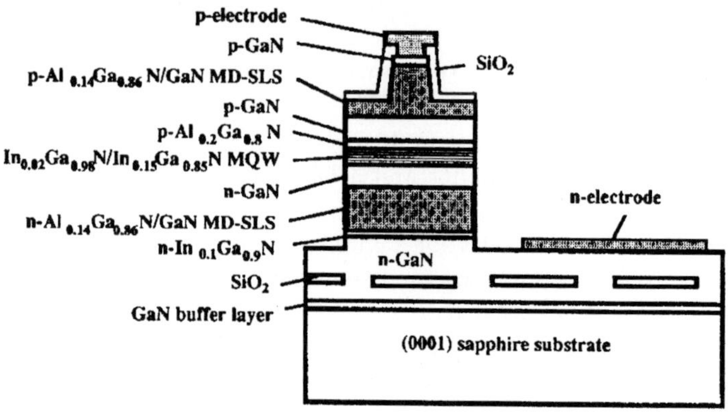

Fig. 14. Schematic of nitride-based LD with modulation-doped SL grown on ELO GaN (from [106]).

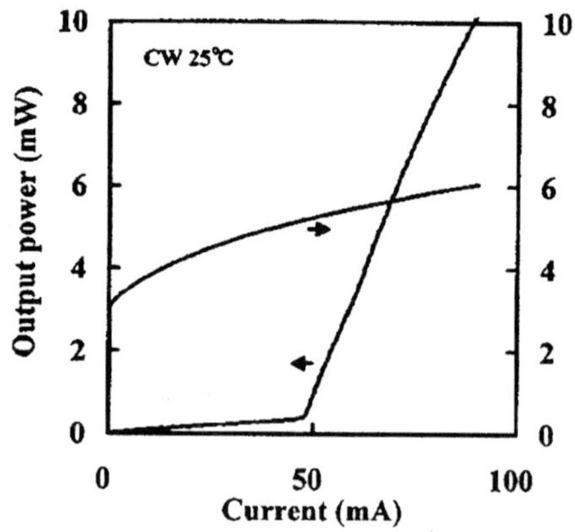

Fig. 15. L-I and I-V curve of nitride-based LD with modulation-doped SL grown on ELO GaN (from [106]).

than LEDs. Also the thick AlGaN cladding layers were prone to cracking due to a large strain in these layers. Some of the lasing modes in these early LDs were related to cavities formed by accidental cracks rather than the actual laser cavity. To reduce strain and cracking in the LD structure Nakamura et al [105, 106] replaced the AlGaN cladding layers with modulation doped strained-layer GaN/AlGaN SLs illustrated in Figure 14. Modulation doping improved the electrical characteristic of the LD, in particular, the p-side of the devices. To reduce defect density epitaxial lateral overgrown (ELO) GaN was employed. ELO is based on the selective area growth of GaN commonly through a dielectric mask/stripe pattern such as SiO_2 formed on a pre-grown GaN film [20]. Defect densities are reduced by several orders of magnitude in the GaN grown on top of the SiO_2 stripes. This was a significant step forward

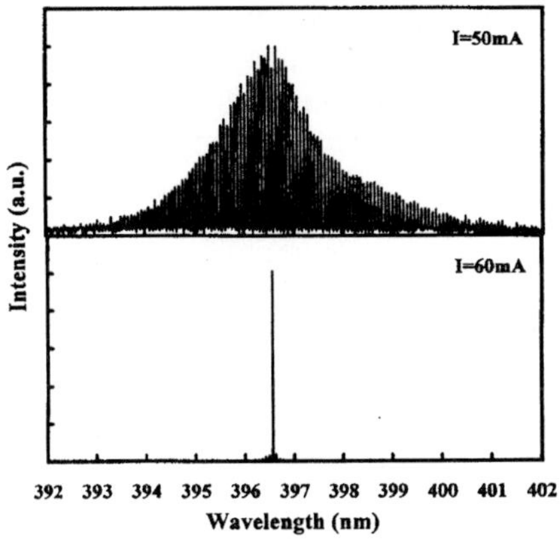

Fig. 16. Spectrum of nitride-based LD with modulation-doped SL grown on ELO GaN operated at two input current. Threshold current is 48 mA (from [106]).

to increase LD lifetimes to 1000 hours at 50°C at an optical output power of 2 mW per facet, with predicted lifetimes of 10000 hours at room temperature. The L-I curve of such an LD is displayed in Fig. 15. Forward threshold voltage was 5 V and threshold current density 4 kA/cm^2 (threshold current of 48 mA). Its emission spectrum is illustrated in Figure 16. Multimode lasing was observed at 50 mA, single mode lasing at 60 mA.

Subsequent developments included the removal of the sapphire substrate by polishing to improve thermal management of the devices. GaN has significantly higher thermal conductivity than sapphire ($\kappa_{\mathrm{sapphire}}$=0.35 Wcm^{-1}K^{-1} compared to κ_{GaN}=1.6 Wcm^{-1}K^{-1}). This resulted in LDs with 30 mW optical output with estimated lifetimes of 15000 hours at a temperature of 60°C [107]. Alternative approaches for good heat sinking of the LDs such as laser lift-off of LD structures from a sapphire substrate and transfer onto a Cu carrier or growth of the LD structure onto SiC substrates with its higher thermal conductivity of κ_{SiC}=3–4 Wcm^{-1}K^{-1} were also explored [108, 109]. Early LDs showed a rather poor modal structure due to optical leakage into the sapphire substrate. Mainly higher optical modes were observed preventing applications for DVD and projection systems that require a well-defined Gaussian optical output profile. Removal of the sapphire substrate, use of a thick GaN spacer layer or use of a SiC substrate allowed significant improvements in the far field pattern of the LDs. Gaussian optical output profiles as narrow as 9° in lateral and 21° in vertical direction have been demonstrated in optimized device structures [110]. We note that many research laboratories made significant contributions to the development of nitride-based LDs (for example [111–116]).

Initially only the wavelength range of 390 to 420 nm was accessible for nitride-based LDs in continuous wave operation. This is the spectral range where InGaN has the highest quantum efficiency. This spectral range has been extended to the wavelength range up to 450 nm in recent years. LDs with wavelengths as short as 366 nm and as long as 480 nm have also been demonstrated, however, under only in pulsed mode [117]. True blue light emitting (around 480 nm) high-power continuous wave operated LDs are needed for full color projection systems, but deterioration of the InGaN material system at high indium compositions appears to pose challenges for achieving a continuous wave operated device in this wavelength range at present. Threshold current densities in nitride-based LDs today can be as low as $1.2\,\mathrm{kA/cm^2}$ in the 390–420 nm spectral range with operating voltages as low as 4–5 V [18]. Maximum output powers as high as 420 mW have been reported, although at the expense of a reduced lifetime [118]. Kneissl et al [119] reported Q-switched LDs in the blue-UV spectral region using a two-section nitride-based LD consisting of an electroabsorption modulator and amplifier (gain) section. First attempts towards vertical cavity surface emitting lasers (VCSELs) have also been made, however, they are presently still in their infancy. Optical pumped stimulated emission has been demonstrated [120] and there has been one report on electrically induced lasing from a nitride-based VCSEL structure by Zhou et al [121].

First blue-ray disk products, the next generation of large-capacity optical disk recording systems, employing blue nitride-based LDs, should be released in 2003 by Sony, Mitsubishi Chemicals, Hitachi Maxell and by others. We note, however, that although device lifetimes of 15000 hours under 30 mW continuous wave operation at 60°C are achievable today, further extension of reliability under much higher power operation at elevated temperature of 70–80°C is required for operation in double-layer and portable DVD applications in which extreme conditions should be anticipated.

The final question which has so far not be answered is what is the mechanism for lasing, i.e., of optical near-threshold gain, in nitride-based LDs? The question would have been relatively easy to answer if only GaN or AlGaN were involved. Near-threshold gain in GaN has been attributed to electron-hole plasma recombination for temperatures above 150 K since carrier densities are well above the Mott density (the critical carrier density beyond which no exciton can exist). At temperatures below 150 K there are reports attributing gain to exciton-exciton scattering [122] while others find electron-hole plasma recombination the dominant gain mechanism [123]. For AlGaN, near threshold gain has been attributed to electron-hole plasma recombination [124] while for GaN/AlGaN SCH structures exciton-exciton scattering is the dominant gain mechanism [125].

With its compositional disorder which goes well beyond alloy disorder known for other material system, InGaN poses challenging new questions in terms of the lasing mechanism. The question arises not only whether exciton or electron-hole plasma contributes to the gain mechanism, but even more what is the role of the localized states in InGaN for gain formation and whether near-threshold gain in LDs is due to localized or delocalized states. These questions have been controversially debated shortly after the demonstration of the first electrically pumped nitride-based LD. Several different approaches were taken to measure near-threshold optical gain in active and passive device structures. Frankowsky et al [126] employed optical pumping of a DH structure that contained an 200 Å-thick InGaN layer using the variable stripe-length method. In this method the sample is excited from the top using a pulsed laser source focused as stripe and amplified spontaneous emission is detected from the side as function of stripe length. The optical gain spectrum showed a single maximum close to the band gap of InGaN, accompanied with a blue-shift of the gain maximum with increasing pump power. The gain spectrum could be fitted using a model that considered band-to-band transitions in an electron-hole plasma, with discrepancies attributed to excitonic enhancements.

Kuball et al [127] employed a large electrical forward bias to generate high electron-hole pair densities in a structure that contained five 15 Å-thick InGaN QWs with 45 Å-thick GaN barriers within an AlGaN/GaN pn-junction and a tunable optical source as probe for transmission changes induced by the electrical injection. The absorption coefficient of the InGaN QWs was determined from photocurrent spectroscopy measurements to determine gain from the electrical pump-optical probe measurements. At a current density of $3.5\,\mathrm{kA/cm^2}$, a single-pass total gain of about $5000\,\mathrm{cm^{-1}}$ was determined for the 125 Å of active InGaN layers. This value was consistent with estimates of Nakamura et al [128] for LD threshold gain at comparable current densities assuming an optical field overlap factor of 0.02. Key result was the rather large linewidth of the gain spectrum in excess of 100 meV and its location in the low energy tail of the absorption spectrum of the InGaN QWs. A scenario was proposed where electron-hole pair states under low injection are subject to localization in InGaN due to In clustering. With increasing current density these states fill up until optical gain develops. Nakamura et al [129] determined gain spectra on an electrical pumped InGaN LD by the Hakki-Paoli method. In this method the LD is operated just below threshold and the gain spectrum is determined from interference fringes in the emission spectrum. A broad gain spectrum was also measured.

Song et al [130] used the Henry method on electrically pumped InGaN QW LDs grown on SiC. This method is based on the fundamental relationship between spontaneous emission, stimulated emission and absorption. The gain spectrum was determined from the measurement of the spontaneous emission spectrum that was recorded through the SiC substrate. Figure 17 shows typical gain spectra obtained using the Henry method at different input currents. Threshold current was 300 mA. Similar to Kuball et al [127] the main result of significance was the pronounced extension of the gain spectra as much as 200 meV below its peak position, indicative of the participation of a corresponding large range of states. The gain peak position blue-shifts

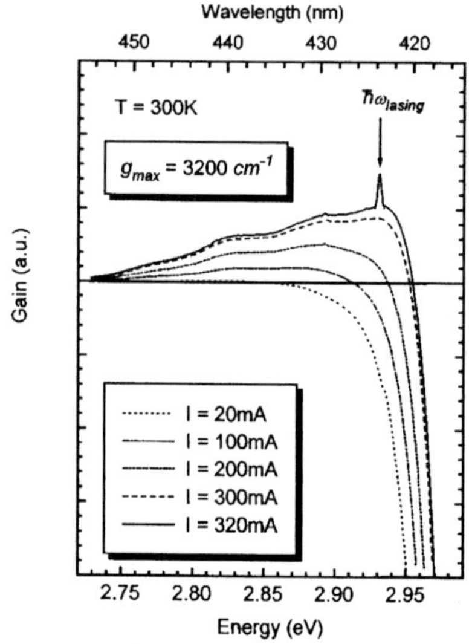

Fig. 17. Gain spectra of electrically pumped InGaN QW LD at different input currents obtained using the Henry method (from [130]).

with higher injection levels, but considerably less than anticipated from a one electron state-filling picture. Transparency condition was reached rather easily, at levels of injection which were not very different from that of the conventional LED regime. Peak gain at threshold was estimated to be about $3200\,\mathrm{cm}^{-1}$. These observations provided further evidence that filling of the localized electron-hole pair states at the lowest interband transition in the InGaN QW is a necessary prerequisite prior to the built-up of a sufficient population inversion for threshold gain. Further evidence for localized states to contribute to the gain formation in this material system was provided by Schmidt et al [131] and Deguchi et al [132] which investigated in detail InGaN multiple QWs using the variable stripe-length method. The question, however, remained whether the states responsible for lasing are localized or delocalized at the carrier densities needed for lasing.

Recent time-resolved studies on InGaN LDs provided further evidence on the nature of the states responsible for lasing in nitride-based LDs. Kawakami et al [133] studied $In_{0.1}Ga_{0.9}N$ near-UV (390 nm), $In_{0.2}Ga_{0.8}N$ violet-blue (420 nm) and $In_{0.3}Ga_{0.7}N$ blue (440 nm) LDs using time-resolved photoluminescence and pump-probe-spectroscopy. They confirmed the importance of localized states for gain formation in this material system. Photo-generated carriers transfer rapidly (in less than 1 ps) to the lowest quantized levels within the QWs, and then relax to the localized tail within a timescale of a few ps, giving rise to optical gain. Optical gain was found to be due to nearly delocalized states (the lowest quantized level within the QW) in the violet LDs, while it was from highly localized states with respect to the low-

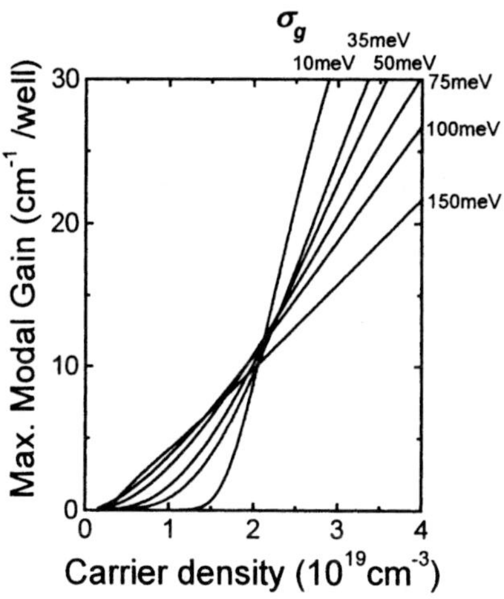

Fig. 18. Maximum modal gain per QW as function of carrier density for InGaN QW LDs for different degrees on indium compositional fluctuations. σ denotes the standard deviation for the fluctuated potential related to In compositional variations in the QW (from [135]).

est quantized level in violet-blue (localization energy of 250 meV) and in blue (localization energy of 500 meV) LDs.

While initial theoretical models of gain in the InGaN QW system did not consider the possibility of carrier localization due to a spatially inhomogeneous band gap recent models that include indium clustering by Chow et al [134] suggest that this clustering results in a deterioration of the LD characteristic. This was confirmed by Yamaguchi et al [135] if band gap fluctuations are on the micron or larger scale. However, assuming submicron spatial variations in indium composition it was found to have positive effects for specific laser operations as illustrated in Fig. 18. Indium compositional fluctuations result in a reduction of the threshold carrier density, although, at the expense of a smaller differential gain. In certain LDs it might therefore be beneficial of having a certain degree of indium compositional fluctuations. A larger indium clustering in the early nitride-based LDs might even have been a crucial factor for achieving lasing at that time. This theoretical result was confirmed by comparing differential gain in LD structures with a different degree of compositional fluctuations. The unusual nature of the InGaN alloy has significant effects on the lasing mechanism in nitride-based LDs.

6 Conclusions

Optical properties of nitrides and the current status of nitride-based LEDs and LDs were reviewed. Nitride research established more than 60 years ago with the first growth of GaN crystals and the achievement of epitaxial growth of both n- and p-type nitrides later on resulted in the development of a large variety of devices in a relatively short time-span in recent years for applications ranging from LEDs for traffic lights, automotive applications and general illumination to LDs for DVD and projection systems. Significant scientific and technological challenges, however, remain. Apart from increasing optical output power and lifetime of nitride-based LEDs and LDs, extension of the wavelength range into the UV and the longer wavelength range is required. For example true blue LDs operated at high power around 480 nm are needed for full color projection systems. Improved luminescence conversion concepts are required to yield higher luminous efficiency white LEDs for general illumination applications. Both polarization effects as well as a high point defect density reduce quantum efficiency in AlGaN containing heterostructures needed to access the UV/near-UV spectral range. Reductions in defect density and possibly the use of the quarternary InAlGaN alloy to induce carrier localization will be required on the pathway of achieving efficient UV light emitters, together with improvements in their electrical properties and new thermal management concepts.

Acknowledgment

I would like to thank my present and former postgraduate students (M. Benyoucef, J.M. Hayes, J. Pomeroy, S. Rajasingam) and post-docs (M. Lin, A. Sarua) at the University of Bristol and the many present and past collaborators. I would like to acknowledge financial support by EPSRC, Renishaw plc, QinetiQ Ltd, British Council, Royal Society and the Office of Naval Research International Field Office London, and thank the Alexander-von-Humboldt Foundation for support during my stay at Brown University. Finally I would like to thank Claus Klingshirn for the opportunity to work with him during my Diploma thesis in 1991/92 at the University of Kaiserslautern, and for sharing his enthusiasm for the optical properties of semiconductors.

References

1. W.C. Johnson, J.B. Parsons, M.C. Crew: J. Phys. Chem. **234**, 2651 (1938)
2. R. Juza, E. Hahn, Z. Anorg. Allgem. Chem. **234**: 282 (1938)
3. H. Grimmeiss, H. Koelmans: Z. Naturfg. **14a**, 264 (1959)
4. H.P. Maruska, J.J. Tietjen: Appl. Phys. Lett. **15**, 367 (1969)

5. H.P. Maruska, L.J. Anderson, D.A. Stevenson: J. Electrochem. Soc. **121**, 1202 (1974)
6. W. Seifert, R. Franzheld, E. Buttler, H. Sobotta, V. Riedle: Crystal. Res. Technol. **18**, 383 (1983)
7. J.I. Pankove, E.A. Miller, J.E. Berkeyheiser: J. Luminescence **5**, 84 (1972)
8. H.P. Maruska, D.A. Stephenson, J.I. Pankove: Appl. Phys. Lett. **22**, 303 (1973)
9. I. Akasaki, T. Kozowa, K. Hiramatsu, N. Sawak, K. Ikeda, Y. Ishii: J. Luminescence **40-41**, 121 (1988)
10. H. Amano, M. Kito, K. Hiramatsu, I. Akasaki: Jpn. J. Appl. Phys. **28**, L2112 (1989)
11. I. Akasaki, H. Amano, M. Kito, K. Hiramatsu: J. Luminescence **48-49**, 666 (1991)
12. S. Nakamura, N. Iwasa, M. Senoh, T. Mukai: Jpn. J. Appl. Phys. **31**, 1258 (1992)
13. S. Nakamura: J. Cryst. Growth, **145**, 911 (1994)
14. S. Nakamura, T. Mukai, M. Senoh: Appl. Phys. Lett. **64**, 1687 (1994)
15. S. Nakamura, M. Senoh, N. Iwasa, S. Nagahama: Appl. Phys. Lett. **67**, 1868 (1995)
16. I. Akasaki, H. Amano, S. Sota, H. Sakai, T. Tanaka, M. Koide: Jpn. J. Appl. Phys. **34**, 1517 (1995)
17. S. Nakamura, M. Senoh, S. Nagahama, N. Iwasa, T. Yamada, T. Matsushita, H. Kiyoku, Y. Sugimoto: Jap. J. Appl. Phys. **35**, L74 (1996)
18. S. Nagahama, T. Yanamoto, M. Sano, T. Mukai: Phys. Stat. Sol. (a) **190**, 235 (2002)
19. D.J. As, K. Lischka: Phys. Stat. Sol. (a) **176**, 475 (1999)
20. O.H. Nam, M.D. Bremser, T.S. Zheleva, R.F. Davis: Appl. Phys. Lett. **71**, 2638 (1997)
21. B. Monemar: Mat. Sci. Eng. B **59**, 122 (1999) and references therein
22. S. Chichibu, H. Okumura, S. Nakamura, G. Feuillet, T. Azuhata, T. Sota, S. Yoshida: Jpn. J. Appl. Phys. **36**, 1976 (1997)
23. A.K. Viswanath, J.I. Lee, S. Yu, D. Kim, Y. Choi, C.H. Hong: J. Appl. Phys. **84**, 3848 (1998)
24. J. Wu, W. Walukiewicz, K.M. Yu, J.W. Ager III, E.E. Haller, H. Lu, W.J. Schaff: Appl. Phys. Lett. **80**, 4741 (2002)
25. F. Yun, M.A. Reshnikov, L. He, T. King, H. Morkoç, S.W. Novak, L. Wei: J. Appl. Phys. **92**, 4837 (2002) and reference therein
26. Y.K. Kuo, W.W. Lin: Jpn. J. Appl. Phys. **41** 5557 (2002)
27. K. Kornitzer, T. Ebner, K. Thonke, R. Sauer, C. Kirchner, V. Schwegler, M. Kamp, M. Leszczynski, I. Grzegory, S. Porowski, Phys. Rev. B **60**, 1471 (1999)
28. B. Monemar. P.P. Paskov, T. Paskova, J.P. Bergman, G. Pozina, W.M. Chen, P.N. Hai, I.A. Buyanova, H. Amano, I. Akasaki: Mat. Sci. Eng. B **93**, 112 (2002)
29. J.A. Freitas Jr., W.J. Moore, B.V. Shanabrook, G.C.B. Braga, S.K. Lee, S.S. Park, J.Y. Han: Phys. Rev. B **66**, (2002) 233311
30. G. Martinez-Criado, M. Kuball, M. Benyoucef, A. Sarua, E. Frayssinet, B. Beaumont, P. Gibart, C. R. Miskys, M. Stutzmann: J. Cryst. Growth **255**, 277 (2003)
31. M. Leroux, B. Beaumont, N. Granjean, P. Lorenzini, S. Haffouz, P. Vennegues, J. Massies, P. Gibart: Mater. Sci. Eng. B **50**, 97 (1997)

32. M.A.L. Johnson, Z. Yu, C. Boney, W.C. Hughes, J.J.W. Cook, J.F. Schetzina, H. Zhao, B.J. Skromme, J.A. Edmond: Mater. Res. Soc. Symp. Proc. **449**, 215 (1997)
33. S. Fischer, C. Wetzel, E.E. Haller, B.K. Meyer: Appl. Phys. Lett. **67**, 1298 (1995)
34. M.A. Reshchikov, G.C. Yi, B.W. Wessels: Phys. Rev. B **59**, 13176 (1999)
35. E. Kuokstis, J. Zhang, Q. Fareed, J.W. Yang, G. Simin, M.A. Khan, R. Gaska, M. Shur, C. Rojo, L. Schowalter: Appl. Phys. Lett. **81**, 2755 (2002)
36. J. Li, K.B. Nam, M.L. Nakarmi, J.Y. Lin, H.X. Jiang: Appl. Phys. Lett. **81**, 3365 (2002)
37. T. Onuma, S.F. Chichibu, T. Sota, K. Asai, S. Sumiya, T. Shibata, M. Tanaka: Appl. Phys. Lett. **81**, 652 (2002)
38. R.A. Youngman, J.H. Harris: J. Am. Ceram. Soc **73**, 3238 (1990)
39. G.A. Slack, L.J. Schowalter, D. Morelli, J.A. Freitas, Jr.: J. Cryst. Growth **246**, 287 (2002)
40. T.L. Tansley, C.P. Foley: Electron. Lett. **20**, 1066 (1984)
41. V.Yu. Davydov, A.A. Klochikhin, V.V. Emtsev, D.A. Kurdyukov, S.V. Ivanov, V.A. Vekshin, F. Bechstedt, J. Furthmüller, J. Aderhold, J. Graul, A.V. Mudryi, H. Harima, A. Hashimoto, A. Yamamoto, E.E. Haller: Phys. Stat. Sol. (b) **234**, 787 (2002)
42. K.S.A. Butcher, M. Wintrebert-Fouquet, Motlan, S.K. Shrestha, H. Timmers, K.E. Prince, T.L. Tansley: Mat. Res. Soc. Symp. Proc. **743**, 707 (2003)
43. T .Saito, Y. Arakawa, Phys. Rev. B **60**, 1701 (1999)
44. Y. Nurukawa, Y. Kawakami, Sz. Fujiita, Sg. Fulita, S. Nakamura: Phys. Rev. B **55**, R1938 (1997)
45. K.P. O'Donnell, P.G. Middleton, C. Trager-Cowan, C. Young, S.C. Bayliss, I. Fletcher, W. Van der Stricht, I. Moerman, P. Demeester: Appl. Phys. Lett. **73**, 3273 (1998)
46. M. Smith, G.D. Chen, J.Y. Lin, H.X. Jiang, M.A. Khan, Q. Chen: Appl. Phys. Lett. **69**, 2837 (1996)
47. J.S. Im, V. Härle, F. Scholz, A. Hangleiter: MRS Internet J. Nitride Semicond. Res. **1**, 37 (1996)
48. E.S. Jeon, V. Kozlov, Y.-K. Song, A. Vertikov, M. Kuball, A.V. Nurmikko, H. Liu, C. Chen, R.S. Kern, C.P. Kuo, M.G. Craford: Appl. Phys. Lett. **69**, 4194 (1996)
49. A. Zunger: Phys. Stat. Sol. (b) **216**, 117 (1999)
50. F. Bernadini, V. Fiorentini, D. Vanderbilt: Phys. Rev. B **56**, R10024 (1997)
51. N. Grandjean, B. Damilano, J. Massies: Mat. Res. Soc. Symp. **639**, G.12.1.1 (2001)
52. H. Jiang, G.Y. Zhao, H. Ishikawa, T. Egawa, T. Jimbo, M. Umeno: J. Appl. Phs. **89**, 1046 (2001)
53. M. Shatalov, J. Zhang, A.S. Chitnis, V. Adivarahan, J. Yang, G. Simin, M.A. Khan: IEEE J. Sel. Top. Quantum Electron. **8**, 302 (2002)
54. H. Hirayama, A. Kinoshita, M. Ainoya, T. Yamabi, T. Yamanaka, A. Hirata, Y. Aoyagi: Inst. Phys. Conf. Ser. **170**, 195 (2002)
55. J. Li, K.B. Nam, K.H. Kim, J.Y. Lin, H.X. Jiang: Appl. Phys. Lett. **78**, 61 (2001)
56. E. Monroy, N. Gogneau, D. Jalabert, F. Enjalabert, E. Bellet-Amalric, Y. Hori, L.S. Dang, B. Daudin: Mat. Res. Soc. Symp, Proc. **743**, 305 (2003)

57. J. Han, J.J. Figiel, G.A. Petersen, S.M. Myers, M.H. Crawford, M.A. Banas, J.J. Hearne: MRS Internet J. Nitride Semicon. Res. **5**, W6.2 (2000)
58. H. Teisseyre, T. Susksi, S.P. Lepkowski, S. Anceau, P. Perlin, P. Lefebvre, L. Konczewicz, H. Hiramyama, Y. Aoyagi: Appl. Phys. Lett. **82**, 1541 (2003)
59. M.R. Krames, J. Bhat, D. Collins, N.F. Gardner, W. Götz, C.H. Lowery, M. Ludowise, P.S. Martin, G. Mueller, R. Mueller-Mach, S. Rudaz, D.A. Steigerwald, S.A. Stockman, J.J. Wierer: Phys. Stat. Sol. (a) **192**, 237 (2002)
60. A. Hangleiter: MRS Bull. **28**, 350 (2003)
61. T. Mukai, S. Nagahama, N. Iwasa, M. Senoh, T. Yamada: J. Phys. Cond. Matter **13**, 7089 (2001)
62. H.X. Jiang, S.X. Jin, J. Li, J. Shakya, J.Y. Lin: Appl. Phys. Lett. **78**, 1303 (2001)
63. I. Ozden, M. Diagne, A.V. Nurmikko, J. Han, T. Takeuchi: Phys. Stat. Sol. (a) **188**, 139 (2002)
64. M. Osinski, J. Zeller, P.-C. Chiu, B.S. Phillips, D.L. Barton: Appl. Phys. Lett. **69**, 898 (1996)
65. G. Meneghesso, S. Levada, E. Zanoni, S. Podda, G. Mura, M. Vanzi, A. Cavallini, A. Castaldini, S. Du, I. Eliasshevich: Phys. Stat. Sol. (a) **194**, 389 (2002)
66. S. Nakamura: Sol. State Commun. **102**, 237 (1997)
67. S. Nagahama, N. Iwasa, M. Senoh, T. Matsushita, Y. Sugimoto, H. Kiyoku, T. Kozaki, M. Sano, H. Matsumura, H. Umemoto, K. Chocho, T. Yanamoto, T. Mukai: Phys. Stat. Sol. (a) **188**, 1 (2001)
68. S. Chichibu, T. Azuhata, T. Soda, S. Nakamura: Appl. Phys. Lett. **69**, 4188 (1996)
69. J. Baur, B. Hahn, M. Fehrer, D. Eisert, W. Stein, A. Pössl, F. Kühn, H. Zull, M. Winter, V. Härle: Phys. Stat. Sol. (a) **194**, 399 (2002)
70. I. Akasaki, H. Amano: Extended Abstracts of the 1992 International Conf. Solid State Devices and Materials, Aug. 1992, p.327
71. J. Han, M.H. Crawford, R.J. Shul, J.J. Figiel, M. Banas, L. Zhang, Y.K. Song, H. Zhou, A.V. Nurmikko: Appl. Phys. Lett. **73**, 1688 (1998)
72. A. Kinoshita, H. Hirayama, A. Ainoya, Y. Aoyagi, A. Hirata: Appl. Phys. Lett. **77**, 175 (2000)
73. A.Y. Polyakov, N.B. Smirnov, A.V. Govorkov, J.M. Redwing: Solid-State Electron. **42**, 831 (1998)
74. T. Nishida, H. Saito, N. Kobayashi: Appl. Phys. Lett. **70**, 711 (2001)
75. S. Kamiyama, M. Iwaya, H. Amano, I. Akasaki: Phys. Stat. Sol. (a) **194**, 393 (2002)
76. A. Chitnis, J.P. Zhang, V. Adivarahan, W. Shuai, J. Sun, M. Shatalov, J.W. Yang, G. Simin, M.A. Khan: Jpn. J. Appl. Phys. **41**, L450 (2002)
77. A. Chitnis, J.P. Zhang, V. Adivarahan, M. Shatalov, S. Wu, R. Pachipulusu, V. Mandavilli, M.A. Khan: Appl. Phys. Lett. **82**, 2565 (2003)
78. J.P. Zhang, A. Chitnis, V. Adivarahan, S. Wu, V. Mandavilli, R. Pachipulusu, M. Shatalov, G. Simin, J.W. Yang, M.A. Khan: Appl. Phys. Lett. **81**, 4910 (2002)
79. A. Hanlon, P.M. Pattison, J.F. Kaeding, R. Sharma, P. Fini, S. Nakamura: Jap. J. Appl. Phys. **42**, L628 (2003)
80. A. Sarua, M. Kuball, M.J. Uren, A. Chitnis, J.P. Zhang, V. Adivarahan, M. Shatalov, M.A. Khan: Mat. Res. Soc. Symp. Proc. **743**, 493 (2003)

81. A. Chitnis, J. Sun, V. Mandavilli, R. Pachipulusu, S. Wu, M. Gaevski, V. Adivarahan, J.P. Zhang, M.A. Khan, S. Sarua, M. Kuball: Appl. Phys. Lett. **81**, 3491 (2002)
82. Y. Taniyasu, M. Kasu, N. Kobayashi: Appl. Phys. Lett. **81**, 1255 (2002)
83. M. Pophristic, S.P. Guo, B. Peres: Appl. Phys. Lett. **82**, 4289 (2003)
84. R. Zeisel, M.W. Bayerl, S.T.B. Goennenwein, R. Dimitrov, O. Ambacher, M.S. Brandt, M. Stutzmann: Phys. Rev. B **61**, R16283 (2000)
85. P. Kozodoy, M. Hansen, S.P. DenBaars, U.K. Mishra: Appl. Phys. Lett. **74**, 3681 (1999)
86. A. Yasan, M. Razeghi: Sol. State Electron. **47**, 303 (2003)
87. A. Chitnis, V. Adivarahan, M. Shatalov, J.P. Zhang, M. Gaevski, W. Shuai, P. Pachipulusu, J. Sun, K. Simin, G. Simin, J. Yang M.A. Khan: Jpn. J. Appl. Phys. **41**, L320 (2002)
88. M.A. Khan, V. Adivarahan, J.P. Zhang, C. Chen, E. Kuokstis, A. Chitnis, M. Shatalov, J.W. Yang, G. Simin: Jpn. J. Appl. Phys. **40**, 1308 (2001)
89. T. Wang, Y.H. Liu, Y.B. Lee, J.P. Ao, J. Bai, S. Sakai: Appl. Phys. Lett. **81**, 2508 (2002)
90. S. Muthu, F.J.P. Schuurmans: IEEE J. Sel. Top. Quantum Electron. **8**, 333 (2002)
91. P. Schlotter, R. Schmidt, J. Schneider: Appl. Phys. A **64**, 417 (1997)
92. U. Kauffmann, M. Kunzer, K. Köhler, H. Obloh, W. Pletschen, P. Schlotter, R. Schmidt, J. Wagner, A. Ellens, W. Rossner, M. Knobusch: Phys. Stat. Sol. (a) **188**, 143 (2001)
93. U. Kauffmann, M. Kunzer, K. Köhler, H. Obloh, W. Pletschen, P. Schlotter, J. Wagner, A. Ellens, W. Rossner, M. Kobusch: Phys. Stat. Sol. (a) **192**, 246 (2002)
94. M.R. Krames: private communication (July 2003)
95. D. Dingle, K.L. Shaklee, R.F. Lehneny, R.B. Zetterstrom: Appl. Phys. Lett. **19**, 5 (1971)
96. H. Amano, T. Asahi, I. Akasaki: Jpn. J. Appl. Phys. **29**, L205 (1990)
97. M.A. Khan, D.T. Olson, J.M. Van Hove, J.N. Kuznia: Appl. Phys. Lett. **58**, 1515 (1991)
98. H. Amano, N. Wanatabe, N. Koide, I. Akasaki: Jpn. J. Appl. Phys. **32**, L1000 (1993)
99. H. Amano, T. Tanaka, Y. Kunii, S.T. Kim, I. Akasaki: Appl. Phys. Lett. **64**, 1377 (1994)
100. M.A. Khan, S. Krishnankutty, R.A. Skogman, J.N. Kuznia, D.T. Olson, R. George: Appl. Phys. Lett. **65**, 520 (1994)
101. S. Nakamura, M. Senoh, S. Nagahama, N. Iwasa, T. Yamada, T. Matsushita, Y. Sugimoto, H. Kiyoku: Appl. Phys. Lett. **69**, 1477 (1996)
102. S. Nakamura, M. Senoh, S. Nagahama, N. Iwasa, T. Yamada, T. Matsushita, Y. Sugimoto, H. Kiyoku: Appl. Phys. Lett. **69**, 3034 (1996)
103. S. Nakamura, M. Senoh, S. Nagahama, N. Iwasa, T. Yamada, T. Matsushita, Y. Sugimoto, H. Kiyoku: Appl. Phys. Lett. **70**, 868 (1997)
104. S. Nakamura, M. Senoh, S. Nagahama, N. Iwasa, T. Yamada, T. Matsushita, Y. Sugimoto, H. Kiyohu: Appl. Phys. Lett. **70**, 1417 (1997)
105. S. Nakamura, M. Senoh, S. Nagahama, N. Iwasa, T. Yamada, T. Matsushita, H. Kiyoku, Y. Sugimoto, T. Kozaki, H. Umenoto, M. Sano, K. Chocho: Jpn. J. Appl. Phys. **36**, L1568 (1997)

106. S. Nakamura, M. Senoh, S. Nagahama, N. Iwasa, T. Yamada, T. Matsushita, H. Kiyoku, Y. Sugimoto, T. Kozaki, H. Umenoto, M. Sano, K. Chocho: Appl. Phys. Lett. **72**, 211 (1998)
107. S. Nagahama, N. Iwasa, M. Senoh. T. Matsushita, Y. Sugimoto, H. Kiyoku, T. Kozaki, M. Sano, H. Matsumura, H. Umemoto, K. Chocho, T. Mukai: Jpn. J. Appl. Phys. **39**, L647 (2000)
108. K. Doverspike, G.E. Bulman, S.T. Sheppard, H.S. Kong, M. Leonard, H. Dieringer, T.W. Weeks, J. Edmond, J.D. Brown, J.T. Swindle, J.F. Schetzina, Y.K. Song, M. Kuball, A. Nurmikko: Mater. Res. Soc. Symp. Proc. **482**, 1169 (1998)
109. W.S. Wong, M. Kneissl, P. Mei, D.W. Treat, M. Teepe, N.M. Johnson: Appl. Phys. Lett. **78**, 1198 (2001)
110. M. Ikeda, S. Uchida: Phys. Stat. Sol. (a) **194**, 407 (2002)
111. A. Kuramata, K. Domen, R. Soejima, K. Horino, S. Kubota, T. Tanahashi: Jpn. J. Appl. Phys. **36**, L1130 (1997)
112. F. Nakamura, T. Kobayashi, T. Asatsuma, K. Funato, K. Yanashima, S. Hashimoto, K. Naganuma, S. Tomioka, T. Miyajima, E. Morita, H. Kawai, M. Ikeda: J. Cryst. Growth **190**, 841 (1998)
113. D.P. Bour, M. Kneissl, L.T. Romano, M.D. McCluskey, C.G. Van deWalle, B.S. Krusor, R.M. Donaldson, J. Walker, C.J. Dunnrowicz, N.M. Johnson: IEEE J. Sel. Top. Quantum Electron. **4**, 498 (1998)
114. M.P. Mack, A.C. Abare, M. Hansen, P. Kozodoy, S. Keller, U. Mishra, L.A. Coldren, S.P. DenBaars: J. Cryst. Growth **190**, 837 (1998)
115. S. Bader, B. Hahn, H.J. Lugauer, A. Lell, A. Weimar, G. Brüderl, J. Baur, D. Eisert, M. Scheubeck, S. Heppel, A. Hangleiter, V. Härle: Phys. Stat. Sol. (a) **180**, 177 (2000)
116. T. Bottcher, C. Zellweger, S. Figge, R. Kroger, C. Petter, H.J. Buhlmann, M. Ilegems, P.L. Ryder, D. Hommel: Phys. Stat. Sol. (a) **191**, R3 (2002)
117. S. Nagahama, T. Yanamoto, M. Sano, T. Mukai: Phys. Stat. Sol. (a) **194**, 423 (2002)
118. S. Nakamura, M. Senoh, S. Nagahama, N. Isawa, T. Yamada, T. Matsushita, H. Kiyoku, Y. Sugimoto, T. Kozaki, H. Umemoto, M. Sano, K. Chocho: Jpn. J. Appl. Phys. **37**, L627 (1998)
119. M. Kneissl, T.L. Paoli, P. Kiesel, D.W. Treat, M. Teepe, N. Miyashita, N.M. Johnson. Appl. Phys. Lett. **80**, 3283 (2002)
120. I.L. Krestnikov, W.V. Lundin, A.V. Sakharov, V.A. Semenov, A.S. Usikov, N.N. Ledentsov, A. Hoffmann, D. Bimberg: Appl. Phys. Lett. **75**, 1192 (1999)
121. H.L. Zhou, M. Diagne, E. Makarona, A.V. Nurmikko, J. Han, K.E. Waldrip, J.J. Figiel: Electron. Lett. **36**, 1777 (2000)
122. M. Smith, J.Y. Lin, H.X. Jiang, M.A. Khan: Appl. Phys. Lett. **71**, 635 (1997)
123. S. Hess, R.A. Taylor, J.F. Ryan, B. Beaumont, P. Gibart: Appl. Phys. Lett. **73**, 199 (1998)
124. J.B. Lam, S. Bidnyk, G.H. Gainer, B.D. Little, J.J. Song, W. Wang: Appl. Phys. Lett. **77**, 4101 (2000)
125. S. Bidnyk, J.B. Lam, B.D. Little, Y.H. Kwon, J.J. Song, G.E. Bulman, H.S. Kong, T.J. Schmidt: Appl. Phys. Lett. **75**, 3905 (1999)
126. G. Frankowsky, F. Steuber, V. Härle, F. Scholz, A. Hangleiter: Appl. Phys. Lett. **68**, 3746 (1996)

127. M. Kuball, E.S. Jeon, Y.K. Song, A.V. Nurmikko, P. Kozodoy, A. Abare, S. Keller, L.A. Coldren, U.K. Mishra, S.P. DenBaars, D.A. Steigerwald: Appl. Phys. Lett. **70**, 2580 (1997)
128. S. Nakamura, M. Senoh, S. Nagahama, N. Iwasa, T. Yamada, T. Matsushita, Y. Sugimoto, H. Kiyoku: Appl. Phys. Lett. **69**, 1568 (1996)
129. S. Nakamura: MRS Internet J. Nitride Semicond. Res. **2**, 5 (1997)
130. Y.K. Song, M. Kuball, A.V. Nurmikko, G.E. Bulman, K. Doverspike, S.T. Sheppard, T.W. Weeks, M. Leonard, H.S. Kong, H. Dieringer, J. Edmonds: Appl. Phys. Lett. **72**, 1418 (1998)
131. T.J. Schmidt, S. Bidnyk, Y.H. Cho, A.J. Fischer, J.J. Song, S. Keller, U.K. Mishra, S.P. DenBaars: Appl. Phys. Lett. **73**, 3689 (1998)
132. T. Deguchi, T. Azuhata, T. Sota, S. Chichibu, M. Arita, N. Nakanishi, S. Nakamura: Semcond. Sci. Technol. **13**, 97 (1998)
133. Y. Kawakami, K. Omae, A. Kaneta, K. Okamoto, Y. Narukawa, T. Mukai, S. Fujita: J. Phys. Condens. Matter **13**, 6993 (2001)
134. W.W. Chow, A.F. Wright, A. Girndt, F. Jahnke, S.W. Koch: Appl. Phys. Lett. **71**, 2608 (1997)
135. A.A. Yamaguchi, M. Kuramoto, M. Nido, M. Mizuta: Phys. Stat. Sol. (a) **188**, 47 (2001)

Thermodynamics of Solar Cells

Peter Würfel

1 Introduction

Solar cells may become the most important option for the energy supply of mankind. It would certainly help if their efficiency could be raised substantially above the present value for commercial panels of 12 to 15%. General thermodynamical arguments give much higher efficiency limits. In order to exploit them with solar cells, future concepts should be based only on necessary requirements. Present day solar cells are based on pn-junctions and their understanding goes back to the beginnings of semiconductor device physics [1]. It will be shown that pn-junctions do not really belong to the necessary requirements. In addition, their understanding is at least misleading when applied to solar cells. In order to be able to find new concepts for more efficient solar cells, wrong ideas must be corrected. Since thermodynamics is independent of particular device aspects, it is particularly well suited to explore the limits of solar cells.

2 Heat Engines Operating with Radiation

Solar cells are able to produce electrical energy from the heat-(radiation) of the sun. As any other engine producing an entropy free form of energy from heat, solar cells have to obey the rules for heat engines. As Fig. 1 shows, the efficiency of a heat engine which does not produce entropy, i.e. a Carnot engine, is given by Carnot's efficiency

$$\eta_{\text{Carnot}} = 1 - \frac{T_0}{T_A}. \tag{1}$$

Since the temperature of the solar radiation is $T_S = 5800$ K and the temperature of the environment is $T_0 = 300$ K, the efficiency is $\eta_{\text{Carnot}} = 0.95$ if the input of the heat engine has a temperature of $T_A = T_S = 5800$ K.

Dealing with heat in the form of radiation, however, imposes some restrictions. Eq. (1) implies that the energy E put into the heat engine is related to the entropy S and the temperature T_A by $E = T_A S$ and that this amount of energy at the input is available for processing by the heat engine. Both

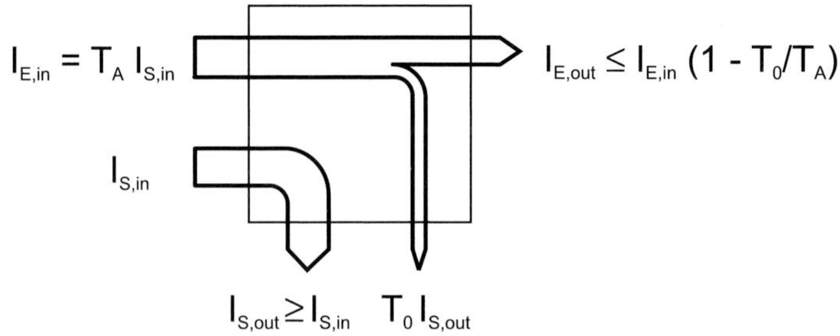

Fig. 1. Entropy and energy currents are processed by a heat engine. Since the desired output (mechanical, chemical or electrical energy) is free of entropy, the entropy from the input must be transferred to a heat dump, the environment, and with it some energy. From the heat energy current $T_A I_{S,\text{in}}$ at the input an amount of $T_0 I_{S,\text{out}}$ is required to dump the entropy. Due to the second principle of thermodynamics, $I_{S,\text{out}} \geq I_{S,\text{in}}$ the efficiency is $(T_A I_{S,\text{in}} - T_0 I_{S,\text{out}})/T_A I_{S,\text{in}} \leq 1 - T_0/T_A$.

assumptions are not valid for radiation. The energy-entropy relation for black-body radiation, a good approximation for solar radiation observed outside of the earth's atmosphere, is $E = 3/4 S T_S$. This higher entropy content of heat radiation is due to the generation of entropy by the emission process [2]. It may be surprising that solar radiation at the location of the earth should still have a temperature of $T_S = 5800$ K although we don't feel it. The reason is that a body is absorbing solar radiation only from the small solid angle of $\Omega_S = 6.8 \cdot 10^{-5}$ which is filled by the sun and mixes it with the cold radiation from galactic space filling the solid angle which the body sees outside the sun. This must be avoided to attain high temperatures. An absorber at the input of the heat engine should, therefore, by means of concentrating optics, only "see" the sun or in a mirror itself. For a planar absorber with a mirror on its backside, solar radiation should be incident from a hemisphere increasing the effective solid angle (considering Lambert's law) to π and the concentration factor to $\pi/\Omega_S = 46200$. Under these conditions of maximal concentration, radiation emitted by the absorber is all directed towards the sun. This ensures that the absorber absorbs solar radiation only. Under open circuit conditions, where no energy is withdrawn from the absorber into the engine, the absorber will be in thermal equilibrium with the sun and have a temperature of $T_A = T_S = 5800$ K at which absorbed and emitted energy and entropy currents balance each other.

2.1 Thermal Heat Engines

The possible presence of thermal equilibrium points to the next and more severe problem: If $T_A = T_S$ then no energy is available for the heat engine to convert into entropy free energy since it is all re-emitted. For a net energy flow into the engine, the input temperature T_A has to be lowered. If the lowering of the temperature by ΔT is small compared to T_S, we find from Stefan-Boltzmanns emission law for the energy current density j_E and the entropy current density j_S emitted by a black body

$$j_E = \sigma T_S^4 \quad \text{and} \quad j_S = 4/3\sigma T_S^3 \tag{2}$$

that the net energy and entropy current densities available to the heat engine are

$$\Delta j_E = 4\sigma T_S^3 \Delta T \quad \text{and} \quad \Delta j_S = 4\sigma T_S^2 \Delta T. \tag{3}$$

For these net current densities Carnot's efficiency in Eq. (1) is found again, but only for the small net energy current density Δj_E which is not re-emitted. This would, however, not be harmful, if we would be reimbursed for the energy given back to the sun (which would in fact extend the sun's lifetime) and if we would only have to pay for the energy actually withdrawn from the sun. But, solar energy is for free, and since there is no reimbursement for re-emission, any emission from the absorber at the input of the heat engine, whether directed towards the sun or not, is considered as a loss and must be minimized.

For an absorber at temperature T_A emitting only towards the sun, i.e. under conditions of maximal concentration of the incident radiation, the absorption efficiency, i.e. the net energy current transferred into the heat engine divided by the incident energy current, is

$$\eta_{\text{abs}} = \frac{\sigma T_S^4 - \sigma T_A^4}{\sigma T_S^4} = 1 - \left(\frac{T_A}{T_S}\right)^4 . \tag{4}$$

Accounting for the internal (Carnot) efficiency of the heat engine, the total efficiency of a black-body thermal engine operating with solar radiation is

$$\eta_{\text{tot}} = \left[1 - \left(\frac{T_A}{T_S}\right)^4\right]\left[1 - \frac{T_0}{T_A}\right] . \tag{5}$$

Fig. 2 shows this total efficiency as a function of the input temperature T_A of a thermal heat engine. The total efficiency is zero at $T_A = T_0 = 300$ K because of zero efficiency of the heat engine, since it needs all the energy to get rid of the entropy, and it is zero at $T_A = T_S = 5800$ K, when all of the incident radiation is re-emitted towards the sun. The maximum value of the total efficiency is 0.85 at an input temperature of the heat engine of $T_A = 2480$ K. This value of the efficiency can even be increased a little to 0.86, as will be explained later, if the incident black-body radiation is spectrally divided into many narrow spectral ranges which are processed separately by as many different heat engines, each absorber operating at its own optimal temperature T_A. While this minute improvement does not justify the complication for thermal engines, it is this principle which leads to large efficiency improvements in solar cells.

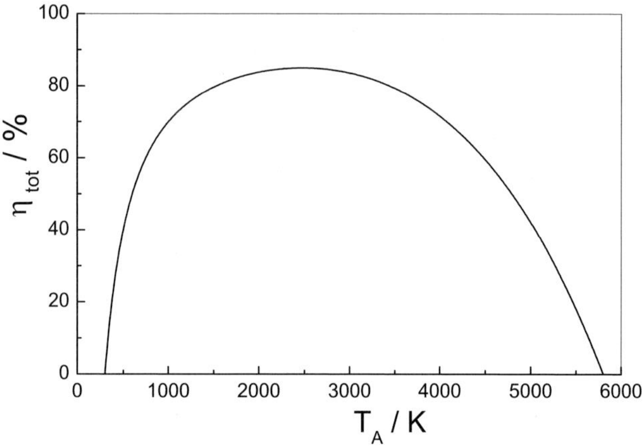

Fig. 2. Efficiency $\eta_{\rm tot}$ of a heat engine processing solar radiation as a function of the temperature T_A at the input of the engine.

2.2 Thermo-Chemical Heat Engines

Thermal equilibrium of an absorber and the sun requires that the absorber sees nothing but the sun or itself (in a mirror). It does not require that the absorber is black and absorbs the full solar spectrum. If photons are exchanged between an emitter and an absorber in a narrow energy interval $d\hbar\omega$ only, Planck's law for the absorbed and the emitted photon current densities per energy interval has to be applied. We use it in a generalized form which, as shown later, covers luminescent radiation, too [3].

$$\frac{dj_\gamma}{d\hbar\omega} = a(\hbar\omega) \cdot \frac{\Omega\,(\hbar\omega)^2}{4\pi\hbar^3 c^2} \cdot \frac{1}{\exp\left(\frac{\hbar\omega - \mu_{eh}}{kT}\right) - 1} \tag{6}$$

$a(\hbar\omega)$ is the absorptivity of the absorber or emitter, which according to Kirchhoff is equal to its emissivity. The second factor is the density of states for photons, multiplied by the light velocity c in the medium, containing the solid angle Ω into which the photons are emitted. The last factor containing the chemical potential μ_{eh} of the electron-hole excitations is the number of photons per mode. For an emitting body in a mirror cavity (or for $a(\hbar\omega) = 1$), there would be chemical equilibrium between electron-hole pairs and photons with $\mu_\gamma = \mu_{eh}$. Equilibrium, however, can only exist for $\mu_{eh} < \hbar\omega$, since the entropy of the photons would be negative for $\mu_\gamma > \hbar\omega$. In contrast, $\mu_{eh} > \hbar\omega$ is not forbidden and is common in lasers which amplify incident photon currents, thereby demonstrating that $\mu_{eh} > \mu_\gamma$ in a non-equilibrium situation.

From (6) we conclude that equilibrium exists between two bodies (1 and 2), in which emitted and absorbed monochromatic energy currents are equal, even if their temperatures are different, as long as

$$\frac{\hbar\omega - \mu_{eh,1}}{kT_1} = \frac{\hbar\omega - \mu_{eh,2}}{kT_2}. \tag{7}$$

This equality requires that one photon is emitted for each absorbed photon. Non-radiative transitions must not be present for this equilibrium to exist. This requirement is not too unrealistic for some III-V materials which exhibit external quantum efficiencies of over 90%. Since in this equilibrium, both the temperature and the chemical potential of electron-hole excitations are involved, we call it a **thermo-chemical equilibrium** in which temperature and chemical potential are coupled as expressed by (7). We are used to other coupled equilibria, e.g. the electro-chemical equilibrium of electrons between the p- and n-parts of a pn-junction, in which their chemical potential and their electrical energy are coupled in a way to yield constant electro-chemical potential.

A real equilibrium requires the equality of absorbed and emitted entropy currents as well. Since the entropy per mode as a function of the number of photons per mode n_γ

$$S_\gamma = k\left[(n_\gamma + 1)\,\ln(n_\gamma + 1) - n_\gamma \ln n_\gamma\right]. \tag{8}$$

depends only on the number of photons per mode, it is conserved in thermo-chemical equilibrium of two bodies which both emit as many photons as they absorb, leaving the number of photons per mode and with it the entropy unchanged.

Applied to the sun with temperature T_S and $\mu_\gamma = 0$ for a thermal emitter and to a body at room temperature T_0 under conditions of maximal concentration, the chemical potential of electron-hole pairs of this body in equilibrium with the sun by exchanging monochromatic radiation with photon energy $\hbar\omega$ is

$$\mu_{eh} = \hbar\omega\left(1 - \frac{T_0}{T_S}\right) \tag{9}$$

Chemical Energy

Equation (9) gives the maximum value of the chemical potential of electron-hole pairs in a body absorbing solar radiation, since the thermo-chemical equilibrium requires that all of the electron-hole pairs generated by solar radiation must recombine radiatively and none are available to be withdrawn carrying their chemical energy μ_{eh} with them. The situation is identical to the thermal equilibrium of two bodies in Sect. 2.1, where no heat could be withdrawn in equilibrium. While non-radiative transitions shall still be excluded, electron-hole pairs shall be extracted. Therefore, the number of emitted photons must be reduced by lowering the chemical potential μ_{eh} of the electron-hole pairs. Due to the reduced value of μ_{eh} compared to the open circuit value in (9), less chemical energy is carried away by both the emitted photons and the withdrawn electron-hole pairs, causing the entropy carried by either the photons or the electron-hole pairs to be larger than in equilibrium. As for the thermal engine, entropy generation seems to be inherent to a net energy conversion.

Although, at this point, we don't know yet, how to withdraw the electrons and holes, it is quite clear that a chemical energy current dj_μ

$$dj_\mu = \mu_{eh} \cdot dj_{eh} = \mu_{eh} \left(dj_{\gamma,\text{abs}} - dj_{\gamma,\text{emit}} \right) \tag{10}$$

is withdrawn together with the electron-hole current dj_{eh} which is given by the difference of absorbed and emitted photon currents, since non-radiative transitions are excluded. Fig. 3 shows the emitted photon current $dj_{\gamma,\text{emit}}$ as a function of μ_{eh} and the absorbed photon current $dj_{\gamma,\text{abs}}$ which does not depend on μ_{eh}. Both consist of mono-energetic photons with energy between $\hbar\omega$ and $\hbar\omega + d\hbar\omega$. The absorbed energy current is $j_{\gamma,\text{abs}}$ multiplied by the photon energy $\hbar\omega$ and indicated in Fig. 3 by the lightly shaded area. Although the electron-hole current which is withdrawn is maximal at $\mu_{eh,\text{sc}} = 0$ the current of chemical energy j_μ is zero because the electron-hole pairs don't carry chemical energy ($\mu_{eh,\text{sc}} = 0$). Similarly, no chemical energy is withdrawn if μ_{eh} has its maximum value $\mu_{eh,\text{oc}}$ given by (9), when all electron-hole pairs recombine radiatively. The current of chemical energy becomes maximal at some intermediate value $\mu_{eh,\text{mp}}$ indicated by the area of the dark rectangle in Fig. 3. The ratio of the dark area and the larger gray area represents the efficiency with which chemical energy is produced. This efficiency is shown as a function of $\hbar\omega$ in Fig. 4.

Large efficiencies are achieved for all photon energies if the incident monochromatic radiation is fully concentrated. For non-concentrated sun light, photons with energy $\hbar\omega < 0.3$ eV are useless. At these energies, a

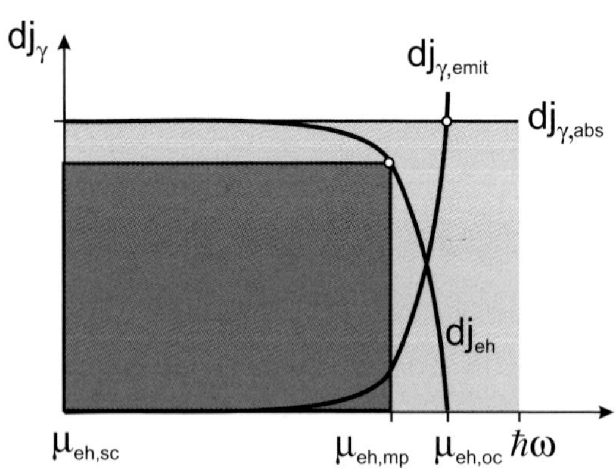

Fig. 3. Absorbed monochromatic photon current $dj_{\gamma,\text{abs}}$, emitted monochromatic photon current $dj_{\gamma,\text{emit}}$ and electron-hole current dj_{eh}, both as a function of the chemical potential μ_{eh} of the electron-hole pairs. The maximal current of chemical energy available for use outside the absorber, $j_\mu = \mu_{eh,\text{mp}} \cdot j_{eh}$ is given by the area of the dark rectangle.

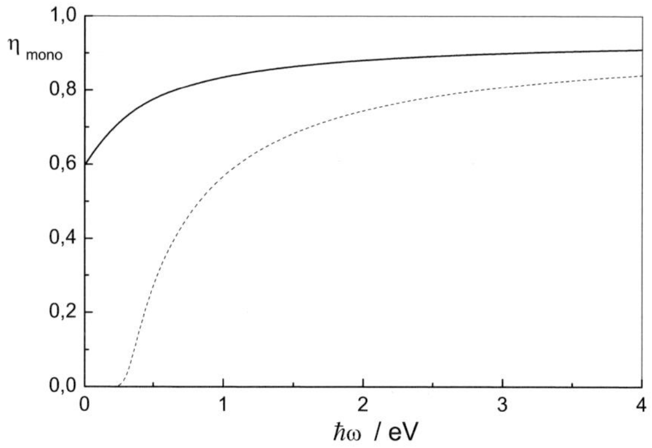

Fig. 4. Efficiency of the conversion of monochromatic radiation into chemical energy as a function of the photon energy $\hbar\omega$. The full curve is for maximal concentration, the broken curve is for non-concentrated sun light.

monochromatic absorber at 300 K and with $\mu_{eh} > 0$ would emit more photons than are incident from non-concentrated sun light. Each absorber covers only the narrow energy interval $d\hbar\omega$. To make proper use of the broad solar spectrum, it must be divided into many narrow energy ranges, each for a different absorber. The simplest way to do this, is to place these many absorbers one behind the other, with the absorber for the largest photon energy in front facing the incident radiation. For infinitely many absorbers, each absorbing monochromatic radiation, the efficiency for converting the whole spectrum into chemical energy is given by

$$\eta_{\rm tot} = \frac{\int_0^\infty \eta_{\rm mono} \cdot dj_\gamma(\hbar\omega) d\hbar\omega}{\int_0^\infty dj_\gamma(\hbar\omega) d\hbar\omega}, \tag{11}$$

where $\eta_{\rm mono}$ is taken from Fig. 4 and $dj_\gamma(\hbar\omega)$ is given by Eq. (6). For fully concentrated radiation, the efficiency is $\eta_{\rm tot} = 0.86$ and for non-concentrated radiation it is $\eta_{\rm tot} = 0.67$ [4].

3 Semiconductors

The conversion of solar heat into chemical energy of electron-hole pairs occurs in semiconductors without any further provision. The principle is the same as in any other heat engine. Heat radiation is absorbed by a working gas which is cooled down by thermodynamic processes to a temperature at which the entropy is discharged with less energy than when it was received. The working gas in a semiconductor is made up by the electrons and holes generated by the absorbed photons. Before these experience any interaction, their energy distribution reflects the energy distribution of the broad solar spectrum and the high temperature of solar radiation. If the incident radiation is fully concentrated, this energy distribution would not even be changed by scattering of electrons and holes among each other. The electron and hole temperatures would be equal to the sun's temperature and the chemical potential of the electron-hole pairs would be $\mu_{eh} = 0$. If radiative recombination would occur right away, the emitted spectrum would be identical to the absorbed spectrum. In reality, however, electrons and holes experience a strong interaction with phonons, which cools them down to the lattice temperature of $T_0 = 300$ K. Fig. 5 shows the situation before and after cooling.

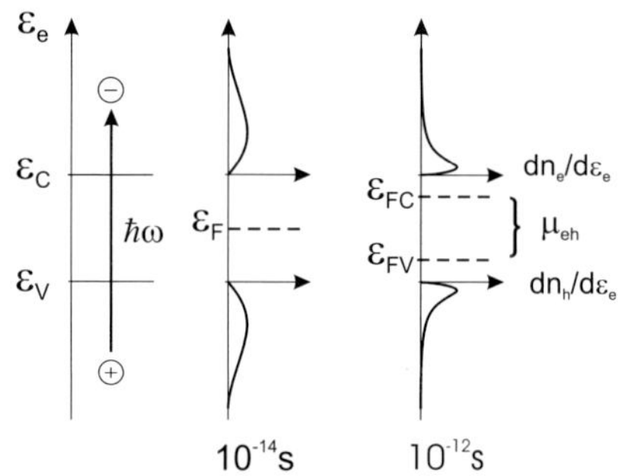

Fig. 5. Energy distribution of electrons and holes right after their generation by incident photons (middle) and after their interaction with phonons (right), when their temperature is $T_0 = 300$ K and their chemical potential is $\mu_{eh} = \epsilon_{FC} - \epsilon_{FV} > 0$.

Since the cooling proceeds at constant concentrations, the resulting energy distribution in Fig. 5 contains much more electrons and holes than would be present in equilibrium with black-body radiation of $T_0 = 300$ K, and the product of their concentrations is $n_e n_h > n_i^2$. Due to scattering with phonons the electron and hole temperature is well defined to be $T_0 = 300$ K and the electrons and holes are each distributed according to a Fermi-distribution. In contrast to the equilibrium with 300 K radiation, however, these Fermi-distributions have different Fermi-energies, ϵ_{FC} for the electron distribution in the conduction band and ϵ_{FV} for the electron distribution in the valence band. The difference of the Fermi-energies is recognized as the chemical potential μ_{eh} of the electron-hole pairs.

The conversion of solar heat carried by all photons with energy $\hbar\omega \geq \epsilon_C - \epsilon_V$ into chemical energy μ_{eh} by cooling of the generated electrons and holes, as shown in Fig. 5, is in no way ideal. A lot of entropy is generated and a lot of energy is lost by the generation of phonons. The efficiency of the production of chemical energy as the chemical energy current available from a semiconductor divided by the incident solar energy current was first calculated by W. Shockley and H.-J. Queisser [5] under the assumption of

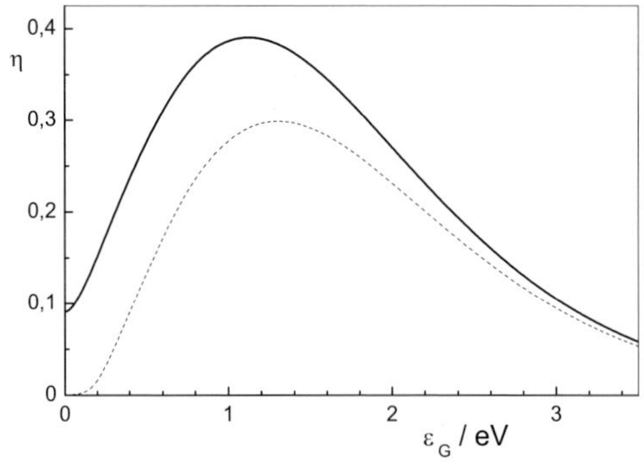

Fig. 6. Efficiency η of the conversion of solar energy into extractable chemical energy in semiconductors with radiative transitions only, as a function of their band gap ϵ_G for full concentration (solid line) and for non-concentrated sun light (broken line).

only radiative transitions. Its dependence on the band gap $\epsilon_G = \epsilon_C - \epsilon_V$ of the absorbing semiconductor is shown in Fig. 6.

The large losses incurred by processing a broad spectrum by a single semiconductor can be avoided, as was mentioned before, by placing many semiconductors with decreasing band gaps, one behind the other. The energy range of the electron-hole pairs right after their generation is in this configuration restricted to energies between the band gap of the host semiconductor and the band gap of the semiconductor in front. Since the energy distribution of the electrons and holes achieved in this way does not differ much from their energy distribution at $T_0 = 300$ K, the subsequent interaction with phonons is almost without losses. Although semiconductors are broad band absorbers, their configuration in optical series enables each one of them to effectively absorb and emit only monochromatically.

3.1 Quasi Fermi-Distributions

When by the interaction with the phonons the temperature of the electrons and holes is $T_0 = 300$ K as in the right part of Fig. 5, their distributions over energy are given by Fermi-distributions. However, both their concentrations are larger than in equilibrium with the 300 K surrounding radiation and different Fermi-distributions $f(\epsilon)$ (with Fermi-energies ϵ_{FV} for the states with density D_1 at $\epsilon_1 < \epsilon_V$ in the valence band and ϵ_{FC} for the states with density D_2 at $\epsilon_2 > \epsilon_C$ in the conduction band) are required.

$$f(\epsilon_1) = \frac{1}{\exp\left(\frac{\epsilon_1 - \epsilon_{FV}}{kT}\right) + 1} \quad \text{and} \quad f(\epsilon_2) = \frac{1}{\exp\left(\frac{\epsilon_2 - \epsilon_{FC}}{kT}\right) + 1} \tag{12}$$

The concentrations of holes in the valence band dn_h and of electrons in the conduction band dn_e are

$$dn_h(\epsilon_1) = D(\epsilon_1)\,[1 - f(\epsilon_1)]\,d\epsilon = \frac{D(\epsilon_1)\,d\epsilon}{\exp\left(-\frac{\epsilon_1 - \epsilon_{FV}}{kT}\right) + 1} \tag{13}$$

$$dn_e(\epsilon_2) = D(\epsilon_2)\,f(\epsilon_2)\,d\epsilon = \frac{D(\epsilon_2)\,d\epsilon}{\exp\left(\frac{\epsilon_2 - \epsilon_{FC}}{kT}\right) + 1}. \tag{14}$$

The product of the densities n_e of all electrons in the conduction band and n_h of all holes in the valence band can be given only in Boltzmann approximation

$$n_e n_h = N_C N_V \exp\left(-\frac{\epsilon_C - \epsilon_V}{kT}\right) \exp\left(\frac{\epsilon_{FC} - \epsilon_{FV}}{kT}\right)$$

$$n_e n_h = n_i^2 \exp\left(\frac{\epsilon_{FC} - \epsilon_{FV}}{kT}\right). \tag{15}$$

N_C and N_V are the effective densities of states in the conduction and valence bands, respectively, and n_i is the density of electrons and holes if the semiconductor were intrinsic and in equilibrium with the surrounding 300 K radiation.

3.2 Transition Rates and Absorption Coefficient

The generalized Planck law for the emission of photons, on which the analysis in Sect. 2.2 is based is derived in the same way as Planck's original formula, except that the body absorbing or emitting radiation contains electron distributions requiring two different Fermi-functions. This has consequences for the absorption coefficient and the probability of photon emission.

We analyze the absorption and emission of photons by a semiconductor for radiative transitions between states in energy ranges of width $d\epsilon$ at energies ϵ_1 in the valence band and ϵ_2 in the conduction band involving photons with energy in the energy range $\hbar\omega \ldots \hbar\omega + d\hbar\omega$. The rate of upward transitions from states at ϵ_1 to states at ϵ_2 is

$$dr_{\text{up}}(\hbar\omega) = |M|^2\,D_{12} f(\epsilon_1)\,[1 - f(\epsilon_2)]\,\frac{dj_\gamma(\hbar\omega)}{d\hbar\omega}\,d\hbar\omega, \tag{16}$$

where M contains the matrix element for the transition and D_{12} is the combined density of states between which the transition occurs.

The rate of stimulated emission of photons is likewise

$$dr_{\text{stim}}(\hbar\omega) = |M|^2\,D_{12}\,[1 - f(\epsilon_1)]\,f(\epsilon_2)\,\frac{dj_\gamma(\hbar\omega)}{d\hbar\omega}\,d\hbar\omega \tag{17}$$

and the rate of spontaneous emission finally is

$$dr_{\text{spont}}(\hbar\omega) = |M|^2\,\frac{c_0}{n}\,D_\gamma(\hbar\omega) D_{12}\,[1 - f(\epsilon_1)]\,f(\epsilon_2)\,d\hbar\omega, \tag{18}$$

where the density of states for photons in the solid angle Ω in a medium with refractive index n is

$$D_\gamma(\hbar\omega) = \frac{\Omega\, n^3}{4\pi\hbar^3 c_0^3}\,(\hbar\omega)^2, \tag{19}$$

and c_0 is the vacuum velocity of light.

Stimulated emission is a process by which a photon is duplicated, producing one additional photon in exactly the same state as the incident photon initiating the transition. These photons, therefore, cannot be distinguished from non-absorbed photons. The net absorption rate at which photons disappear and which defines the absorption coefficient $\alpha(\hbar\omega)$ then is

$$\begin{aligned} dr_{\text{abs}}(\hbar\omega) &= dr_{\text{up}}(\hbar\omega) - dr_{\text{stim}}(\hbar\omega) \\ &= |M|^2\, D_{12}\,[f(\epsilon_1) - f(\epsilon_2)]\, dj_\gamma(\hbar\omega) = \alpha(\hbar\omega)\, dj_\gamma(\hbar\omega). \end{aligned} \tag{20}$$

resulting in an absorption coefficient

$$\alpha(\hbar\omega) = |M|^2\, D_{12}\,[f(\epsilon_1) - f(\epsilon_2)] \tag{21}$$

This important relation allows to replace the matrix element and the combined density of states in the spontaneous emission rate by the absorption coefficient $\alpha(\hbar\omega)$ and makes this treatment applicable to real materials for which only the absorption coefficient is known.

$$dr_{\text{spont}}(\hbar\omega) = \alpha(\hbar\omega)\frac{c_0}{n}D_\gamma(\hbar\omega)\frac{[1-f(\epsilon_1)]\,f(\epsilon_2)}{f(\epsilon_1)-f(\epsilon_2)}\,d\hbar\omega. \tag{22}$$

By inserting $f(\epsilon_1)$ and $f(\epsilon_2)$ from (12) and remembering that the energy difference $\epsilon_2 - \epsilon_1$ over which the transitions occur is equal to the photon energy $\hbar\omega$, (22) becomes

$$dr_{\text{spont}}(\hbar\omega) = \alpha(\hbar\omega)\frac{c_0}{n}D_\gamma(\hbar\omega)\frac{d\hbar\omega}{\exp\left(\frac{\hbar\omega-(\epsilon_{FC}-\epsilon_{FV})}{kT}\right)-1}. \tag{23}$$

Since (23) does not depend on the occupation probabilities $f(\epsilon_1)$ and $f(\epsilon_2)$ directly, each pair of states with energies ϵ_j in the conduction band and ϵ_i in the valence band such that $\epsilon_j - \epsilon_i = \hbar\omega$ contributes to (23) in the same way and is contained in (23) if $\alpha(\hbar\omega)$ is the absorption coefficient for all these transitions.

Equation (23) is a generalization of Kirchhoff's and Planck's laws and is valid for materials, which are neither black nor have a single Fermi-distribution over all states. It reduces to Planck's original emission law for a thermal emitter, in which all electrons belong to a single Fermi-distribution and for which $\epsilon_{FC} - \epsilon_{FV} = 0$. The difference of the Fermi-energies $\mu_{eh} = \epsilon_{FC} - \epsilon_{FV}$ is the Free Energy per electron-hole pair, also called the chemical potential of electron-hole pairs. It is free of entropy and we may therefore hope to transfer it into electrical energy without losses. If electron-hole pairs are not allowed to leave the absorber, i.e. under open circuit conditions, they have to recombine and photons are emitted. These photons carry the free energy of the electron-hole pairs, and $\mu_\gamma = \mu_{eh} = \epsilon_{FC} - \epsilon_{FV}$ is recognized as the chemical potential of the photons.

The above treatment is valid quite generally, even if $\epsilon_{FC} - \epsilon_{FV} > \hbar\omega$, where the denominator in (23) is negative. Under the same conditions, the absorption coefficient in (21) is negative, too and the spontaneous emission rate in (22) remains positive. When the absorption coefficient is negative, stimulated emission overcompensates the rate of upward transitions and the semiconductor amplifies the incident light as in a laser. $\epsilon_{FC} - \epsilon_{FV} > \hbar\omega$ is also known as the condition for lasing.

Equation (23) describes the rate of radiative transitions. In a real material, there are also non-radiative transitions occurring in parallel and in addition. This does not invalidate (23), as long as the assumptions are justified that two separate Fermi-distributions describe the occupation of states in the two energy ranges between which the transitions occur.

4 Semiconductor Devices

Semiconductor devices are based on the property of semiconductors to harbor two different kinds of electrons, electrons in the conduction band and electrons in the valence band which are each distributed over energy according to a separate Fermi-distribution. This is different from metals because in semiconductors thermalization times after an injection or generation of a carrier are short compared to recombination times after which a single Fermi-distribution over all energies is reestablished.

The pn-junction is the basic building block of most semiconductor devices. Its understanding is the basis for understanding LEDs and solar cells.

Fig. 7 shows a pn-junction in the dark, under forward bias V. Going back to the original interpretation of W. Shockley [1], the current through a pn-junction is thought of being composed of a field current and a diffusion current. This interpretation acknowledges the presence of a gradient of the electrical potential φ and of a gradient of the concentrations. At zero bias, field and diffusion currents cancel each other and the resulting current is zero. A positive current for forward bias is dominated by the diffusion current, whereas a negative current is dominated by the field current. The exponential current-voltage characteristic of a pn-junction is, since Shockley, interpreted as being caused by the barrier of the electrical potential between p- and n-type regions.

These interpretations are at least questionable and, especially for solar cells, lead to wrong conclusions.

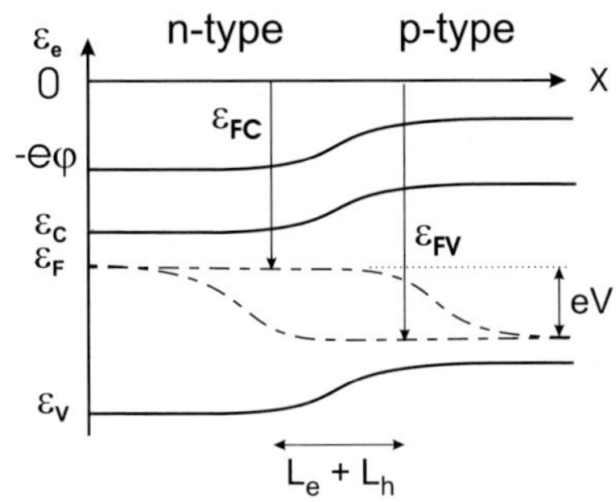

Fig. 7. Energy diagram of a pn-junction in the dark, under forward bias V. Separate Fermi-energies $\epsilon_{FC} \neq \epsilon_{FV}$ exist over a distance equal to the sum of the diffusion lengths $L_e + L_h$.

4.1 Field and Diffusion Currents

Field and diffusion currents are in conflict with physical laws and with logic. While they add up to zero in a pn-junction for zero bias, their entropy generation (Joule heat), being proportional to the square of the currents, does not add to zero as it should in electro-chemical equilibrium (zero bias). Furthermore, they have the strange property that all electrons and holes contribute to the field current and at the same time all of them contribute to the diffusion current into the opposite direction, as well. For solar cells, the negative current at maximum power is interpreted as a field current and the conclusion is that an electrical potential gradient driving the current must exist in any solar cell structure and only those structures in which an electrical potential gradient exists are suitable for a solar cell. As a result of these ideas, the pn-junction appears to be the ideal structure for solar cells. This view shall be criticized in the following.

From irreversible thermodynamics we know that a current of a certain type of charged particles is driven by the gradient of its electro-chemical potential. For electrons and holes the electro-chemical potentials are

$$\begin{aligned} \eta_e &= \mu_e - e\varphi = \epsilon_{FC} && \text{for electrons} \\ \eta_h &= \mu_h + e\varphi = -\epsilon_{FV} && \text{for holes} \end{aligned} \tag{24}$$

The electro-chemical potential η consists of the chemical potential μ and of the potential energy of a charged particle in an electrical potential φ. The gradient of the chemical potential is the driving force behind diffusion and the gradient of the electrical potential energy is the field force. If both forces are present as in semiconductor devices, these forces have to be added first to find the resulting force, the gradient of the electrochemical potential, before a charge current as the result of a combination of both forces is determined. Applied to electrons and holes the charge current is

$$j_Q = \frac{\sigma_e}{e}\,\text{grad}\,\epsilon_{FC} + \frac{\sigma_h}{e}\,\text{grad}\,\epsilon_{FV} \tag{25}$$

The current j_Q in (25) is proportional to the conductivities σ and represents the motion of electrons and holes as being caused by the combination of the forces. Field currents and diffusion currents arising from the components of the resulting force, separately, do not exist.

For solar cells the conclusion from these considerations is that the presence of gradients of the electro-chemical potentials or of the Fermi-energies is all that matters for a charge current. The presence of an electric field as in a pn-junction is not a necessary property of a solar cell structure.

4.2 Current-Voltage Characteristic of the pn-Junction

Another misconception of semiconductor devices is that the current-voltage characteristic of a pn-junction, as shown by the continuous line in Fig. 8 is caused by the potential barrier in Fig. 7 which electrons (from left to right) and holes (from right to left) have to surmount for a current in forward direction. This misconception is the result of looking for a transport resistance as the only cause which limits the current as is true for a homogeneous material. The presence of a transport resistance would require a large driving force and would, according to (25) show up in a large gradient of the Fermi-energies. In fact, in Fig. 7 the gradient of ϵ_{FC} driving the electrons is negligibly small in the n-region and the gradient of ϵ_{FV} driving the holes is negligibly small in the p-region, reasonable mobilities assured. In particular, both gradients are small right at the junction where the barrier is thought to cause a transport resistance.

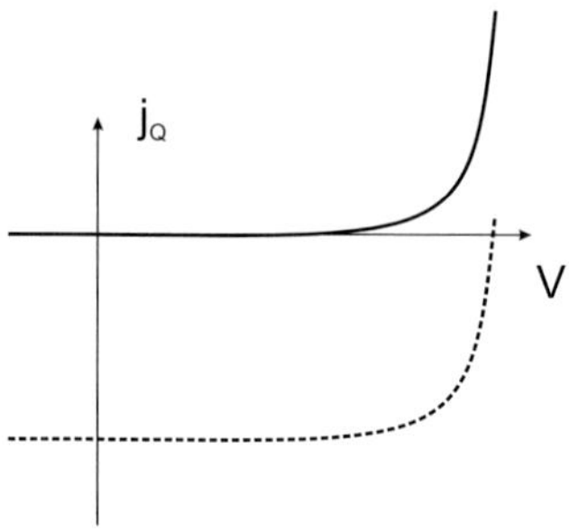

Fig. 8. Charge current j_Q as a function of the voltage V through a pn-junction in the dark (continuous line) and under illumination (broken line).

A close look at Fig. 7 reveals the following. Outside a diffusion length from the pn-junction, the charge current is to a very good approximation carried only by electrons in the n-type region and only by holes in the p-type region. In forward direction, electrons moving from the n-region towards the junction, will, therefore not arrive at the contact to the p-region and holes moving from the p-region towards the junction will not arrive at the contact to the n-region. Conversely, in reverse direction, electrons move away from the junction towards the contact to the n-region and holes move towards the contact to the p-region. But the electrons don't come from the contact to the p-region, since there, the current is a pure hole current and by the same reasoning, the holes don't come from the n-region. Outside a diffusion length from the junction, the n-region acts as a semi-permeable membrane allowing only the passage of electrons, but blocking the passage of holes. The p-region acts as a semi-permeable membrane for the holes.

Clearly, in forward direction ($j_Q > 0$ in Fig. 8) electrons and holes must be annihilated near the junction by recombining with each other. Recombination is a chemical reaction of electrons and holes by which photons and phonons are created. Only so many electrons and holes can flow towards the junction in steady state as are consumed in the chemical reaction $e+h \rightarrow \gamma, \Gamma$ which produces photons and phonons. The rate of recombination reactions is proportional to the product of the electron and hole densities. According to Eq. (15) the recombination rate rises exponentially with the difference $\epsilon_{FC} - \epsilon_{FV}$, which is equal to eV, as in Fig. 7, if the gradients of ϵ_{FC} and of ϵ_{FV} are small in the n- and p-regions, respectively.

In reverse direction, electrons and holes contributing to the current are generated by the same chemical reaction running in reverse, consuming phonons and photons to create electron-hole pairs ($e+h \leftarrow \Gamma, \gamma$). The current in reverse direction ($j_Q < 0$ in Fig. 8) is so small in a pn-junction in the dark because there are not many photons γ available in the 300 K surrounding radiation which can generate electron-hole pairs in common semiconductors and a band-band transition by phonons Γ is not a probable event since many phonons are required simultaneously. In reverse direction, it is simply not possible to extract more electrons and holes than are generated in the region within a diffusion length on both sides of the junction and as a result, the reverse current is small and almost independent of the voltage.

From these considerations, the current voltage relation of a pn-junction is easily derived. For perfect membranes the current density is given by the difference between generation rate G and recombination rate R inside a volume extending from a diffusion length of the holes L_h away from the junction in the n-region to a diffusion length of the electrons L_e from the junction in the p-region.

$$j_Q = -e\,(L_e + L_h)(G - R) \tag{26}$$

The recombination rate R_0 in the dark with no bias must be equal to the generation rate G_0 in the dark. For band to band recombination the recombination rate is proportional to the product of electron and hole concentrations $n_e\, n_h$. With (15) we have

$$j_Q = e\,(L_e + L_h)\, G_0 \left[\exp\left(\frac{eV}{kT}\right) - 1\right] - e\,(L_e + L_h)\Delta G \tag{27}$$

where the generation rate $G = G_0 + \Delta G$ consists of contributions from the 300 K background radiation and from additional illumination as from the sun.

Instead of a transport resistance which would require large gradients of the Fermi-energies as driving forces, a "resistance" of the chemical reaction $e + h \to \Gamma, \gamma$ is limiting the current in forward direction. The driving force which determines the rate of this reaction is given by the difference of the Fermi-energies $\epsilon_{FC} - \epsilon_{FV}$. The difference of generation and recombination determines the current-voltage characteristic of a pn-junction.

For many semiconductor devices, mobilities of electrons and holes as large as possible are very desirable. With increasing mobilities the diffusion lengths are increasing as well. However, the current-voltage characteristic of the pn-junction is lost, if the diffusion lengths become larger than the distance of the junction from the contacts, where recombination and generation of electron-hole pairs is less impeded. A better structure is discussed in the following paragraph.

4.3 Solar Cells

While the production of chemical energy from solar heat in an absorber is useful for subsequent chemical reactions as in photosynthesis, solar cells are to produce electrical energy. They have to convert chemical into electrical energy. In order to produce an electrical current, electrons and holes generated within the semiconductor have to leave it through two different ports or contacts. One should transmit electrons and block holes and the other should transmit holes and block electrons. Fig. 9 shows a principle arrangement of absorber and selective contacts which act as semi-permeable membranes.

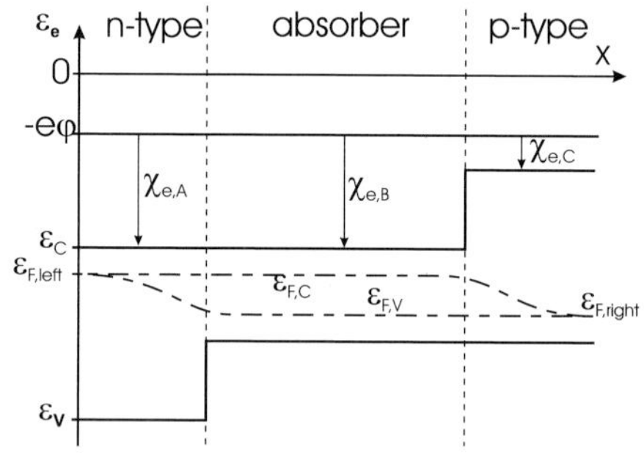

Fig. 9. Ideal structure of solar cells. The semiconductor absorber is placed in between n-type and p-type materials which selectively allow electrons to flow to the left and holes to flow to the right, thereby generating a charge current to flow from left to right.

A material which selectively conducts electrons is an n-type semiconductor, which has a much higher conductivity for electrons than for holes. If the thickness of this material, however is less than a diffusion length of holes, holes may be injected and increase the hole conductivity. In order to avoid this, a contact material was selected in Fig. 9 which has a larger band gap than the absorber and thus presents a barrier for the holes in the valence band which prevents their injection. By the same principle, the semi-permeable membrane for holes is a p-type semiconductor with a large band gap and a barrier in the conduction band preventing the injection of electrons. One additional and necessary property of a semipermeable membrane must be that it does not in any way impede the transport of the carriers which it should transmit. Its conductivity for the carriers to be transmitted must, therefore, be large enough to allow their transport without undue voltage drop, otherwise giving rise to a bad fill factor of the current-voltage characteristic of the solar cell.

The solar cell structure in Fig. 9 is ideal, if the corresponding materials are available and if the interface between them is not contaminated with too many interface states causing non-radiative recombination. If three materials differing in band gap and/or electron affinity with low interface state densities are not available, the n- and p- type membranes may be made from the same material as the absorber. In fact, both membranes or one of them may even act as the absorber. This brings us to the standard crystalline silicon cell, with a very thin highly doped n-type membrane facing the incident light, followed by a rather thick (200–300 μm) p-type absorber and membrane, a pn-junction. Although the p-type absorber is so thick it does not effectively block the transport of electrons towards the back side (in the wrong direction) because of the large diffusion length of electrons of about 1 mm in good quality silicon. In order to avoid recombination, i.e. loss, of too many electron-hole pairs at the metallic back contact, the contact area is reduced to the absolute minimum, the rest of the back surface being passivated by $Si0_2$, producing an interface with a low interface state density.

The solar cell structure in Fig. 9 allows the electrons to flow out of the absorber to the left and the holes to the right. This corresponds to a charge current flowing to the right. The question is now, how well the chemical energy stored in the electron-hole pairs is transformed into electrical energy. If it is assumed that the electron conductivity on the way to the left and the hole conductivity on the way to the right are large, the driving forces to support the currents are small and the gradients of the electron Fermi-energy ϵ_{FC} from the absorber to the left and of the hole Fermi-energy ϵ_{FV} to the right are negligible. As a result, the difference of the Fermi-energies in the left and the right contact is very nearly equal to $\epsilon_{FC} - \epsilon_{FV} = \mu_{eh}$, the chemical potential of the electron-hole pairs. The structure in Fig. 9 is, therefore, able to completely convert chemical energy into electrical energy. And this is true not only for this idealized structure, but to a good approximation for real solar cells as well. In conclusion, the efficiency of good solar cells is limited by the ability of converting solar heat into chemical energy as has been discussed in Sect. 3, while the subsequent conversion into electrical energy is close to 100% efficient.

Solar Cells Without pn-Junctions

The illuminated solar cell structure shown in Fig. 9 has a constant electrical potential φ everywhere. Without illumination, or even with illumination, but at a smaller voltage, however, a potential barrier exists between the p- and n-membranes. Although our derivation of the current voltage characteristic in Sect. 4.2 does not make use of a barrier, it would be nice, if a structure without a barrier would have a similar characteristic. Such a structure is the dye solar cell [6] .

In the dye solar cell the absorber is a monomolecular dye layer, in which excitons are produced upon illumination. The dye is sandwiched between an n- and a p-type membrane. The n-membrane is a sponge-like structure of n-type TiO_2 nano-particles in contact with one of the electrodes. The p-membrane is an ionic redox system in an electrolyte in contact with the other electrode. Due to the highly conducting electrolyte filling every pore in the TiO_2 sponge, all electrical fields are so effectively screened that no potential barriers can exist between the membranes. Nevertheless, a current-voltage characteristic as in a pn-junction is observed in the dye solar cell. A similar case is the plastic solar cell, in which p- and n-type membranes are also intermixed on a nanometer scale [7].

Tandem Cells

A solar cell with an absorber absorbing all photons with energy $\hbar\omega \geq \epsilon_G$, the band gap, is subject to the thermodynamic limits shown in Fig. 6. Higher efficiencies require absorbers in which the range of electron energies available for thermalization to a Fermi-distribution is reduced. As a result more than 2 Fermi-distributions must be generated from the broad solar spectrum and incident photons must be diverted to different transitions according to their energy.

From Sect. 2.2 we know that monochromatic operation of each solar cell avoids entropy generation and thermalization losses. This mode of operation is realized by placing (infinitely) many solar cells with decreasing band gap one behind the other, starting with the largest band gap facing the sun. Every absorber absorbs photons in an energy rage extending from its own band gap to the band gap of the preceding absorber. In tandem cells, the optical series connection lends itself to an electrical series connection. As is obvious from Fig. 10, charge has to be passed from the holes flowing out of the p-membrane of one cell (on the right in Fig. 10) to the electrons flowing out of

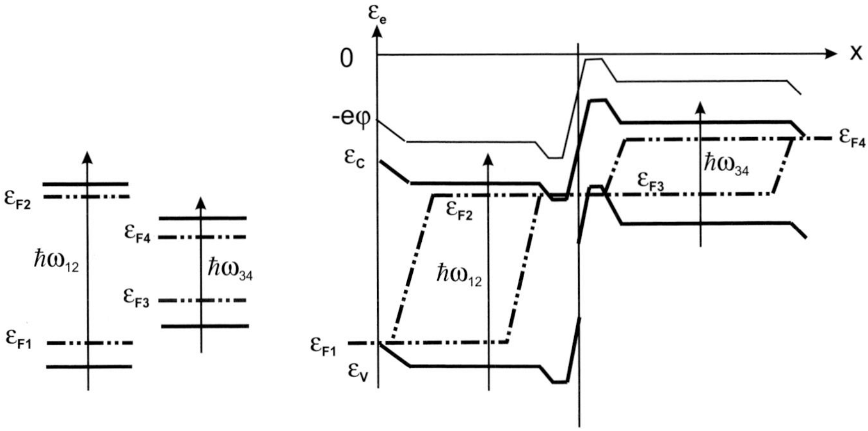

Fig. 10. In a tandem of two solar cells in optical and electrical series, the one with the higher band gap is facing the sun. Thermalization occurs in two different energy ranges, from the smaller band gap to the higher band gap in the cell on the right and from the higher band gap to ∞ in the cell on the left, leading to three different Fermi-distributions with Fermi-energies ϵ_{F1}, $\epsilon_{F2} = \epsilon_{F3}$ and ϵ_{F4}. The lossless connection between the n-membrane of the cell on the left and the p-membrane of the cell on the right requires a tunnel junction between them.

the n-membrane of the adjacent cell. In order not to lose free energy (μ_{eh}) the recombination reaction of electrons and holes at the interface must proceed within a single Fermi-distribution. This is achieved by a tunnel junction as shown in Fig. 10.

Although the efficiency of 86% derived in Sect. 2.2 for an infinite tandem does not take into account the additional constraint of equal currents in all cells imposed by the series connection, this is only a problem for tandems of a few cells, which do not reach their unconstrained efficiencies and which are susceptible to changes of the incident spectrum.

Hot Carrier Cells

A totally different approach to high efficiency solar cells is opened if it is possible to extract electrons and holes from the absorber through appropriate membranes before they are thermalized inside the absorber [8]. In order to avoid thermalization losses in the membranes or contacts, electrons (and holes) are allowed to pass through the membrane in only a narrow energy range. In a narrow energy range electron-phonon interaction cannot cause an energy loss and entropy generation is largely avoided.

For a hot carrier cell, the membranes have not only to be selective with respect to the type of carrier, they have to be selective with respect to the energy range as well. Carriers extracted from the absorber in these narrow energy ranges are replenished by carrier-carrier scattering. If carrier-carrier scat-

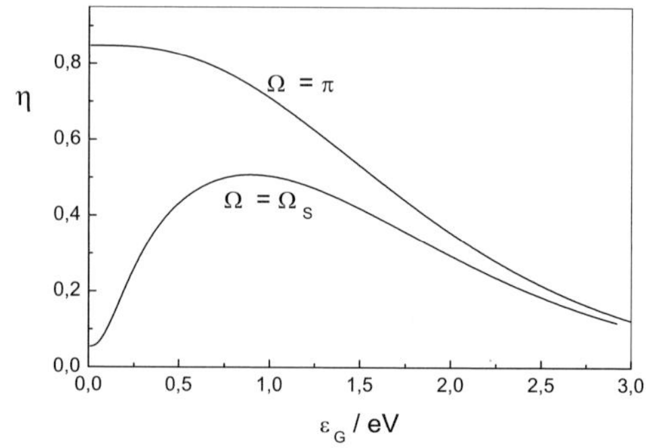

Fig. 11. In a hot carrier solar cell the absorber is placed between membranes which allow transport of carriers in only a narrow energy range, thereby avoiding energy losses caused by thermalization. The efficiency is shown as a function of the band gap ϵ_G of the absorber for maximally concentrated and non-concentrated sun light.

tering includes impact ionization and Auger recombination, processes which do not conserve the number of carriers, an energy distribution results which is a single Fermi-distribution $\mu_{eh} = 0$ with a high temperature in the absorber. The efficiency for hot carrier energy conversion under the assumption of only radiative recombination in the absorber is shown in Fig. 11.

5 Conclusions

Because of the high temperature of solar radiation rather high efficiencies for solar energy conversion are thermodynamically feasible. The thermodynamical limitations concern only the conversion of solar heat into chemical energy of electron-hole pairs in an absorber. The subsequent conversion into electrical energy is not limited by thermodynamics, since both forms of energy are free of entropy. The generation of chemical energy is based on the thermalization of the electron-hole gas. Thermalization can proceed almost without entropy generation if it is restricted to narrow energy ranges. In contrast, in a single semiconductor absorber, illuminated by the broad solar spectrum, representing present day solar cells, entropy generation and energy loss by thermalization are substantial. Tandem cell configurations are the conventional way to avoid thermalization losses by employing several different absorber materials. If, however, carrier extraction can be achieved before thermalization occurs in the absorber, high efficiencies could be achieved with only a single absorber.

References

1. W. Shockley: *Electrons and Holes in Semiconductors* (van Nostrand, New York 1954)
2. P. Würfel: J. Phys. Chem. Solids **49**, 721 (1988)
3. G. Lasher, F. Stern: Phys. Rev. **133**, A553 (1964), P. Würfel: J. Phys.: Condens. Matter **15**, 3967 (1982), P. Würfel, S. Finkbeiner, E. Daub: Appl. Phys. A **60**, 67 (1995)
4. A. de Vos: *Proc. 5.E.C. Photovoltaic Solar Energy Conf.*, Athens 1983, p 186
5. W. Shockley, H.J. Queisser: J. Appl. Phys. **32**, 510 (1961)
6. B. O'Reagan, M. Graetzel: Nature **353**, 737 (1991)
7. C.J. Brabec, N.S. Sariciftci, J.C. Hummelen: Advanced Functional Materials **11**, 15 (2001)
8. P. Würfel: Solar Energy Materials and Solar Cells **46**, 43 (1997)

Index

Springer Series in Solid-State Sciences

Editors: M. Cardona P. Fulde K. von Klitzing H.-J. Queisser

1 **Principles of Magnetic Resonance**
3rd Edition By C. P. Slichter

2 **Introduction to Solid-State Theory**
By O. Madelung

3 **Dynamical Scattering of X-Rays in Crystals** By Z. G. Pinsker

4 **Inelastic Electron Tunneling Spectroscopy**
Editor: T. Wolfram

5 **Fundamentals of Crystal Growth I**
Macroscopic Equilibrium and Transport Concepts
By F. E. Rosenberger

6 **Magnetic Flux Structures in Superconductors**
2nd Edition By R. P. Huebener

7 **Green's Functions in Quantum Physics**
2nd Edition By E. N. Economou

8 **Solitons and Condensed Matter Physics**
Editors: A. R. Bishop and T. Schneider

9 **Photoferroelectrics** By V. M. Fridkin

10 **Phonon Dispersion Relations in Insulators** By H. Bilz and W. Kress

11 **Electron Transport in Compound Semiconductors** By B. R. Nag

12 **The Physics of Elementary Excitations**
By S. Nakajima, Y. Toyozawa, and R. Abe

13 **The Physics of Selenium and Tellurium**
Editors: E. Gerlach and P. Grosse

14 **Magnetic Bubble Technology** 2nd Edition
By A. H. Eschenfelder

15 **Modern Crystallography I**
Fundamentals of Crystals
Symmetry, and Methods of Structural Crystallography
2nd Edition
By B. K. Vainshtein

16 **Organic Molecular Crystals**
Their Electronic States By E. A. Silinsh

17 **The Theory of Magnetism I**
Statics and Dynamics
By D. C. Mattis

18 **Relaxation of Elementary Excitations**
Editors: R. Kubo and E. Hanamura

19 **Solitons** Mathematical Methods for Physicists
By. G. Eilenberger

20 **Theory of Nonlinear Lattices**
2nd Edition By M. Toda

21 **Modern Crystallography II**
Structure of Crystals 2nd Edition
By B. K. Vainshtein, V. L. Indenbom, and V. M. Fridkin

22 **Point Defects in Semiconductors I**
Theoretical Aspects
By M. Lannoo and J. Bourgoin

23 **Physics in One Dimension**
Editors: J. Bernasconi and T. Schneider

24 **Physics in High Magnetics Fields**
Editors: S. Chikazumi and N. Miura

25 **Fundamental Physics of Amorphous Semiconductors** Editor: F. Yonezawa

26 **Elastic Media with Microstructure I**
One-Dimensional Models By I. A. Kunin

27 **Superconductivity of Transition Metals**
Their Alloys and Compounds
By S. V. Vonsovsky, Yu. A. Izyumov, and E. Z. Kurmaev

28 **The Structure and Properties of Matter**
Editor: T. Matsubara

29 **Electron Correlation and Magnetism in Narrow-Band Systems** Editor: T. Moriya

30 **Statistical Physics I** Equilibrium Statistical Mechanics 2nd Edition
By M. Toda, R. Kubo, N. Saito

31 **Statistical Physics II** Nonequilibrium Statistical Mechanics 2nd Edition
By R. Kubo, M. Toda, N. Hashitsume

32 **Quantum Theory of Magnetism**
2nd Edition By R. M. White

33 **Mixed Crystals** By A. I. Kitaigorodsky

34 **Phonons: Theory and Experiments I**
Lattice Dynamics and Models of Interatomic Forces By P. Brüesch

35 **Point Defects in Semiconductors II**
Experimental Aspects
By J. Bourgoin and M. Lannoo

36 **Modern Crystallography III**
Crystal Growth
By A. A. Chernov

37 **Modern Chrystallography IV**
Physical Properties of Crystals
Editor: L. A. Shuvalov

38 **Physics of Intercalation Compounds**
Editors: L. Pietronero and E. Tosatti

39 **Anderson Localization**
Editors: Y. Nagaoka and H. Fukuyama

40 **Semiconductor Physics** An Introduction
6th Edition By K. Seeger

41 **The LMTO Method**
Muffin-Tin Orbitals and Electronic Structure
By H. L. Skriver

42 **Crystal Optics with Spatial Dispersion, and Excitons** 2nd Edition
By V. M. Agranovich and V. L. Ginzburg

43 **Structure Analysis of Point Defects in Solids**
An Introduction to Multiple Magnetic Resonance Spectroscopy
By J.-M. Spaeth, J. R. Niklas, and R. H. Bartram

44 **Elastic Media with Microstructure II**
Three-Dimensional Models By I. A. Kunin

45 **Electronic Properties of Doped Semiconductors**
By B. I. Shklovskii and A. L. Efros

46 **Topological Disorder in Condensed Matter**
Editors: F. Yonezawa and T. Ninomiya

Springer Series in Solid-State Sciences

Editors: M. Cardona P. Fulde K. von Klitzing H.-J. Queisser

47 **Statics and Dynamics of Nonlinear Systems**
Editors: G. Benedek, H. Bilz, and R. Zeyher

48 **Magnetic Phase Transitions**
Editors: M. Ausloos and R. J. Elliott

49 **Organic Molecular Aggregates**
Electronic Excitation and Interaction Processes
Editors: P. Reineker, H. Haken, and H. C. Wolf

50 **Multiple Diffraction of X-Rays in Crystals**
By Shih-Lin Chang

51 **Phonon Scattering in Condensed Matter**
Editors: W. Eisenmenger, K. Laßmann, and S. Döttinger

52 **Superconductivity in Magnetic and Exotic Materials** Editors: T. Matsubara and A. Kotani

53 **Two-Dimensional Systems, Heterostructures, and Superlattices**
Editors: G. Bauer, F. Kuchar, and H. Heinrich

54 **Magnetic Excitations and Fluctuations**
Editors: S. W. Lovesey, U. Balucani, F. Borsa, and V. Tognetti

55 **The Theory of Magnetism II** Thermodynamics and Statistical Mechanics By D. C. Mattis

56 **Spin Fluctuations in Itinerant Electron Magnetism** By T. Moriya

57 **Polycrystalline Semiconductors**
Physical Properties and Applications
Editor: G. Harbeke

58 **The Recursion Method and Its Applications**
Editors: D. G. Pettifor and D. L. Weaire

59 **Dynamical Processes and Ordering on Solid Surfaces** Editors: A. Yoshimori and M. Tsukada

60 **Excitonic Processes in Solids**
By M. Ueta, H. Kanzaki, K. Kobayashi, Y. Toyozawa, and E. Hanamura

61 **Localization, Interaction, and Transport Phenomena** Editors: B. Kramer, G. Bergmann, and Y. Bruynseraede

62 **Theory of Heavy Fermions and Valence Fluctuations** Editors: T. Kasuya and T. Saso

63 **Electronic Properties of Polymers and Related Compounds**
Editors: H. Kuzmany, M. Mehring, and S. Roth

64 **Symmetries in Physics** Group Theory Applied to Physical Problems 2nd Edition
By W. Ludwig and C. Falter

65 **Phonons: Theory and Experiments II**
Experiments and Interpretation of Experimental Results By P. Brüesch

66 **Phonons: Theory and Experiments III**
Phenomena Related to Phonons
By P. Brüesch

67 **Two-Dimensional Systems: Physics and New Devices**
Editors: G. Bauer, F. Kuchar, and H. Heinrich

68 **Phonon Scattering in Condensed Matter V**
Editors: A. C. Anderson and J. P. Wolfe

69 **Nonlinearity in Condensed Matter**
Editors: A. R. Bishop, D. K. Campbell, P. Kumar, and S. E. Trullinger

70 **From Hamiltonians to Phase Diagrams**
The Electronic and Statistical-Mechanical Theory of sp-Bonded Metals and Alloys By J. Hafner

71 **High Magnetic Fields in Semiconductor Physics**
Editor: G. Landwehr

72 **One-Dimensional Conductors**
By S. Kagoshima, H. Nagasawa, and T. Sambongi

73 **Quantum Solid-State Physics**
Editors: S. V. Vonsovsky and M. I. Katsnelson

74 **Quantum Monte Carlo Methods in Equilibrium and Nonequilibrium Systems** Editor: M. Suzuki

75 **Electronic Structure and Optical Properties of Semiconductors** 2nd Edition
By M. L. Cohen and J. R. Chelikowsky

76 **Electronic Properties of Conjugated Polymers**
Editors: H. Kuzmany, M. Mehring, and S. Roth

77 **Fermi Surface Effects**
Editors: J. Kondo and A. Yoshimori

78 **Group Theory and Its Applications in Physics**
2nd Edition
By T. Inui, Y. Tanabe, and Y. Onodera

79 **Elementary Excitations in Quantum Fluids**
Editors: K. Ohbayashi and M. Watabe

80 **Monte Carlo Simulation in Statistical Physics**
An Introduction 4th Edition
By K. Binder and D. W. Heermann

81 **Core-Level Spectroscopy in Condensed Systems**
Editors: J. Kanamori and A. Kotani

82 **Photoelectron Spectroscopy**
Principle and Applications 2nd Edition
By S. Hüfner

83 **Physics and Technology of Submicron Structures**
Editors: H. Heinrich, G. Bauer, and F. Kuchar

84 **Beyond the Crystalline State** An Emerging Perspective By G. Venkataraman, D. Sahoo, and V. Balakrishnan

85 **The Quantum Hall Effects**
Fractional and Integral 2nd Edition
By T. Chakraborty and P. Pietiläinen

86 **The Quantum Statistics of Dynamic Processes**
By E. Fick and G. Sauermann

87 **High Magnetic Fields in Semiconductor Physics II**
Transport and Optics Editor: G. Landwehr

88 **Organic Superconductors** 2nd Edition
By T. Ishiguro, K. Yamaji, and G. Saito

89 **Strong Correlation and Superconductivity**
Editors: H. Fukuyama, S. Maekawa, and A. P. Malozemoff

Springer Series in Solid-State Sciences

Editors: M. Cardona P. Fulde K. von Klitzing H.-J. Queisser

90 **Earlier and Recent Aspects of Superconductivity**
Editors: J. G. Bednorz and K. A. Müller

91 **Electronic Properties of Conjugated Polymers III** Basic Models and Applications
Editors: H. Kuzmany, M. Mehring, and S. Roth

92 **Physics and Engineering Applications of Magnetism** Editors: Y. Ishikawa and N. Miura

93 **Quasicrystals** Editors: T. Fujiwara and T. Ogawa

94 **Electronic Conduction in Oxides** 2nd Edition
By N. Tsuda, K. Nasu, F. Atsushi, and K. Siratori

95 **Electronic Materials**
A New Era in Materials Science
Editors: J. R. Chelikowsky and A. Franciosi

96 **Electron Liquids** 2nd Edition By A. Isihara

97 **Localization and Confinement of Electrons in Semiconductors**
Editors: F. Kuchar, H. Heinrich, and G. Bauer

98 **Magnetism and the Electronic Structure of Crystals** By V.A. Gubanov, A.I. Liechtenstein, and A.V. Postnikov

99 **Electronic Properties of High-T_c Superconductors and Related Compounds**
Editors: H. Kuzmany, M. Mehring, and J. Fink

100 **Electron Correlations in Molecules and Solids** 3rd Edition By P. Fulde

101 **High Magnetic Fields in Semiconductor Physics III** Quantum Hall Effect, Transport and Optics By G. Landwehr

102 **Conjugated Conducting Polymers**
Editor: H. Kiess

103 **Molecular Dynamics Simulations**
Editor: F. Yonezawa

104 **Products of Random Matrices** in Statistical Physics By A. Crisanti, G. Paladin, and A. Vulpiani

105 **Self-Trapped Excitons**
2nd Edition By K. S. Song and R. T. Williams

106 **Physics of High-Temperature Superconductors**
Editors: S. Maekawa and M. Sato

107 **Electronic Properties of Polymers**
Orientation and Dimensionality of Conjugated Systems Editors: H. Kuzmany, M. Mehring, and S. Roth

108 **Site Symmetry in Crystals**
Theory and Applications 2nd Edition
By R. A. Evarestov and V. P. Smirnov

109 **Transport Phenomena in Mesoscopic Systems** Editors: H. Fukuyama and T. Ando

110 **Superlattices and Other Heterostructures**
Symmetry and Optical Phenomena 2nd Edition
By E. L. Ivchenko and G. E. Pikus

111 **Low-Dimensional Electronic Systems**
New Concepts
Editors: G. Bauer, F. Kuchar, and H. Heinrich

112 **Phonon Scattering in Condensed Matter VII**
Editors: M. Meissner and R. O. Pohl

113 **Electronic Properties of High-T_c Superconductors**
Editors: H. Kuzmany, M. Mehring, and J. Fink

114 **Interatomic Potential and Structural Stability**
Editors: K. Terakura and H. Akai

115 **Ultrafast Spectroscopy of Semiconductors and Semiconductor Nanostructures**
2nd Edition By J. Shah

116 **Electron Spectrum of Gapless Semiconductors**
By J. M. Tsidilkovski

117 **Electronic Properties of Fullerenes**
Editors: H. Kuzmany, J. Fink, M. Mehring, and S. Roth

118 **Correlation Effects in Low-Dimensional Electron Systems**
Editors: A. Okiji and N. Kawakami

119 **Spectroscopy of Mott Insulators and Correlated Metals**
Editors: A. Fujimori and Y. Tokura

120 **Optical Properties of III–V Semiconductors**
The Influence of Multi-Valley Band Structures
By H. Kalt

121 **Elementary Processes in Excitations and Reactions on Solid Surfaces**
Editors: A. Okiji, H. Kasai, and K. Makoshi

122 **Theory of Magnetism**
By K. Yosida

123 **Quantum Kinetics in Transport and Optics of Semiconductors**
By H. Haug and A.-P. Jauho

124 **Relaxations of Excited States and Photo-Induced Structural Phase Transitions**
Editor: K. Nasu

125 **Physics and Chemistry of Transition-Metal Oxides**
Editors: H. Fukuyama and N. Nagaosa